ATTACHMENT DEVICES OF INSECT CUTICLE

Attachment Devices of Insect Cuticle

by

Stanislav Gorb

KLUWER ACADEMIC PUBLISHERS
DORDRECHT / BOSTON / LONDON

A C.I.P. Catalogue record for this book is available from the Library of Congress.

ISBN 0-7923-7153-4

Published by Kluwer Academic Publishers,
P.O. Box 17, 3300 AA Dordrecht, The Netherlands.

Sold and distributed in North, Central and South America
by Kluwer Academic Publishers,
101 Philip Drive, Norwell, MA 02061, U.S.A.

In all other countries, sold and distributed
by Kluwer Academic Publishers,
P.O. Box 322, 3300 AH Dordrecht, The Netherlands.

Cover illustration: Drosophila attachment pad by courtesy of Jürgen Berger

Printed on acid-free paper

Printed in the Netherlands.

Contents

Foreword

In 1974 when I published my book, *Biological Mechanism of Attachment*, not many pages were required to report on the attachment devices of insect cuticles. As in most fields of research, our knowledge on this specific subject has simply exploded. Dr. Stanislav N. Gorb now describes the present day level of our knowledge, to which he has personally contributed so much, and a research team working on biological microtribology has gradually developed, also.

With modern methods of measurement it is possible to enter the structure – function relationship much more deeply, even down to a molecular level, which was not possible two and a half decades ago. It is a well known fact that, in biology, the more sophisticated the measuring method, the greater the achievement of biological fundamental research, and its resulting evidence. Our knowledge remains at a certain level until new methods once more permit a forward leap. Biological knowledge develops in the form of a stepped curve rather than linear, as reflected in the studies carried out on the attachment devices of insect cuticles.

I had great pleasure in reading this work, which is exceptionally rich in substance, well arranged, and superbly documented. I also have learned many new facts from it. I hope the book will be widely distributed. I also believe that it contains a great deal of stimulation for extended technical creativity. It is also entirely in agreement with Technical Biology and Bionics that we operate in Saarbrücken, which aims to promote the knowledge gained from fundamental research in biology, and help it to find its way into technical applications.

Prof. Dr. Werner Nachtigall

Preface

During the last 20-30 years, scanning electron microscopy (SEM) has been established as a standard method to study biological surfaces. Using SEM, much of information has been collected in the literature on the design and function of the animal integument. A large proportion of works have been devoted to attachment systems occurring in arthropods, animals bearing a cuticle, or a kind of outer skeleton. This book gives an overview of a wide variety of devices of the insect integument, adapted for attachment.

Because of limited space, the book is mainly focused on so-called frictional systems, which are a kind of attachment device containing tiny outgrowths on the contact surfaces. The presence of fields of such protuberances on the insect body increases frictional forces in the region of contact, and may result in fixation between these surfaces. Some of these systems employ secreted substances, modulating forces in the contact area.

To show different principles of morphology and ultrastructure of frictional systems, this book describes some examples, and outlines general rules of interrelationships between the construction and function of such devices. It summarises twelve years of my ultrastructural and experimental studies. Several examples are described in detail which allow me to discuss the combination of morphological, ultrastructural, and physiological data. Since comparative studies on several attachment systems include a wide variety of taxa, questions on the evolution of frictional devices are discussed. Nearly 400 references on cuticle microsculpture, ultrastructure, and attachment devices provide a reasonable basis for further interdisciplinary studies of biological attachment.

Lectures on biological attachment, which the author has given to biologists, material scientists and engineers, were quite successful, judging by the feedback received. It is believed that this book will also help non-biologists to get an impression of how the systems of biological attachment are designed and how they work. These design principles may inspire new ideas on biomimetics of new surface-active materials.

Different parts of my original studies were done at the Department of Zoology, University of Kiev; Schmalhausen Institute of Zoology, Kiev; Sechenov Institute of Evolutionary Physiology and Biochemistry, St. Petersburg; Department of Neurobiology, University of Vienna; Institute of Systematic Zoology and Evolutionary Biology, University of Jena, and Max-Planck-Institute of Developmental Biology, Tübingen. The author is most grateful to his teachers L.I. Frantsevich (Institute of Zoology, Kiev) and G.I. Shcherbak (University of Kiev), who initially started and supported studies on biological attachment systems. F.G. Barth (University of Vienna) and U. Schwarz (Max-Planck-Institute of Developmental Biology, Tübingen) provided valuable insight during my postdoctoral work in their laboratories. I have greatly benefited from collaboration with M. Scherge (Institut für Angewandte Verschleissforschung GmbH, Karlsruhe), who provided much advice on the measurements of small forces. Collaboration with R. Beutel (University of Jena) inspired me with evolutionary questions.

I wish to thank D.B. Tsarichkova, A.G. Kononko, N.N. Iljenko (Department of Zoology, University of Kiev), V.A. Krivonos, V.A. Krivosheev (Kholodny Institute of Botany, Kiev), F. G. Gribakin, A.D. Poljanovskij (Sechenov Institute of Evolutionary Physiology and Biochemistry, St. Petersburg), T. Losert (University of Vienna), H. Schwarz, J. Berger, B. Lattemann, I. Zimmermann, B. Sailer, G. Scheer (Max-Planck-Institute of Developmental Biology, Tübingen), Ch. Ziegler, M. Mondon, H. Stadler, and J. Wallentine (University of Kaiserslautern) for their kind help with diverse microscopical procedures. Members of the Biological Microtribology Group at Max-Planck-Institute of Developmental Biology, F. Baur, O. Gorb, Y. Jiao, V. Kastner, S. Niederegger, and V. Vötsch provided unpublished data for this book. Discussions with L.N. Pritykina (Paleontological Institute, Moscow), G. Bechly (Museum für Naturkunde, Stuttgart) on odonate phylogeny, and W. Federle (University of Würzburg) on insect attachment, are greatly acknowledged.

The author is thankful to colleagues who provided their original figures to illustrate the text (H. Krenn, H. Rischer, J. Berger, R. Beutel, W. Federle, A. Schnittger, M. Mondon, H. Stadler) or specimens for microscopy (L.N. Pritykina, B.F. Belyshev, A.Yu. Haritonov, S.N. Borisov, R.S. Pavljuk, V.N. Fursov, L.A. Zhiltsova, N.N. Shcherbak, J. DeMarmels, M. Samways, M. Sugimura, C. Cook, S.W. Dunkle, F. Haas, H. Rischer). I would particularly like to acknowledge the help of those colleagues who have commented either on particular chapters (L. Gaume-Vial, Université

Montpellier; A. Schnittger, University of Tübingen; U. Schwarz, MPI Tübingen; H. Silyn-Roberts, University of Auckland), or on the entire book (R. Beutel, University of Jena; L. Frantsevich, Institute of Zoology Kiev; F. Haas, University of Ulm; V. Kastner, MPI Tübingen; W. Nachtigall, University of Saarbrücken).

Permission to reproduce figures in the text was generously given by Georg Thieme Verlag (Stuttgart), Birkhäuser Verlag AG (Basel), and Company of Biologists Ltd. I am grateful for the financial support provided by the Federal Ministry of Education and Research of Germany (BMBF grant Biofuture 0311851).

Finally, I thank my parents, Nikolaj and Evdokia, my wife Olena, and my daughter Katheryna, who always gave me so much support.

Stanislav Gorb

Chapter 1

Friction and adhesion: fundamentals

1. DEALING WITH FRICTION

Any movement involving contact between two surfaces or between a surface and a medium, deals with the resistance of the surfaces or medium. This resistance we will call *friction*, a phenomenon which has a complex nature, as we will see in the next few paragraphs. This resistance may have a great influence on the design of biomechanical structures which arose during the process of evolution. Living creatures have developed fascinating systems for decreasing friction (*anti-frictional systems*), and vice versa, for increasing it (*frictional systems*). Interestingly, in both cases the purpose of such a system is to save energy. One always needs friction to generate the force for overcoming the, drag caused by friction elsewhere. Optimization then, strangely, becomes the exercise of minimizing friction at one end of the system, while maximizing it at the other (Radhakrishnan, 1998). For example, in the case of terrestrial locomotion, for effective propulsive movements a high friction is necessary for contact of the limbs with the substratum, and a lower friction within the joints of the limbs.

Surface microsculpture, mechanical properties of biological materials, and the resulting friction seem to be important factors for the understanding of (1) design of the living motion systems, (2) functional loads of numerous biological structures, and (3) evolution of these structures. This chapter will give basic information, collected in technical literature, about the nature of the friction and behaviour between the contacting surfaces. Only cases, which are relevant to biological frictional devices and which will be needed further to discuss observed phenomena, have been selected.

2. BASIC LAWS OF FRICTION

The first experimental tests of frictional forces were done by Leonardo da Vinci, who established the first two primary laws of friction. These laws are as follows: the friction force is (1) dependent on the applied force (*force dependence*) and (2) independent of the area of the contacting surfaces (*area independence*). There is static friction, the force required to start sliding, and kinetic friction, the force required to maintain this sliding. (3) Kinetic friction is considerably lower than static friction and is independent of the sliding speed (*speed independence*). This rule is often defined as the third law of friction (Bowden and Tabor, 1986). The fourth and fifth basic laws of friction are that (4) frictional forces are dependent on the material of the contacting surfaces (*material dependence*), and (5) on their texture, more precisely, on the surface profile (*surface-profile dependence*).

3. AREA OF CONTACT BETWEEN SOLIDS

Indeed, it is very difficult to prepare surfaces that are really flat. Even on carefully polished surfaces there are hills and valleys. Dimensions of such microsculptures lie in the micrometer-range. Thus, the surface-profile dependence of the frictional force is due to the interaction between contact areas of surfaces. If two solids are placed in contact, the upper surface will be supported by the summits of the irregularities of both surfaces, and a large area of the surface will be separated by a great distance, compared with the molecular range of action (Bowden and Tabor, 1986). This means that the friction is due to the interlocking of the surface asperities. Two solid bodies are supported on the hills of the highest irregularities, so the area of intimate contact is very small. The area of contact is thus independent of the surface area and is determined by load. The force applied to the contact points results in plastic deformation and flow, which can sufficiently increase the area of actual contact between solids. The plastic deformation and flow continue until the contact area is great enough to support the load. Generally, the friction is often defined as the work of lifting the load over the surface irregularities. If so, the crucial question of the study of the frictional behaviour of any material is, what is the real area of contact between solids which are pressed together? In any case, in most practical applications, for all types and shapes of surface irregularities, the real area of contact will be very nearly proportional to the load.

4. FRICTION AND ADHESION

There is a close relationship between friction and adhesion (Bowden, 1957). They are similar in directionality; they are directed against acting external forces. Additionally, both forces are dependent on the actual area of contact between two surfaces. We speak about the *friction* when an acting force is applied in a parallel manner to contacting surfaces, and results in the sliding of the surfaces. When an acting force is directed perpendicularly to the contacting surfaces and results in breaking of the contact between the surfaces, we describe the force holding the surfaces together as *adhesion*. The differences between adhesion and friction are dependent on the mechanical properties of the contacting surfaces. However, solids with a higher frictional coefficient normally have stronger adhesive properties (Bowden, 1957).

Furthermore, the friction between many surfaces is due to the shearing of junctions, formed by adhesion, at the points of intimate contact. Such junctions are usually formed when water or another liquid film, between the surfaces, is present. This example of friction is called *wet friction*. Experiments show that adhesion is independent from the thickness of the liquid film. If the surfaces are exposed to air no adhesion is observed (*dry friction*). In studying frictional surfaces it is, therefore, very important to examine the presence and nature of liquid films and their effect on adhesion and friction.

5. MECHANISMS OF ADHESION

The various types of forces, occurring in the area between the adhesive substance and the substrate may contribute to the adhesive bond. These forces are often called the mechanisms of adhesion (Naldrett, 1992). There are several adhesion mechanisms, namely (1) mechanical interlocking, (2) diffusion, (3) electrostatic forces, and (4) adsorption. Adhesive substances may interlock with the surface irregularities of the substrate to produce adhesion. The diffusion theory postulates that the molecular diffusion across the interface aids attraction of the polymers to themselves, and to each other. It has also been suggested that electrostatic forces, which arise from the contact of two materials with different charges, may contribute significantly to intrinsic adhesion. Unlike these three mechanisms, which presumably make only a minor contribution to the overall adhesion, the adsorption mechanism is generally accepted as the primary way by which adhesive substances adhere to the substrate.

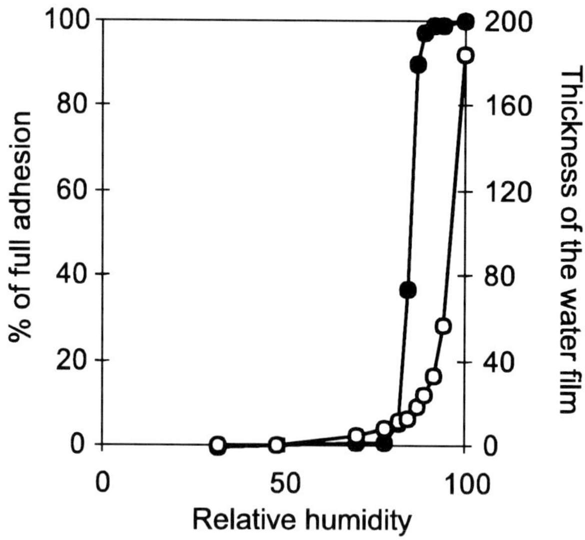

Fig. 1.1. Adhesion of glass surfaces as a function of the humidity of the surrounding atmosphere. The adhesion (•) rises rapidly at humidity levels exceeding 80%. There is a close parallel to the results, regarding thickness of the water film adsorbed on glass surfaces (○). Data from Bowden and Tabor (1986).

6. SURFACE TEXTURE, LIQUID FILMS AND ADHESION

As mentioned above, there is no capillary adhesion between dry, clean surfaces in a dry atmosphere, independent of the surface texture. Such an adhesion type first occurs, when an atmosphere is saturated with water vapour (Fig. 1.1) (Bowden and Tabor, 1986). Adhesion produced in such an atmosphere, decreases with increasing roughness of contacting surfaces. However, a thin water-film applied to the rough surfaces may drastically increase the adhesion. This indicates that the adhesion depends on the height of the surface peaks, and on the thickness of the absorbed layer of liquid. A decrease in adhesion would occur, when the height of the irregularities is comparable to the thickness of the absorbed film.

7. INTERDEPENDENCE OF THE ADHESION, SURFACE TENSION AND VISCOSITY OF THE LIQUID FILM

Adhesion is partly produced by cohesive forces in the liquid film. However, the pull force required to separate the surfaces, rapidly increases when the film thickness decreases at some level. This fact leads us to the conclusion that the adhesion observed, with very thin films, is not simply due to the surface tension of the liquid.

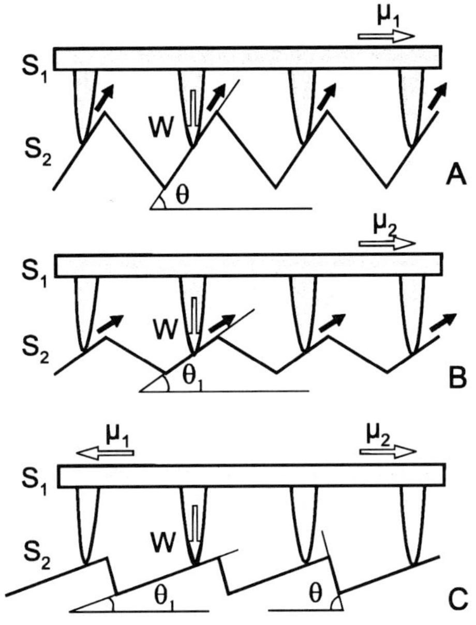

Fig. 1.2. Friction is dependent not only on the height of the surface irregularities, but also on their shape. The diagram shows why friction will depend on the angle of cone-shaped irregularities. S_1, S_2, contacting surfaces; *black arrows*, directions of movement of the S_1 on the structures of the S_2; μ_1, μ_2, friction dependent on the slope angle θ and θ_1, respectively; W, weight or normal force of the S_1. Friction may be dependent on the direction, when the shape of the asperities is asymmetrical (C).

If the space between contacting surfaces is completely surrounded by the liquid, the surface tension forces should be zero. In this case, therefore, the surfaces may be separated by the slightest normal force, when the separation is carried out very slowly. If the rate of separation is rapid, the viscosity of the liquid may become a very important factor. Bowden and Tabor (1986) provided a calculation showing that two round disks (R=1.27 cm) separated by oil, with the film density $h_1=10^{-5}$ cm, required nearly 10^5 N of pulling force to separate them against the viscous force, in a time of 10 seconds (Bowden and Tabor, 1986).

8. GLUE

When adhesive film can be polymerised we have glue. When glue is polymerised, it holds two surfaces together due to the frictional forces. The friction coefficient of glues is very high, due to the optimal actual area of contact with both surfaces. Glues can effectively hold two surfaces together, if (1) the actual contacting area is as large as possible, (2) there are no surface films present, and (3) the tension in the surrounding area is largely reduced (Bowden, 1957).

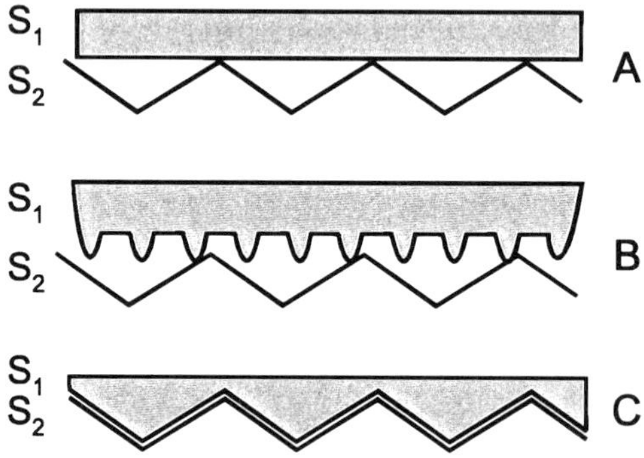

Fig. 1.3. Friction depends on the profile of both hard, contacting surfaces. A. A rough surface, S_2, has minimum contacting points with a smooth surface, S_1; friction will be minimal. When irregularities of the two surfaces have different dimensions (B), friction will be considerably lower than in the case of two co-opted surfaces (C), having similar dimensions of irregularities.

9. SURFACE IRREGULARITIES AND FRICTION

Friction is dependent not only on the height of the surface irregularities, but also on their shape. Let us imagine two hard surfaces sliding against each other. In the two-dimensional diagram of the cone-shaped irregularities (Fig. 1.2), the friction will depend on the angle θ. If we try to slide the upper surface there will be, in addition to the friction of the sloping faces, the component of weight, since we have to raise the upper surface in order to allow sliding. Surfaces whose irregularities have smaller θ will have lower friction.

Another important parameter is the microsculpture of the corresponding surface. The surface with high irregularities will slide on smooth surfaces with low friction (this is a good example of an anti-frictional system) (Fig. 1.3 A). If both surfaces have irregularities that fit well into each other, the friction will be higher (Fig. 1.3 C), as compared to a situation of the irregularities of one surface having a different dimension of those of the other (Fig. 1.3 B).

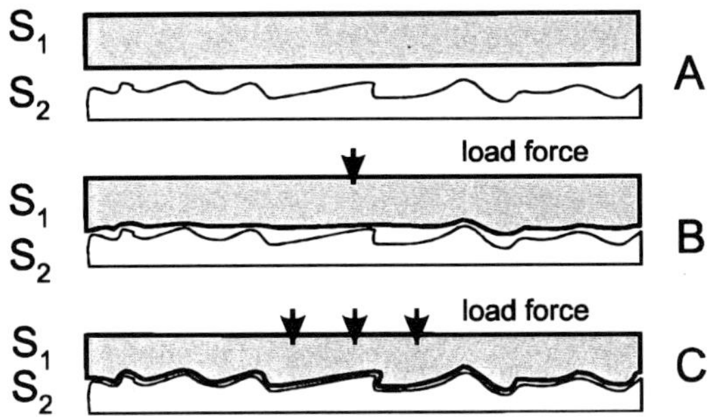

Fig. 1.4. Friction is dependent on the material properties of contacting surfaces. In the case of soft (S1) and hard (S2) contacting surfaces, flows of the flexible material will cover asperities of the hard material. As a result, the frictional force will increase.

10. DIRECTIONALITY OF FRICTION

When surfaces have cone-shaped or spherical irregularities, the frictional force measured during surface sliding in different directions, will be similar. A similar result would be obtained for surfaces with randomly-shaped microsculpture. Often, technical surfaces have, however, asymmetrical irregularities. This is also the case for many biological surfaces, as we will see below. In the two-dimensional diagram of the irregularities shaped like sloped cones, the friction, which depends on the angle θ, will be lower during sliding to the left (Fig. 1.2). Sometimes microsculpture, in the form of plates, may allow sliding with low friction in only one direction.

11. FRICTION AND MATERIAL PROPERTIES OF CONTACTING SURFACES

The examples given in the three subsections above are based on the surfaces of hard material. In the case where one material is hard and the other flexible or elastic, the frictional behaviour of the surfaces, during sliding, will differ from the examples given above. If the flows of the flexible material are so that the asperities of the hard material are completely covered by the deformations of the flexible one, it is clear that the frictional force will be greater, compared with the sliding of two hard materials with similar surface microsculptures (Fig. 1.4). Dependence of the friction on the load is greater in flexible materials than in hard ones.

Chapter 2

Insect cuticle: structure and properties of the material

1. FUNCTIONS OF THE CUTICLE

The insect exoskeleton is usually called a cuticle. Being an interface between a living animal and an environment, the cuticle of an insect serves many functions. (1) It limits the dimensions of an exoskeleton and is a basis for muscle insertions (mechanical function and function of locomotion). (2) It is an important element in organism defence against a variety of external factors, such as mechanical stresses, dry, wet, cold or hot environments. (3) It takes part in the transport of diverse epidermal secretions, and serves as a chemical reservoir for the storage of metabolic waste products. (4) A variety of cuticular structures are parts of mechano- and chemoreceptors. (5) The cuticle, its coloration pattern, and chemical components are important for thermoregulation, and are often involved in diverse communication systems. (6) Specialised cuticular protuberances may serve a variety of functions, such as oxygen retention, food grinding, body cleaning (grooming), etc. (see Chapter 3).

The cuticle is subdivided into two main layers epicuticle, and procuticle. Each layer consists of several layers. Additionally, there are two layers comprising the epidermis: (1) a layer of epidermal cells producing the cuticle, and other secretions, and (2) a basement membrane underlying the epidermal cells.

2. CUTICLE DESIGN

2.1 Basement membrane

The basement membrane is connected to the layer of epidermal cells (Richards, 1951). The membrane consists of collagen fibres, as has been shown for *Locusta migratoria* (Rinterknecht and Levi, 1966), and for the beetle, *Hydrocyrius columbiae* (Neville, 1975). It has been hypothesised that the membrane derives from plasmatocytes in the haemolymph (Wigglesworth, 1973).

2.2 Epicuticle

The epicuticle is the very thin, outermost layer of the cuticle. However, it is complex and consists of several sublayers, namely 1) cement sublayer, 2) wax sublayer, 3) outer epicuticle, and 4) inner epicuticle (Fig. 2.1). Each sublayer is of unique composition and properties. This composite structure does not contain chitin, and overlies the chitin-bearing procuticle. The epicuticle covers the surface of an insect. However, some exceptions are known: it is missing, for example, in chemoreceptors (Slifer, 1961) and midgut (Bertram and Bird, 1961).

Functions of the epicuticle are water homeostasis, setting limits on the dimensions of an exoskeleton, and transport of moulting fluids and epidermal secretions. The epicuticle may also serve as a chemical reservoir for the storage of metabolic waste products, or for the brief retention of juvenile hormones or defensive secretions (Filshie and Waterhouse, 1969). The high extension ability of the recently ecdysed cuticle, and specialised extending cuticles, such as, in fluid-sucking insects, is a result of the mechanical properties of the epicuticle (see review by Hepburn (1985)).

The outermost layer is a *cement sublayer* that is subtended by a wax layer. The cement layer is the combined secretory product of dermal glands. It consists of proteins and lipids stabilised by various polyphenolic substances (Beament, 1955). The distribution and thickness of this layer varies even in the same insect species. Some insects, such as the honey-bee, completely lack the cement sublayer (Locke, 1964). The primary function of the sublayer is to protect the lipid monolayer and outer epicuticle from possible abrasion, or impact damage.

The *wax sublayer* is located below the cement sublayer. The chemical composition of the waxes varies, even between individuals of the same species (Warthen et al., 1981; Gilby, 1980). The wax is usually a combination of long-chain, saturated alcohols and fatty acids. Such a chemical composition may reflect their properties: usually saturated and esterified fatty acids are associated with hardness in commercial waxes (Warth, 1956). Several variations of the structural arrangement of wax in the epicuticle are known: a) the lipid monolayer, located above the outer epicuticle; b) wax combined with the cement layer (Fig. 2.1); c) wax extending above the cement layer in the form of wax blooms (Gorb, 1995b). The lipids of the wax layer are secreted by epidermal cells. Presumably, the lipids are initially released in a mobile liquid state, from the cells, into a pore canal system, which is terminated by thin terminal, or wax canals. It has been previously hypothesised that the final wax substance is actually synthesised in the pore canal filaments (Locke, 1961). During transport to the surface, esterases modify wax precursors coming from the epidermal cells.

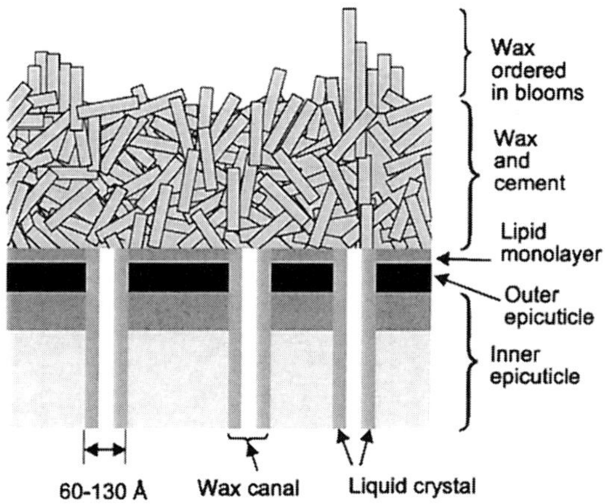

Fig. 2.1. The structure of the epicuticle (diagram). Wax is associated with the cement layer (sublayer), and wax also occurs in a bloom above the cement layer.

The *outer epicuticle,* often called cuticulin, occurs in all types of cuticle containing epicuticle. It has been described as a trilaminar layer of about 12-18 nm thick, which presumably consists of polythene-like polymers and quinone-tanned protein (Locke, 1965; 1966; 1976). The outer epicuticle is derived from specialised areas of the apical membrane of the epidermal cells, called plasma membrane plaques (Fig. 2.2) (Filshie and Waterhouse, 1969). During development, rather discrete plates appear above the membrane and gradually form a continuous layer (Filshie and Waterhouse,

1969; Locke, 1976). When the outer epicuticle is formed, the shape of the cuticle surface will be repeated by underlying cuticular layers. This mechanism of surface shape formation is responsible for the diversity of surface microsculpture described and discussed below. The surface patterns of insects serving as plastrons, diffraction grating, and frictional surfaces are formed at the same time as the outer epicuticle (Locke, 1964). Although the morphological significance of the outer epicuticle is great (Hepburn, 1985), the basic questions about the cellular mechanism of surface pattern formation, still remain unresolved. To what extent is the outer epicuticle itself the determinant of final exoskeleton shape? What are the mechanisms of formation of the microvilli and plaques, of the apical plasma membrane?

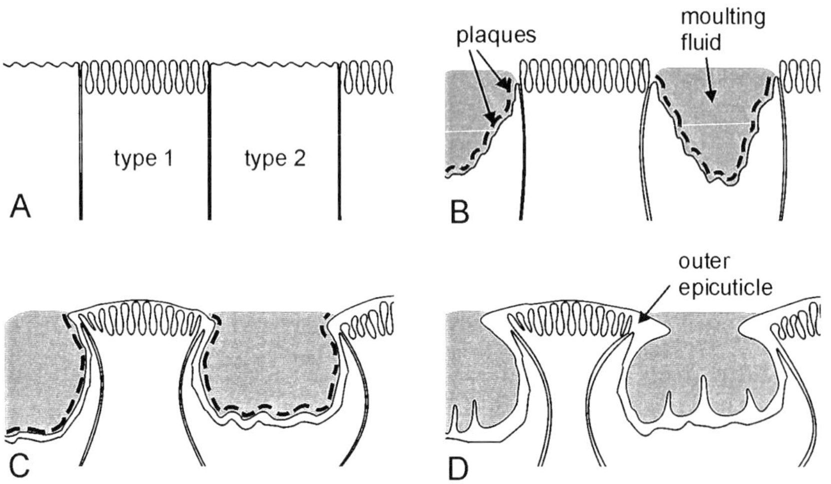

Fig. 2.2. Formation of the surface pattern of the outer epicuticle of the bug, *Nezara viridula* (Hemiptera: Pentatomidae). A. Differentiation of epidermal cells prior to secretion of the moulting fluid. B. Moulting fluid associated with type-2 cells. The formation of the outer epicuticle occurs in small patches above the microvilli. C. Outer epicuticle formation proceeds at a greater rate over type-1 cells, laterally displacing the forming cuticle. D. The expanding outer epicuticle results in the "surface folding" of the sheeted material.

The *inner epicuticle* is an optically isotropic layer consisting of polymerised lipoprotein stabilised by quinones (Hepburn, 1985). It serves as the reservoir for extracellular substances, such as, extracellular enzymes.

2.3 Procuticle

The *procuticle* is the chitin-bearing part of the cuticle, and it represents secretory products of the epidermis. It is generally known that it consists of two principal components, chitin and protein. Chitin is a structural polymer, which occurs in insects in crystalline form. Structures of chitin molecules and their arrangement are well studied, but the structure of chitin-protein assemblages still remains unknown. There are three morphs of chitin, differing in the molecule packing and polarities of adjacent chains: α-, β-, and γ-chitin (Rudall, 1963). The arrangement of molecules usually differs within the different morphs of chitin (Fig. 2.3 A). All three morphs of chitin are classified as visco-elastic polymers (Hepburn and Chandler, 1978). Chitin crystals always occur in insects as bundles of microfibrils, of 25-30 Å in diameter. Microfibrils are always associated with protein in a chitin-protein complex.

2.4 Cuticle as fibrous composite material

Chitin bearing cuticular microfibrils have a complex pattern of orientation in the three-dimensional space of the procuticle. The question arises as to whether the resulting regular geometry represents a self-assembly crystallisation phenomena, or whether the observed geometry is both produced and controlled by the synthesising epidermal cells (Hepburn, 1985). It is generally accepted that chitin monomers are secreted into apolysial spaces where they are then polymerised, and added to a growing chain by the enzyme chitin synthetase. This enzyme is presumably, somehow, attached to the plasma membrane. This point is crucial to the understanding of the 3D orientation of chitin microfibrils, but unfortunately, to date, still remains unresolved. The parabolic pattern of the microfibril arrangement, first described by M. Locke (1960b), from transmission electron micrographs has been explained by several models of the ultrastructural architecture of the procuticle.

2.5 Spatial orientation of the cuticle fibres

If we do not concentrate on some particularly specialised cases of the cuticle ultrastructure, the main observation on the cuticle architecture is that the fibers within sheets are non-randomly orientated.

2.5.1 Bouligand «helicoid» model

Bouligand (Bouligand, 1965; 1972) proposed a geometrical interpretation to explain such parabolic patterns (Fig. 2.3 D). His model is as follows: 1) within the cuticle the chitin microfibrils occur in layers (sheets); 2) within any sheet, all of the microfibrils are arranged in a format parallel to the surface plane; 3) neighbouring sheets of microfibrils are rotated in relation to each other, in the same direction. The angle of rotation may vary or remain constant (Neville, 1975). A set of sheets that has rotated approximately 180° is usually called lamella (Fig. 2.3 C). The thickness of lamella is dependent on the thickness of the microfibrils and rotation angle. Such a model very well explains the parabolic pattern as an optical artefact. Since it is impossible to obtain an ideally parallel section, the rotational angle of the microfibrils in successive layers of the lamella, will compose the parabolic pattern. However, there are numerous exceptions to be seen, also vertical orientation of microfibrils (Hepburn and Ball, 1973; Hepburn and Chandler, 1976; Dalingwater, 1975a; b; Mutvei, 1974). They have shown that this model cannot be universally applicable to all types of insect procuticle. Additionally, the helicoidal model could not successfully accommodate spiralling bow-curves occurring in the corneal lens of the cuticle (Neville, 1975; Chu et al., 1975).

2.5.2 Weis-Fogh «screw-carpet» model

Numerous exceptions have yet to be satisfactorily explained by the Bouligand model. The so called «screw-carpet» model overcame this difficulty (Weis-Fogh, 1970). This model asserts that the microfibrils are continuously secreted and the sheets of microfibrils gradually change position, vertically. Following such a principle, we can conclude that a single microfibril is a helicoid which can contribute to several sheets of microfibrils (Fig. 2.3 B). Interestingly, this model is analogous to crystal structure in dome synthetic plastics. However, our results on the ultrastructural architecture of the cuticle of the head-arresting system of dragonflies, did not support this model (Gorb, 1997b; d).

2.5.3 Filshie & Smith «perpendicular» model

Other authors supported the idea of the actual existence of vertical fibres (Filshie and Smith, 1980). Vertically arranged structures have been explained by including regular rows of presumably proteinous particles, whih oriented perpendicularly to the chitin microfibrils within lamella. Other authors have doubted the universality of this model (Mutvei, 1974; Dalingwater, 1975a).

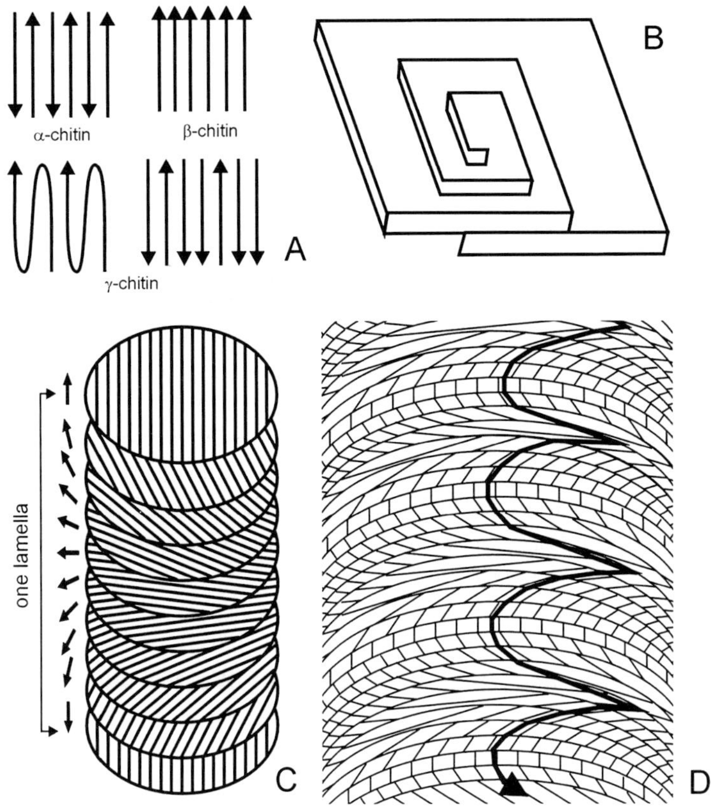

Fig. 2.3. Spatial orientation of the chitin fibres. A. Diagram of chitin chains in the unit cells of α- β- and γ-chitin (Rudall, 1963). B-D. Models of orientation of microfibrils in the procuticle. B. «Screw-carpet» model to account for the occurrence of vertically orientated microfibrils, within helicoidally orientated sheets of microfibrils (Weis-Fogh, 1970). C. Diagram representing a few sheets of chitin microfibrils rotating 180°, and defining single lamella (the arrows indicate the orientation of the microfibrils within each of the sheets) (Hepburn, 1985). D. Bouligan's pattern, created using the same rules as in C, to demonstrate how the parabolic pattern of arcs (bold curve) arise in sloped sections of cuticle, viewed in a transmission electron microscope (Bouligand, 1965).

2.5.4 Neville & Luke «integrative two system» model

Neville and Luke (1969a) basically used a helicoid model (Bouligand, 1965; 1972) with one difference; that there are two basic types of the microfibril orientation within lamella: helicoidal and preferred. Their «two-system» model stated that (1) sheets of chitin microfibrils are helicoidally arranged within the protein matrix, and (2) neighbouring sheets of microfibrils may be orientated in the same direction, and thus form a «preferred» layer, or be helicoidally rotated. (A) Entirely helicoidal microfibrillar sheets have been reported on Lepidoptera, Diptera, Coleoptera, Collembola, Diplura and Thysanura. (B) Entirely preferred, or unidirectional sheets, occur in the femoral apodeme of grasshoppers *Eutropidacris*. (C) Alternating helicoidal and preferred layers in which preferred layers lie in the same orientation, have been found in Orthoptera, Cursoria and Mantodea. Circadian clocks are responsible for the alternation of layers (Neville, 1965). (D) Alternating helicoidal and preferred layers in which the preferred layers change direction with respect to one another, (*plywood structure*) occur in Heteroptera, Odonata, some Coleoptera, Phasmida and some Mecoptera.

3. MECHANICAL PROPERTIES OF THE CUTICLE

3.1 Epicuticle

The epicuticle is stronger in compression than in tension (Joffe et al., 1975), and therefore it is presumed that it is in some ways analogous to a lacquer.

3.2 Procuticle

Macroscopic samples of different morphs of chitin vary in their rates of deformation under loads, depending on the inter-chain bonding characteristics. γ-chitin appears to deform, visco-elastically, by neighbouring chains slipping past one another. The same behaviour is presumed for α- and β-chitin, when the chains of both morphs slip past one another. The deformations of β-chitin chains are consistent with molecular conformational changes (Hepburn and Chandler, 1978; Rudall, 1963).

The cuticle consists of a laminate arrangement of five components (excluding epidermis and epicuticle) (Jensen and Weis-Fogh, 1962). (1) A system of chitin lamellae parallel to the surface (Neville, 1975; Hepburn, 1976), which are of considerable tensile strength (Hepburn and Joffe, 1976). (2) A rubber-like proteinaceous glue between lamellae, cross-linked to different extents, but poor in tyrosine, and not tanned. (3) Additional resin-like protein that is rich in tyrosine when tanned (Andersen, 1979). (4) Rubber-like proteins (Weis-Fogh, 1960; 1961). (5) Water (Hepburn and Chandler, 1976; Hepburn and Joffe, 1976).

The interplay of these components results in the wide spectrum of properties. Cuticles can be functionally grouped into (1) the so-called «solid» cuticle, typical for sclerites, apodems, phragmata, mandibles and claws; (2) thin arhrodial membranous cuticle, as well as, the cuticle of caterpillars; and (3) resilin-containing cuticle, from tendons of the flight muscles, springs of mouthparts, and wing articulations (Weis-Fogh, 1960; 1961).

3.3 Solid cuticle

The lack of flexibility of the solid cuticle is highly-correlated to the amount of NaOH non-extractable protein there in, which is responsible for cross-linking in the matrix (Andersen and Barrett, 1971). As the amount of non-extractable protein increases, the cuticle's stiffness increases (Fig. 2.4) (Hepburn, 1985). Additionally, the stiffness depends on the total cross-sectional area of the cuticle specimen.

The extensibility of the cuticles (*breaking strain*) is negatively correlated to the stiffness. This is because, as the extent of sclerotisation proceeds, the ability of proteins to slip past one another, or to be stretched, is greatly reduced due to cross-linking of the matrix. Where the matrix is not cross-linked, the material can flow plastically (Hepburn, 1985).

3.4 Arhrodial membranes

The external skeleton of Insecta contains hard sclerites connected to each other by highly-flexible regions, called arthrodial membranes (Hepburn, 1985). The term «intersegmental membranes» covers only a small number of examples (Snodgrass, 1935). The membranes exist in the movable joints of legs, antennae, organs of copulation, ovipositor, mouth-parts, and between segments of the body. During movement, the cuticular membranes can stretch quite rapidly to a great extent. Thus, these types of cuticles are grouped together, because of their mechanical properties and reactions to histological stains (Wigglesworth, 1956; Richards, 1967). Two main structural properties of arthrodial membranes are as follows: (1) relatively little crystalline order in the cuticle, and (2) ability of the chitin crystallites to orient, when suitably loaded in compression or tension (Fraenkel and Rudall, 1940; 1947).

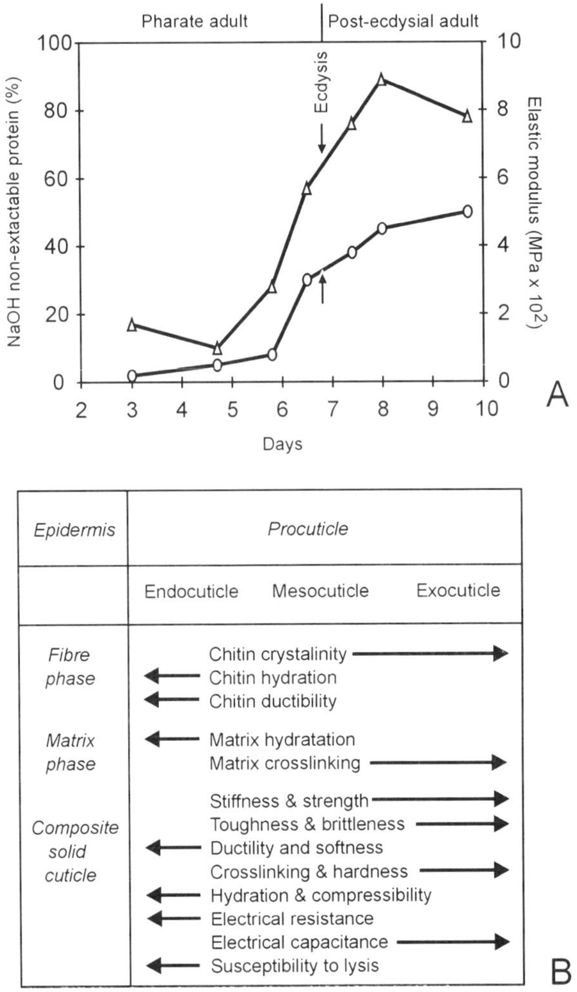

Fig. 2.4. Changes in the physical properties of insect cuticle during the course of differentiation and sclerotisation. A. Changes in the amounts of NaOH non-extractable protein (circles) and elastic modulus (triangles) during post-ecdysial development of the third abdominal tergite of the honey-bee, *Apis mellifera*. B. A summary of the direction of changes.

Generally, arthrodial and caterpillar-like cuticles are materials of great flexibility, great distensibility and reasonable strength. The interaction between chitin fibre and protein matrix is very loose, as has been demonstrated in deformation and fractographic studies. Under tension, these cuticles are visco-plastic and show so called «necking», as do the steel specimens under the same conditions (Hepburn and Chandler, 1976). However, there is remarkable variation regarding the extent to which such cuticles can be extended (Vincent and Wood, 1972).

There are two types of membrane design found in insects. 1) Highly extensible membranes that can extend more than 1000% of their length, are found in the locust abdomen (Vincent, 1981). This cuticle is highly specialised at the molecular level due to its protein composition (Hackman, 1975; Hackman and Goldberg, 1987). The explanation for such an extreme extension ability is the fact that chitin fibres are arranged normally to the longitudinal axis of the animal, and one can presume that the protein matrix is not completely bound in itself and/or to the chitin fraction. 2) Folding laminated membranous cuticle with a somewhat lower degree of extension. This cuticle, such as that found in the abdominal membranes of the tsetse fly, *Glossina morsitans*, is often armored by parallel rows of microtrichia (Hackman and Goldberg, 1987).

3.5 Resilin

Resilin is a protein that has molecular structure in a three-dimensional network of chains, nearly free of one another; thermally agitated, randomly kinked, and fixed within a network of only a few stable cross links (Andersen, 1963; 1964; 1966). These necessary covalent cross-links are fluorescent amino-acids, dityrosine and tertyrosine, derived from tyrosine (Andersen and Weis-Fogh, 1964). Amino acid composition of resilin from different species is quite similar (Hepburn, 1985).

Resilin was initially described in the prealar arm and wing-hinge ligaments of the locust, *Schistocerca gregaria*, and in the pleuro-subalar muscle tendon of the dragonfly, *Aeshna grandis*. These preparations showed the recoverable mechanical deformability, so that the resilin-containing cuticle has been called rubber-like cuticle, in contrast to «solid» cuticle. Resilin can be reversibly distended over 200% of its initial length. Its principal physical properties are its long-range elasticity, and lack of either creep or stress relaxation (Weis-Fogh, 1960; 1961; Andersen and Weis-Fogh, 1964). The resilin-containing structures work as mechanical springs under tension and compression (Frazier et al., 1999).

Chapter 3

Cuticular protuberances of insects

1. DIVERSITY OF CUTICULAR PROTUBERANCES AND PRINCIPLES OF THEIR CLASSIFICATION

The entire spectrum of microstructures in Arthropoda is composed of two external layers: epicuticle (1-2 μm) and exocuticle (a few hundred μm). The cuticular surface usually has a polygonal microsculpture, originating from the shape of underlying epidermal cells. However, the literature contains a huge amount of data on the invaginated and evaginated processes of cuticle. Some of them were already described in the 19th century (Dewitz, 1883; 1884; Rombouts, 1884; Simmermacher, 1884). New possibilities for studying the cuticular microstructure arose with the application of electron microscopy (TEM and SEM). The remarkable diversity of microsculptures of the cuticular surface of Arthropoda, has been studied with the aid of this device. In taxonomic literature, the following terms are used for the diverse cuticular protuberances: cornuti, signa, spines, chaetae, bristles, spicules, darts, denticles, hairs, tubercles, scobinate patches, graniculi, virga, pile, pubescence, tomentum, dust, pollen, etc. (Tuxen, 1970). What are the differences between them?

There are several basic principles for classification of these structures. An initial principle of classification has been based on the presence and type of innervation (non-innervated, single and multiple innervation) (Snodgrass, 1935). If one takes functional principle into account, for example mobility, cuticular protuberances may be separated into two groups: mobile and immobile. The first attempt to classify cuticular processes according to a developmental principle was undertaken by A.G. Richards (1951).

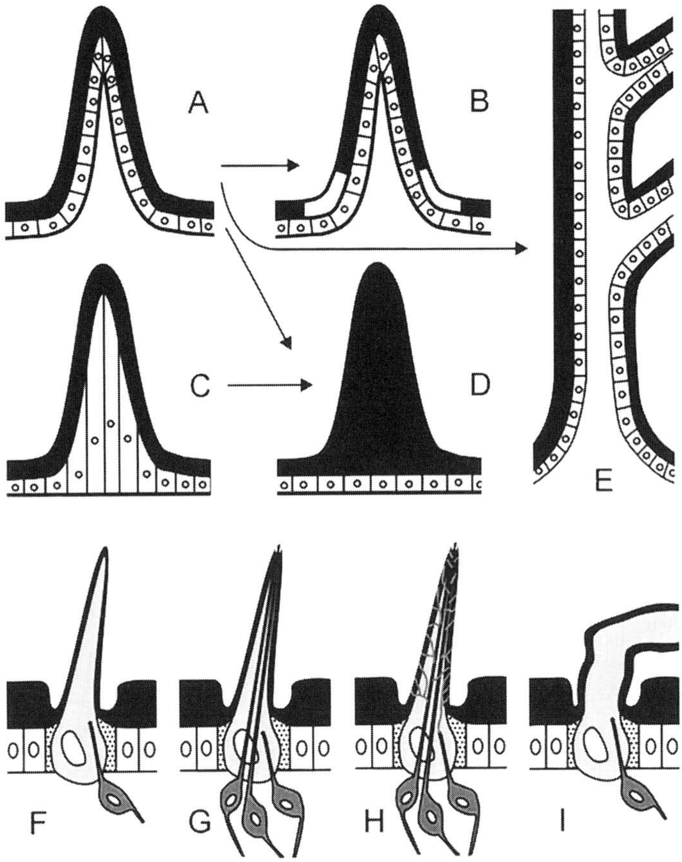

Fig. 3.1. Diagrams of multicellular projections. The secreted cuticle is solid black, cytoplasm is light-grey, and the tormogen cells are dotted. A. A simple multicellular spine. B. Modification of a simple spine with a membranous ring at the base. C, D. Unusual solid spine, termed Brunner's organ. F-I. Major types of protruding trichoid sensilla.

Later, A.G. Richards and P.A. Richards accumulated data primarily from transmission electron microscopy of ultrathin sections, and suggested a general classification and evolutionary scheme, into which a variety of modifications fits quite well. Their principle is based on the origin of such processes (Richards and Richards, 1979). There are four main types of processes: (1) the multicellular processes, in which the epidermal cells are not differentiated and are simply underlying the process; (2) the multicellular processes, in which the epidermal cells are differentiated into several specialised cells; (3) the unicellular processes; and (4) the subcellular processes, where there is more than one projection per cell. Smooth cuticle, as an additional principle, may also be taken into account.

2. SPINES AND HORNS

Spines and horns are multicellular processes without differentiation of cells. In other words, a single protuberance originates from cells which are similar in appearance to cells of the remainder of the epidermis (Fig. 3.1 A-E) (Richards and Richards, 1979). Such multicellular processes may be mobile due to an unsclerotised basal region, as in claws, and can be extended to include eversible osmeteria of caterpillars, or different types of glands of chrysomelid beetles, honey-bees, and male Mecoptera (Richards and Richards, 1979; Hepburn, 1985). Other cases of such projections are given in Table 3.1.

Table 3.1. Muticellular processes without differentiation of specialised cells

Structures	Insects	Source
Large teeth in the proventriculi	Acrididae (Orthoptera)	(Richards, 1951)
Osmeteria and other eversible glands	larvae of Lepidoptera	(Pliske and Salpeter, 1971)
Ptilinum of the head	Cyclorrhapha (Diptera)	(Henning, 1974)
Endophallus	most insects	(Tuxen, 1970)
Large flanges around the proventriculi	Nematocera (Diptera)	(Richards and Seilheimer, 1977; Richards et al., 1977)
«Felt chamber» of the caudal spiracle	Brachycera (Diptera)	(Clausen and Richards, 1951)
Brunner's organ of the jumping legs	Acrididae (Orthoptera)	(Slifer and Uvarov, 1938)

3. BRISTLES, CHAETAE, SETAE, AND HAIRS

These multicellular processes originate from specially differentiated cells: *the trichogen cell*, which secretes the cuticle of the process; *the tormogen cell*, which is responsible for the formation of the setal socket and associated cuticle; and often *the sensory cell*, which is differentiated into a bipolar primary sense cell body, with peripheral dendrites and an axonal connection to the central nervous system (Fig. 3.1 F-I). The fourth cell, *a dendritic sheath cell*, has also been reported (Ernst, 1969; 1972; Schmidt and Gnatzy, 1971). In multimodal sensila, there are a few sensory cells responsible for the detection of different signals. Such functional differentiation of cells, producing multicellular processes, has been supported by data from some gene-controlled mutants in *Drosophila* (Lees and Waddington, 1942; Stern, 1954). These protuberances usually serve as a part of the mechano- or chemoreceptor (Dethier, 1963). Tactile setae are often called trichoid sensilla. They are covered with solid cuticle, whereas chemoreceptive hairs usually contain pores on the side of the hair shaft or its tip (Cribb and Jones, 1995). The pores arise during cuticle secretion from the trichogen cell (Ernst, 1969; 1972), which usually makes the actual protuberance and secretes cuticle on to it.

Table 3.2. Muticellular processes with differentiation of specialised cells

Structures	Insects	Source
Scales	Lepidoptera, Coleoptera	(Anderson and Richards, 1942a; Ghiradella, 1989; 1994; Karasev, 1989)
Urticating hairs	Lepidoptera	(Pliske and Salpeter, 1971)
Sensory hairs	all Insects and Arthropods	(Lees and Waddington, 1942; Stern, 1954; Ernst, 1969; 1972; Schmidt and Gnatzy, 1971)

Large multicellular processes of this type are often called bristles or chetae. Setae range in length from a few micrometers to several millimetres. Such modifications, such as plumose and fimbriated setae, can be referred to as those containing microtrichia (see below) on the hair shaft. Some derivatives of setae can be very complex indeed. The most elaborate of them are scales found on butterflies, moths (Anderson and Richards, 1942a; Ghiradella, 1989; 1994), and beetles (Karasev, 1989). In the development of the scale, the three typical cells are involved, but usually one cell (assumed to be the sensory cell) degenerates during development. Often both trichogen and tormogen cells, together with neighbouring epidermal cells, may degenerate.

So called, urticating hairs of certain Lepidoptera, contain a gland at the base and a hollow shaft. It has been previously suggested that one of these cells, trichogen or tormogen, after producing the seta, shifts its functional specialisation and begins to secrete another material (Pliske and Salpeter, 1971). Some examples of cuticle protuberances originating from few cells are given in Table 3.2.

4. UNICELLULAR PROCESSES (ACANTHAE)

A single sclerotised protuberance originates from a single cell called, by definition, a trichogen cell (Richards and Richards, 1969). However, the derivation of this cell from a trichoid complex, by loss of both tormogen and sensory cells, is not proven, although it has been suggested (Richards and Richards, 1979). Some special cases of sensory cell degeneration in the trichoid complex were indicated in the previous paragraph.

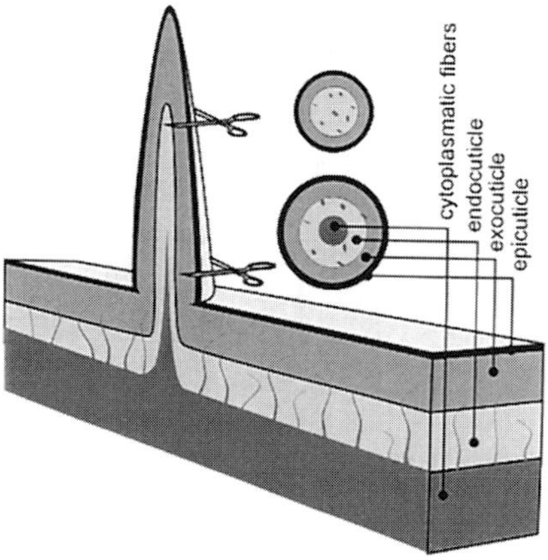

Fig. 3.2. Diagram demonstrating the principal structure of acanthae or microtrichia in longitudinal and cross sections.

Table 3.3. Unicellular processes (acanthae and microtrichia)

Structures	Insects	Source
Ovipositor valvulae	Hymenoptera, Rhaphidioptera	(Smith, 1972; Mickoleit, 1973b; Austin and Browning, 1981; Field and Austin, 1994)
Foregut	Coleoptera, Diptera, Trichoptera, Mecoptera	(Richards, 1965; Gibbs, 1967; Hepburn, 1969; Richards and Richards, 1969; Richards and Seilheimer, 1977)
Reproductive system	Lepidoptera	(Richards and Richards, 1979)
Adhesive pads	Diptera	(Whitten, 1969b; a; 1976)
Comb-like compositions	Siphonaptera	(Rothshild and Traub, 1971)

Acanthae range in length from a few micrometers to at least 0.5 mm, and lack a socket and sensory cell (Richards, 1965; Hepburn, 1969) (Fig. 3.2). Such processes are widespread among insects and are to be found on both internal and external surfaces: they have been reported in the reproductive systems of moths (Richards and Richards, 1979) and in adhesive pads of Diptera (Whitten, 1969a; b; 1976) (Fig. 3.3 A-E). They have various shapes: bifurcated, fimbriate, composed pectines or combs, with straight tips, and with bent tips (Richards and Richards, 1979) (Fig. 3.4).

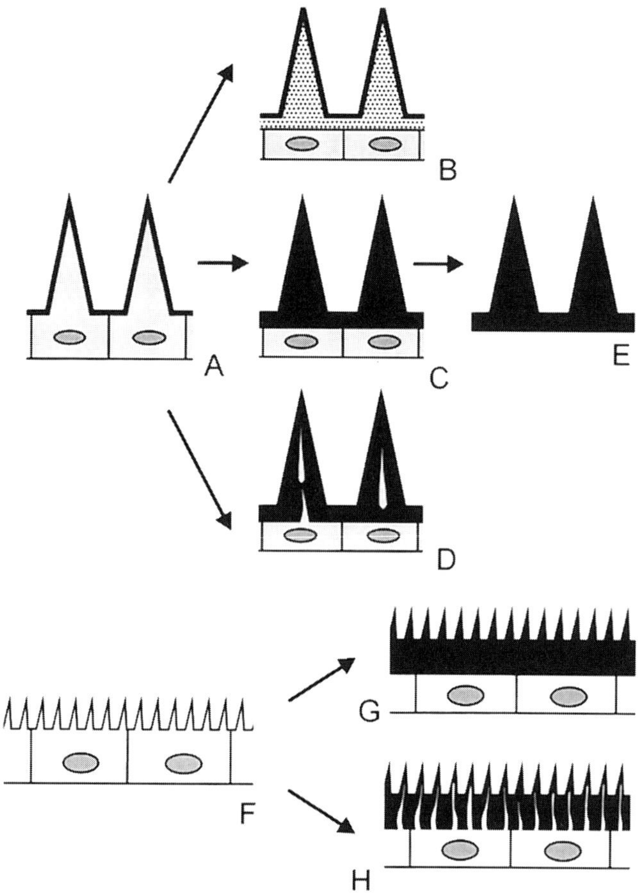

Fig. 3.3. Diagrams of unicellular and subcellular projections (Richards and Richards, 1979). Secreted cuticle, solid black; cytoplasm, light-grey. A-E. Sequences for development of acanthae; A. Early stage of acantha, which is usually modified later in development. B. Cell processes withdrawn from the lumens of acanthae and the space filled with acid mucopolysaccharide, (dotted) as in *Panorpa*. C. Cell process withdrawn, lumens filled with cuticle and processes underlaid with cuticle, as in «cornuti», in the reproductive system of the moth. D. Cell processes cut off by secretion of underlying cuticle, as in proventriculi of certain staphylinid beetles. E. Cell processes withdrawn, lumens filled and underlaid with cuticle, then secreting cells phagocytized by hemocytes, as in tenent hairs of flies. F-H. Sequences for development of microtrichia. F. Early stage of cytoplasmic filaments extending into each microtrichium. G. Solidified microtrichia underlaid by cuticle. H. Microtrichia with relationship to pore canals.

Two subtypes of acanthae have been described: (A) with the process nearly as great as the cross-sectional area of the cell producing it; (B) with the process much smaller, relative to the cell. The acantha is first formed as shown in Figure 3.3, A. Usually, the space within the acantha becomes filled with acid mucopolysaccharide as in a scorpion fly *Panorpa* (Fig. 3.3, B), or filled and underlaid with endocuticle as in the reproductive system of the moth *Dioryctria* (Fig. 3.3, C) (Richards and Richards, 1979).

In the fly pulvillus (*Phormia* and *Sarcophaga*), cells forming acanthae withdraw and are then phagocytised (Fig. 3.3 E) (Whitten, 1969a; b), and the acantha remains filled with endocuticle. However, our data on *Episyrphus* and *Calliphora* show that distal tenent hairs contain lumen, distally connected to the outer space through pores (Chapter 10). In the tenent hairs of the bug *Rhodnius* (Hemiptera: Reduviidae), cells remain in the tenent hairs and become glandular (Gillett and Wigglesworth, 1932).

Usually, acanthae compose specialised fields or combs, as reported on the inner surface of ovipositor valvulae (Austin and Browning, 1981; Field and Austin, 1994; Mickoleit, 1973b; Smith, 1972), but sometimes there may be a single acantha in a large area. Interestingly, acanthae may be characteristic for some structures of large groups of insects. For example, they have been found in the proventriculi of all Mecoptera and Siphonaptera (Richards, 1965) examined. However, they have been reported only for certain representatives of particular coleopteran families (Richards and Richards, 1979). In proventriculi, the acanthae have nearly the same size in any one species studied, but in reproductive ducts a variety of sizes has been observed. The size of the acantha correlates with the size of the corresponding epidermal cell, and its nuclear diameter.

5. SUBCELLULAR PROCESSES (MICROTRICHIA)

Microtrichia are very small hairs, less than 1 μm in diameter and 1-5 μm in length, occurring on diverse surfaces of an insect. More than one such projection corresponds to a single epidermal cell (Fig. 3.3 F-H). These protuberances are usually filled with cuticle. As has been previously mentioned (Richards, 1944; Richards and Korda, 1947; Richards and Richards, 1979), microtrichia are subcellular elements and therefore never get as large as some setae and acanthae. However, they may sometimes be slightly larger than the smallest acanthae. This makes the task of distinguishing between these types of cuticular protuberances quite difficult. Being very small in diameter, they may be very high in density, up to 1,000,000 mm^{-2}, and may cover considerable areas of the body surface, especially on the abdomen. This classification defines plumose setae as microtrichia on the setae (Anderson and Richards, 1942b), and divided tips of acanthae as microtrichia on acanthae (Richards and Seilheimer, 1977). Furthermore, acanthae of type B can also be a type B microtrichia by making, as the main character, the hair size relative to cell size, rather than the number of hairs per cell.

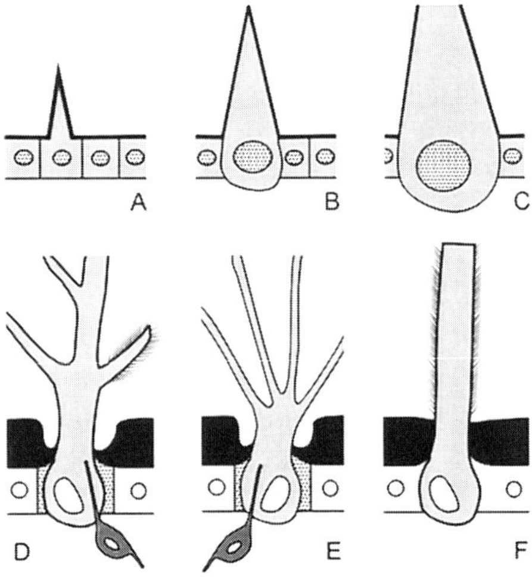

Fig. 3.4. Diagrams of special cases. Secreted cuticle solid black, cytoplasm light-grey. A-C. A series of diagrams of acanthae of various sizes in aedeagus of the moth, *Dioryctria abietella,* showing the relationship between the size of the cell and the size of acantha. Note that this is not a developmental sequence. D-E. Interpretation for plumose and fimbriated setae (tormogen cells are dotted). These could be called «microtrichia on setae». F. Interpretation for plumose acantha. This could be called «microtrichia on acantha».

Table 3.4. Subcellular processes (microtrichia)

Structures	Insects	Source
Part of structural coloration	variety of Insecta	(Hinton, 1970)
Taenidia of trachea	Insecta	(Richards and Anderson, 1942; Richards, 1944)
Coverage of the scopula hairs	Aranei	(Foelix et al., 1984)
Plastron of water bugs	water bugs Nepidae	unpublished data
Body coverage (presumably plastron)	*Nymphula* larva (Lepidoptera)	(Richards and Richards, 1979)
Ommatidia grating	Diptera	(Parker et al., 1998)

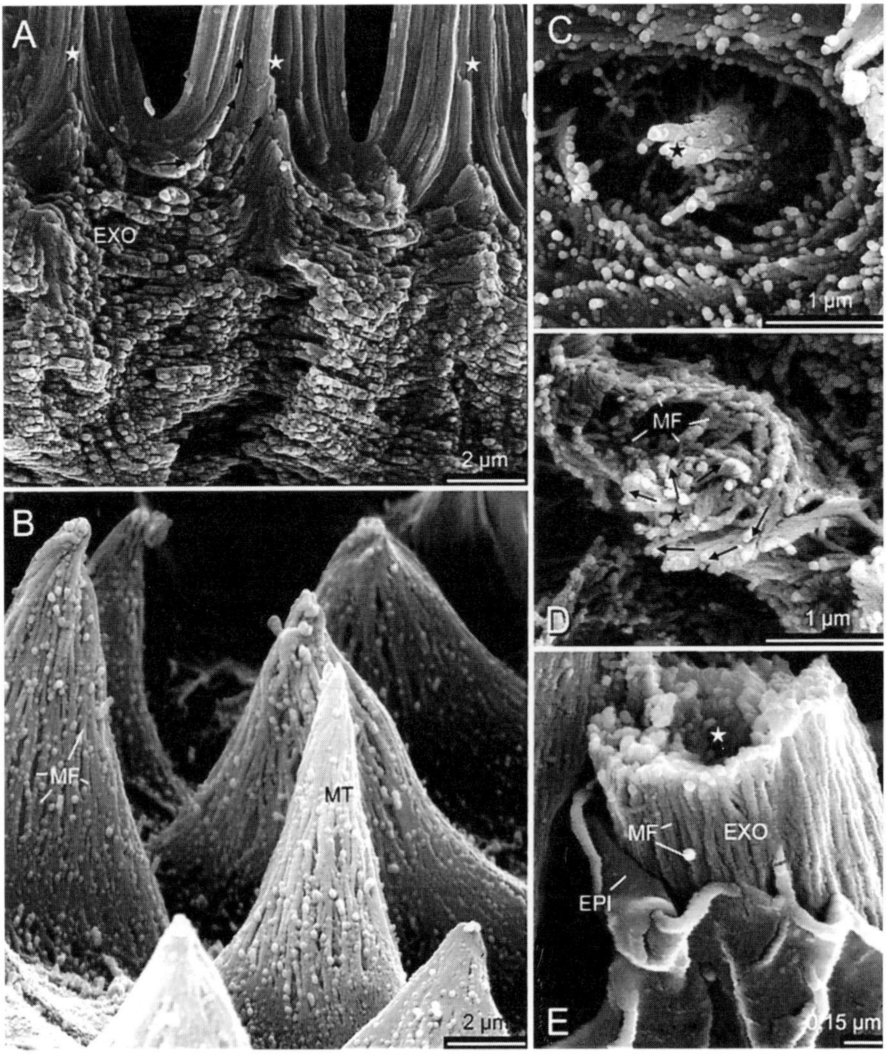

Fig. 3.5. Ultrastructural architecture of the microtrichia (MT) of the dragonfly postcervical sclerite (Gorb, 1997d). A. Cuticle fracture treated by chitinase. B. The outermost lamella of the exocuticle in the microtrichium, after NaOH treatment. C, D. Microfibril orientation within the microtrichium. E. A single fractured microtrichium with partly removed epicuticle. A, B, D, E. *Aeshna mixta*. C. *Sympetrum sanguineum*. *Arrows*, local direction of the fibrils; *asterisks*, axis of the microtrichium; EPI, epicuticle; EXO, exocuticle; MF, microfibrils.

There are three main types of microtrichia in accordance with their internal anatomy: 1) those containing cytoplasmatic fibers (Fig. 3.3 F); 2) those containing pore canals (Fig. 3.3 H); and 3) those with neither cytoplasmatic fibers nor pore canals (Fig. 3.3 G). The presence of the microtrichia results in an increase in the frictional forces in the region of contact, and may be an adaptation for the frictional devices studied here.

The microtrichia are always formed within cytoplasmatic processes. However, cytoplasmatic evaginations, such as microvilli observed during the process of surface pattern formation, are structures substantially smaller than microtrichia (Filshie and Waterhouse, 1969; Locke, 1976). That means that microtrichium is not a result of an evagination of a single cytoplasmic filament. It has been previously suggested that self-assembling processes are likely to be involved in the development of structures of subcellular dimension.

After dissolving the protein coverage of chitin-protein microfibrils with NaOH, or chitin fibres with chitinase, the arrangement of the axial chitin fibres, or protein matrix, can be demonstrated in three dimensions (Fig. 3.5) (Gorb, 1997d). The chitin microfibrils of the outermost exocuticular lamella lie parallel to the slope of the cone surface of the microtrichium, and slightly twist at the top of the microtrichium (Fig. 3.5 B, E). The microfibrils of the external lamella, of the transient area between plain cuticle and raised cuticle of the microtrichium, twisted within one lamella and preferred directions of microtrichia in the successive lamella, also rotating. In the deeper layers of the cuticle, the rotation of both the microfibrils and successive lamella result in the twisted-straw architecture of the microfibril's composition, within the microtrichium (Figs. 3.5 A, C, D; 3.6). The deepest microfibrils, located close to the axis of the microtrichium, form compact clusters of fibrils (1.0-1.5 μm length). The twisted-straw architecture of microfibrils in the MT is derived from principles of development of extracellular fibrous composites (Neville, 1993). This architecture has, presumably, an additional functional significance in strategic design, with particular mechanical properties.

6. FUNCTIONS OF CUTICULAR PROTUBERANCES

Functions of cuticular protuberances are listed in Figure 3.7. Highly specialised, jointless cuticular microstructures, originating from a trichoid complex of epidermal cells, are mainly sensory organs, but in some cases the original sensory function has been lost in evolution (scales). Some setae may also have a defence function, as in the dermestid beetle larva (Nutting, 1963), or the function of pheromone release (Seibt et al., 1972).

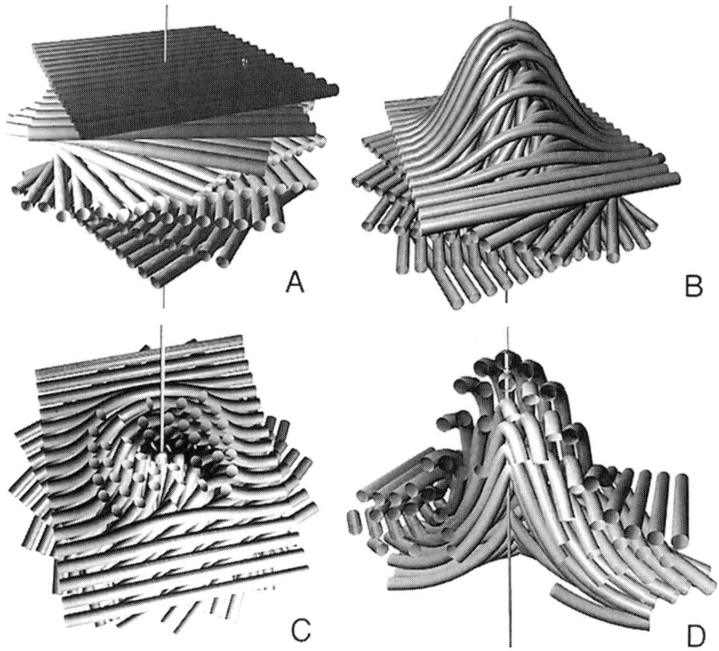

Fig. 3.6. Model of orientation of the microfibrils within lamella in the helicoidal flat cuticle (A), and in the helicoidal cuticle of the microtrichium (B-D) (Gorb, 1997d). B. The entire model of fibre orientation in cone helicoid. C. The cross section of the cone helicoid. D. The longitudinal section of the cone helicoid.

Cuticular microstructures originating from one cell (acanthae), or as a pattern from one epidermal cell (microtrichia), serve diverse mechanical functions or form a coloration pattern. The functions are as follows: grooming, cleaning, oxygen retention, thermoregulation, increasing the aerodynamic activity of the surface, sound generation, food grinding, filtering of food particles, etc. Often, cuticular protuberances are polyfunctional, such as in the case of coloration pattern.

6.1 Body coloration pattern

Structural coloration pattern, which is composed of scales and bristles, is well-known from Lepidoptera (Richards, 1967; Grodnicky, 1988; Ghiradella, 1989), dermestid and cicindellid beetles (Coleoptera) (Schultz and Hadley, 1987) (Fig. 3.7 I). Scales of the scarabaeid beetles from the genus *Hoplia* bear additional microtrichia on their surfaces (Fig. 3.8 C, D).

Coloration pattern serves as species and sex recognition, and also for camouflage and mimicry.

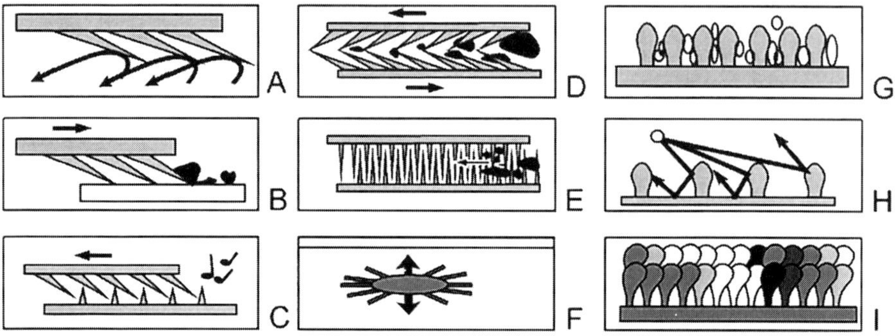

Fig. 3.7. Diagram of functions of cuticular microstructures. A. As aerodynamically active surfaces. B. Grooming. C. Sound generation. D. Food grinding. E. Filtration devices. F. As hydrodynamically active surfaces. G. Oxygen retention. H. Thermoregulation. I. Body coloration pattern.

6.2 Oxygen retention (plastron)

Many aquatic and semiaquatic arthropods have sculptured surfaces responsible for holding air under water (Fig. 3.8 I, J). Such surfaces usually contain microtrichia fields, which are called plastron (Heckmann, 1983). These structures appear convergently in various arthropod taxa, as an adaptation to the aquatic environment: Collembola (Messner, 1986a), Lepidoptera (Messner and Adis, 1987; Messner et al., 1987), Coleoptera (Messner, 1983; Messner and Dieckmann, 1987), Heteroptera (Thorpe and Crisp, 1947), Diptera (Hinton, 1976; Messner, 1986b), Araneae (Messner, 1988), and Diplopoda (Messner and Adis, 1988). Some terrestrial insects, such as Aphididae (Auchenorrhyncha), also bear similar structures in the form of bristles, mushroom-like spines, or stigmal plates, which can protect their surfaces from moisture (Heie, 1987).

6.3 Thermoregulation and prevention of drying

Cuticle protuberances may provide multi-level reflection of sunlight (Fig. 3.7 H). Such an ability of wing scales is suggested to be an adaptation for cooling in adult specimens of Lepidoptera (Grodnicky, 1988; Grodnicky and Kozlov, 1989). In the species of curculionid beetles, of the genus *Tychius* inhabiting arid areas, cuticular scales have been suggested to be a system responsible for maintaining thermal balance (Karasev, 1989). An additional function of such coverage is preventing water loss. This function was suggested for leaf-like bristles at the body margins in Aphididae (Auchenorrhyncha) (Heie, 1987). Body coverage by bristles, scales, and hairs in the honey-bee, *Apis mellifera,* may be used for warming up and influencing metabolism at low temperatures (Southwick, 1985).

6.4 Aero- and hydrodynamic-active surfaces

Small bristles, scales, and microtrichia of the wings of Odonata, Diptera, Hymenoptera, and Lepidoptera (Bocharova-Messner and Dmitriev, 1984; D'Andrea and Carfi, 1988) are aerodynamically-active (Figs. 3.7 A; 3.8 E). The microturbulences, generated around such structures in flight, build a kind of a lubricating layer of air, between an air stream and the insect surface. This can decrease friction during high-speed flight. Very small insects, such as Thysanoptera, can sweep in the air using long setae on the margin of their wings. In water, similar structures have the effect of increased flotation (Fig. 3.7 F). Tiny bristles with this function have been reported for crustaceans *Dosima* (Cirripedia, Lepadomorpha) and *Rhithropanopeus harrisii* (Decapoda, Brachiura) (Memmi and Elfimov, 1987; Morgan, 1989).

6.5 Sound generation

Highly specialised areas of cuticle, responsible for sound generation, usually consist of co-opted patterns of cuticular plates, seldom of microtricha fields (Hinton, 1970). When these surfaces slide over each other, sound is generated (Fig. 3.7 C). Such structures have been previously described from elytra, abdomina and coxae in phylogenetically, far-related arthropod taxa: spiders (Aranei), crustaceans *Trizopagurus*, bugs Cimicomorpha, Pentatomorpha, and beetles *Geotrupes* (Scarabaeoidea) (Gogala, 1984; Starck, 1985; Field et al., 1987; Palestrini et al., 1987; Pirisinu et al., 1988). The sound frequency corresponds to the periodicity of these structures contacting functionally-corresponding surfaces, and the speed of sliding.

6.6 Defence

Long, stiff, sclerotised cuticular spines are often used as defence mechanisms against predators. These structures are particularly wide-spread in the marine arthropods, such as zoea of the crustacean *Rhithropanopeus harrisii* (Decapoda, Brachiura) (Morgan, 1989). A similar function was described for long setae of certain beetle larvae, from the family Dermestidae (Coleoptera) (Nutting, 1963).

6.7 Grooming, sampling, and filtrating

Grooming is a very important function for insects, which sometimes live in extremely dirty environments (Fig. 3.7 B). Their rich sensory equipment has to be kept clean in order to respond adequately to external signals. Many Hymenoptera, such as Formicidae, Tiphidae, Mutillidae (Schönitzer and Lawitzky, 1987; Francouer and Loiselle, 1988) and bees Apoidea (Hymenoptera) (Schönitzer and Penner, 1984; Schönitzer, 1986) bear modified leg spines specialised for the cleaning of antennae. In the aquatic environment, the problem of grooming is even more important because of biofouling. Pinnoterid crabs (Decapoda, Brachiura) use epipodit lobes covered by bristles for cleaning the gills (Pohle, 1989).

The collection of pollen grains and food sampling are functions similar to grooming. They require similar surface structures and generally the same motion. Interestingly, such systems in the mouth-parts' area appeared independently in the evolution of such phylogenetically, far-related animal groups as Insects, Crustaceans, Molluscs and even some Vertebrates (Arens, 1989). An example of such a tooth-like structure is given in Figure 3.8 H, showing the prestomal teeth of the labellum in the calliphorid fly. When food is sampled from rough substrata, such as stones, surfaces of sampling devices show the wear. This is the reason that larval Ephemeroptera may have up to 27 instars. In bees (Apoidea), systems responsible for collecting pollen grains are usually equipped with urticating bristles (Pasteels and Pasteels, 1972; Hesse, 1981).

Filtration systems are usually equipped with long bristles too (Figs. 3.7 E). Such systems are well-known from mouthparts of aquatic insect larvae. The filtering system of insect spiracles is composed of branched acanthae (Fig. 3.8 G). However, in some cases, completely different principles may be involved in the design of these systems. The labellum of a fly (Brachycera, Diptera) bears so-called pseudotrachea (Gracham-Smith, 1930; Dethier, 1976; Zaytsev, 1984; Elzinga and Broce, 1986; Driscoll and Condon, 1994), which are able to change the diameter of the filtration sieve, depending on the size of particles in the food (Fig. 3.8 F). Labellum and pseudotrachea are driven by muscles, resilin springs, and hydraulic pressure. Outgrowths of the pseudotrachea have a complex material design.

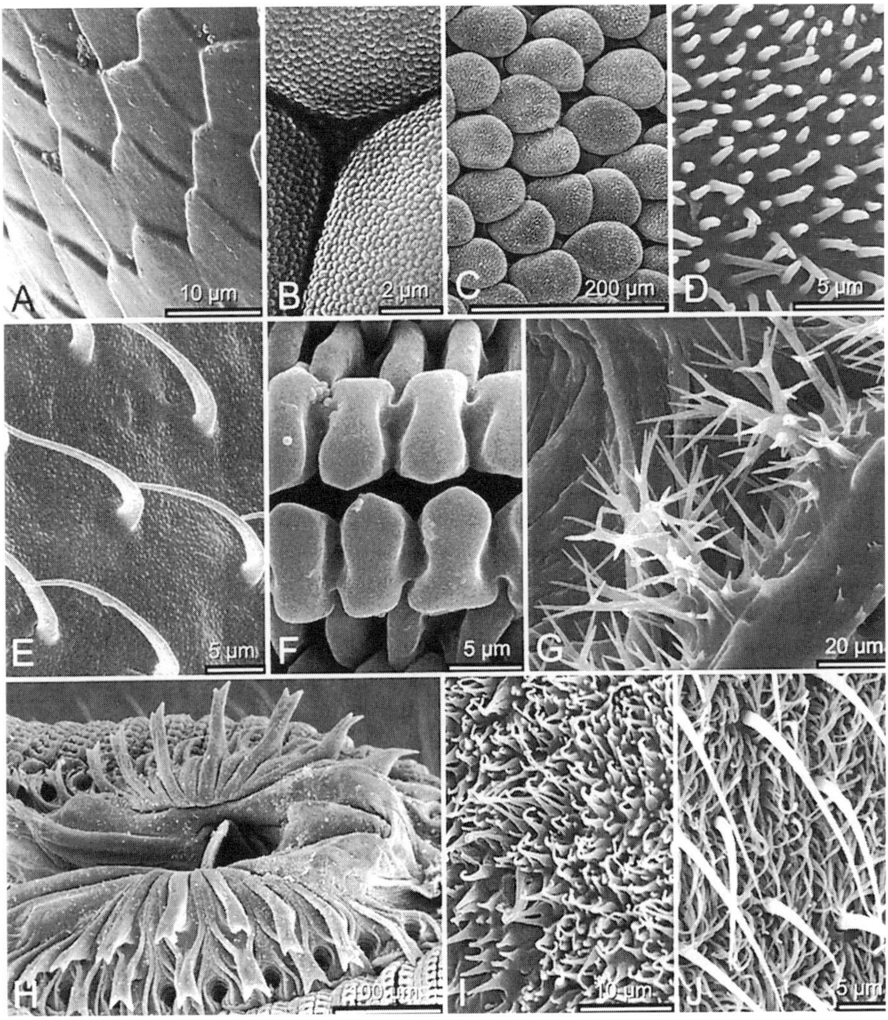

Fig. 3.8. Functional diversity of non-innervated cuticular protuberances in insects. A. Unspecialised polygonal surface on the tarsus of the scarabaeid beetle, *Melolontha melolontha*. B. Ommatidia surface in the calliphorid fly, *Calliphora vicina*. C. Scales on the dorsal surface of the elytron in the scarabaeid beetle, *Hoplia sp*. D. Same, the surface of a single scale. E. Wing surface in the bibionid dipteran, *Bibio ferruginatus*. F. «Pseudotrachea» of the labellum in the *C. vicina*. G. Filter system of the spiracle in the tenebrionid beetle, *Tenebrio molitor*. H. Prestomal teeth in the *C. vicina*. I. Plastron in the nepid bug, *Ranatra linearis*. J. Plastron in the water-strider, *Gerris lacustris*.

6.8 Food grinding

Cuticular protuberances that grind food items occur not only in the insect mouth-parts, but also in one part of the insect digestive system, called proventriculus (Fig. 3.7 B). Spines on the inner surface of the proventriculus have been previously reported in Mecoptera, Odonata, and Siphonaptera (Ferguson, 1955; Richards, 1965; Hepburn, 1969; Richards and Richards, 1969).

6.9 Other functions

There are several functions, which occur relatively seldom in the cuticle protuberances of insects. Long seta of the brush organ of Lepidoptera serve for pheromone dispersal (Seibt et al., 1972). Cuticle outgrowths can be a part of a sensory organ. In addition to the typical trichoid sensilla, there are microtrichia-covered areas on the hair shaft of insect mechanoreceptors. These can influence the force range detected by the receptor. Additionally, rough areas on the hair base may effect hair motion. The shape of bees' trichoid sensilla corresponds to the shape of the microsculpture of flower petals (Kevan and Lane, 1985). This «lock-and-key mechanism» is responsible for the texture recognition of the host plant.

Ommatidia grating are anti-reflective structures on the eyes of insects, especially those which are nocturnally active (Fig. 3.8 B). These protuberances are very small microtrichia (200 nm in diameter), which increase visual efficiency through decreased surface reflection in their density, and increased photon capture for a given stimulus condition (Parker et al., 1998). Such a grating is particularly useful on a curved corneal surface, as it would increase the transmission of incident light through the cornea, compared with a smooth surface. A comparable anti-reflector is employed on solar absorbers. According to the so-called «lotus-effect», newly claimed for many plant surfaces (Barthlott and Neinhuis, 1997; 1998a), hydrophobic surfaces with a microsculpture ranging from 0.1 to 1.0 μm are better protected from contamination than smooth surfaces. A similar self-cleaning function can also be suggested for ommatidia grating.

Chapter 4

Principles of cuticular attachment in Arthropoda

This chapter is a review, classifying attachment systems according to the attachment principle and their functional loads. Most of the information given is about specimens of insects. However, taxa of other arthropod groups are also considered here, since it is impossible to give an overview on principles of cuticular attachment without a broader systematic background. Additional information on this topic can be found in a review by W. Nachtigall (1974).

A functional or technological system is a set of structures (organs) which are dynamically interconnected and adapted to one, or several, complex functions (Dlussky and Fedoseeva, 1988). Attachment devices are functional systems, the purpose of which is either temporary or permanent attachment of an organism to the substrate surface, to another organism, or temporary interconnection of body parts within an organism. Their design varies enormously and is subject to different functional loads (Nachtigall, 1974). There is no doubt that many functional solutions have evolved independently in different lineages.

Almost all species of Arthropoda are supplied with diverse attachment devices, and the morphology depends on the species biology. The evolutionary background, and the habits influence the specific composition of attachment systems in each particular species. Since cuticle and its derivatives play a crucial role in the design of attachment devices in the Arthropoda, these systems are referred to as *cuticular attachment systems*. There are eight fundamental classes of fixation principles: (1) hooks, (2) lock or snap, (3) clamp, (4) spacer, (5) sucker, (6) expansion anchor, (7) adhesive secretions, and (8) friction (Fig. 4.1). However, different combinations of these principles also occur in existing attachment structures.

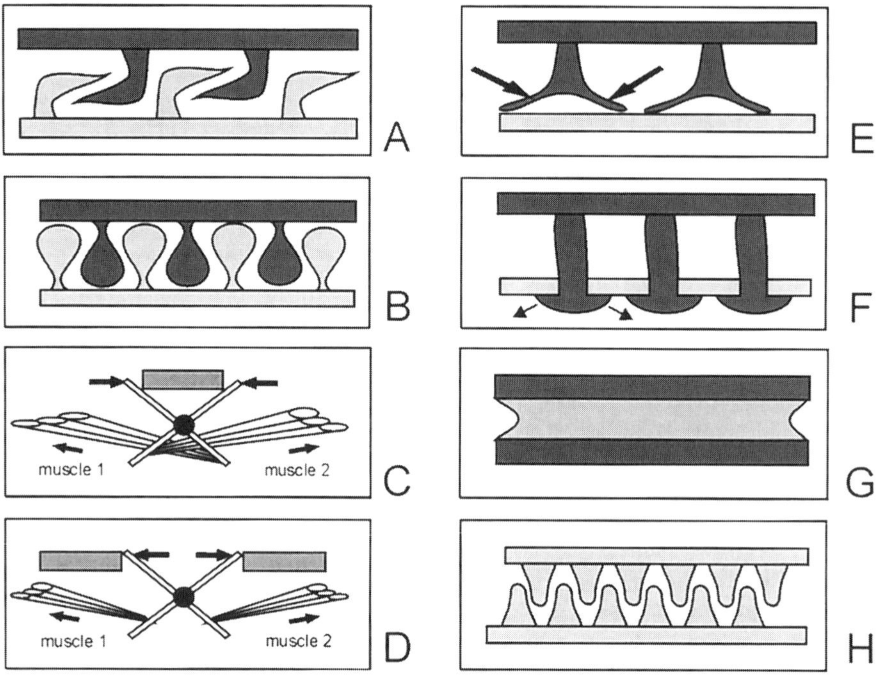

Fig. 4.1. Eight fundamental classes of fixation principles: hooks (A), lock or snap (B), clamp (C), spacer (D), sucker (E), expansion anchor (F), adhesive secretions (G), friction (H).

Each of the eight sections below contains information about attachment systems based on one of these design principles. Within each section, the information is presented in subsections according to functional loads of these devices (e.g. locomotion, attachment to the host, copulation etc.). Attachment of many parasitic insects is comparable with attachment to the substratum, but functional systems of parasitic arthropods are often specialised, to such a degree that a treatment in separate subsections appears appropriate.

1. HOOK

Hooks are a wide-spread attachment principle among arthropods. This occurs mainly in systems adapted for long-term attachment. Numerous specialised, hook-like structures are found in parasitic arthropods, adapted for attachment to particular surfaces of the host body.

1.1 Hooking to the substratum

Terrestrial locomotion is always connected with the attachment of extremities to the substratum. The most common example of the hook-like attachment device is the tarsal claw, which is used to interlock with surface texture during locomotion. Prolegs of butterfly caterpillars bear hooks, or crochets, surrounding the proleg sucker. These hooks are acanthae, because they originate from single epidermal cells (Barbier, 1985). They are not homologous to imaginal claws, but serve a similar function. There are several types of hook arrangements surrounding the sucker. In different species, multiserial, uniordinal, and triordinal arrangements of hooks around the sucker exist (Nielsen and Common, 1991; Hasenfuss, 1999) (Fig. 4.2).

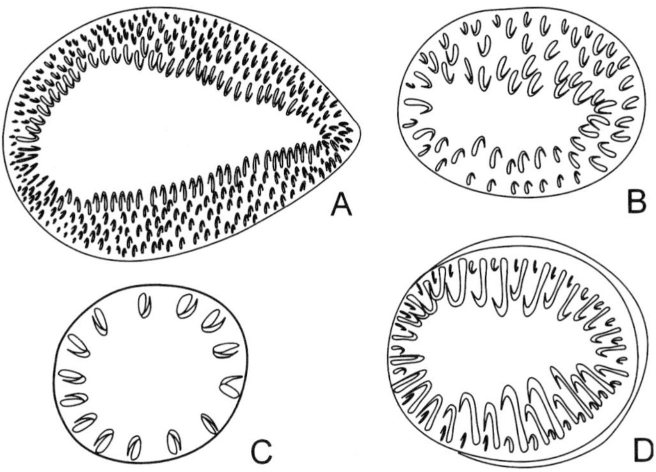

Fig. 4.2. Several examples of hook arrangement on the abdominal prolegs of larval Lepidoptera. A, multiserial (Hepialidae); B, multiserial (Yponomeutidae); C, uniordinal circle (Carposinidae); D, triordinal circle (Pyralidae).

The tip of the tarsus in orb web spiders bears two curved claws, which usually have a comb-like serrate edge (Foelix, 1982) (Fig. 4.3). The hooked design of these claws enables quick grasping onto threads during locomotion within the web. An additional (middle) claw is present in some spiders (*Trionycha*). Serrate bristles located around the claws play an important role in thread interlocking (Fig. 4.3 B, C). Spider claws are controlled by two antagonistic muscles, the *M. levator praetarsi* and the *M. depressor pretarsi*, whereas only one muscle is present in insects.

Posterior abdominal tergites of the pupae of *Thyridopteryx ephemeraeformis* (Lepidoptera, Psychidae), bear rows of anteriorly oriented spines (Neal, 1982). These fine, hook-like outgrowths are sloped at an angle of 60° to the supporting surface. Male pupae bear additional spines on abdominal segments II-VI. These structures prevent pupa from sliding out of the cocoon. In representatives of other lepidopteran families, the spines are mainly posteriorly directed and have a wide variety of shapes from one species to another (Dolynskaja, 1993). In some cases, these structures are involved in the attachment of the pupa to the supporting substratum. They may play an important role in preventing falling of the suspended pupa, during moulting. Similar hook-like structures are known in pupae from several Trichoptera families (Malicky, 1999).

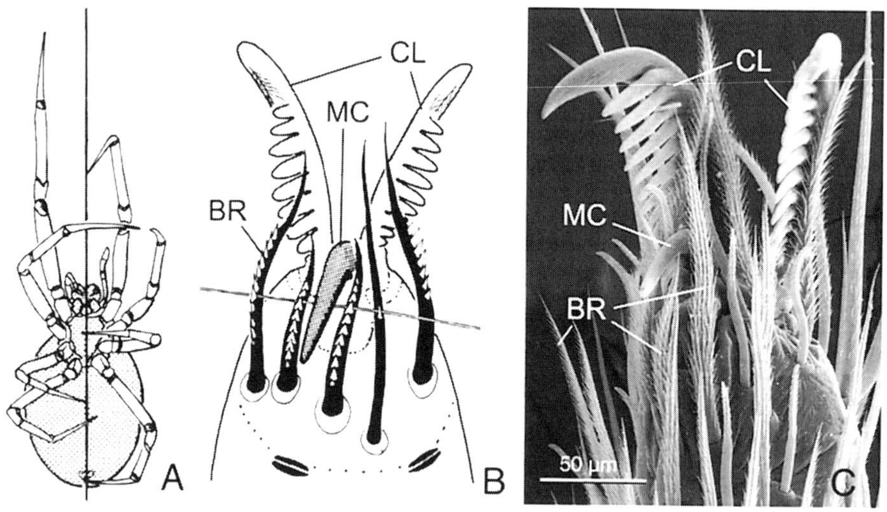

Fig. 4.3. A. The orb web spider *Zygiella* (Araneidae) climbing a vertical thread. B. Interaction of the middle hook and serrated bristles of the pretarsus during the grasping of a thread. C. Pretarsus of the spider *Oxyopes heterophthalmus* (Oxyopidae). BR, serrated bristles; CL, main claws; MC, middle claw. A, B. From Foelix (1982) with permission of Georg Thieme Verlag.

The crustacean species *Latonura rectirostris* (Cladocera, Macrothricidae) uses hook-like attachment structures to hold onto the surface of water plants. Postabdominal armour plates with a system of strong, distally elongated spines serve as hooking devices on soft substrata. Additionally, lancet-like setae on the ventral edge of the carapace, are present as accessory attachment devices.

The attachment of particles to an animal body is a special case of substrate attachment. This phenomenon is known in several arthropod taxa that use camouflage to hide from predators, or to ambush their prey. In many cases, there are specialised hook-like structures to attach different materials to the body surface. For example, the crab *Loxorhynchus crispatus,* densely decorates its carapace with algae, pieces of sponges and bryozoans, hooking them with specialised setae (Wicksten, 1978).

1.2 Animal associations: phcresy, parasitism, predation

A huge variety of hook-like attachment devices has been described for mites (Acari). A number of mesostigmatan mites use the mobility of other arthropods for dispersal (Andreev, 1988). Animal dispersal behaviour using other animals is called *phoresy*. Hooked attachment organs have convergently evolved in many species of phoretic mites (Hunter and Rosario, 1988). An even wider spectrum of hooked attachment devices is present in parasite mites. For example, nymphs of demodicid mites, parasitising on rodents, use scale-like hooking structures on their ventral surface to attach themselves to the host-hairs (Bregetova et al., 1955).

A specific behaviour allows specimens of the mite genus *Michaelicus,* parasitising on the feathers of water birds, to attach themselves to bird feathers. They raise one foreleg at an angle to the body axis, and extend their hook-like tarsus between the barbules (Dubinin, 1951). This kind of attachment behaviour is reflected in the asymmetric design of the legs. An additional seta is present at the base of the left tarsus of *M. javanicus*. This strong, thickened and curved seta works as an additional hook, while the mite is holding onto the host.

Similar attachment systems, and related behaviour is described for the mite species *Alloptes titanopedes,* and *Sulanyssus caput-medusae*. The tarsus of the left leg of the former species is transformed into a tightly compressed, ventrally directed, hook-like structure with two teeth-like structures at its base, and close to the tip. Two cone-shaped outgrowths are present on the ventral side of the femur and trochanter. A strong hook is formed by the trochanteral outgrowth, together with an additional thick and hard plate. While attaching, the mite interlocks itself onto the bird feather between the barbules, using the structures mentioned. The size of these hooks corresponds to the distance between the neighbouring barbules of the feather. Interestingly, hook-like attachment devices exist only in mites parasitising on the stiff part of the feather. Mites living in the softer parts of the feather, or in the feather itself, have evolved attachment systems of the clamp-type (see below).

Larvae of meloid beetles use hook-like claws to attach themselves to setae on the body of solitary bees. Small parasitic wasps of the genus *Rielia* (Sceleonidae) use a similar system to attach themselves to mantids (Lopatin, 1980).

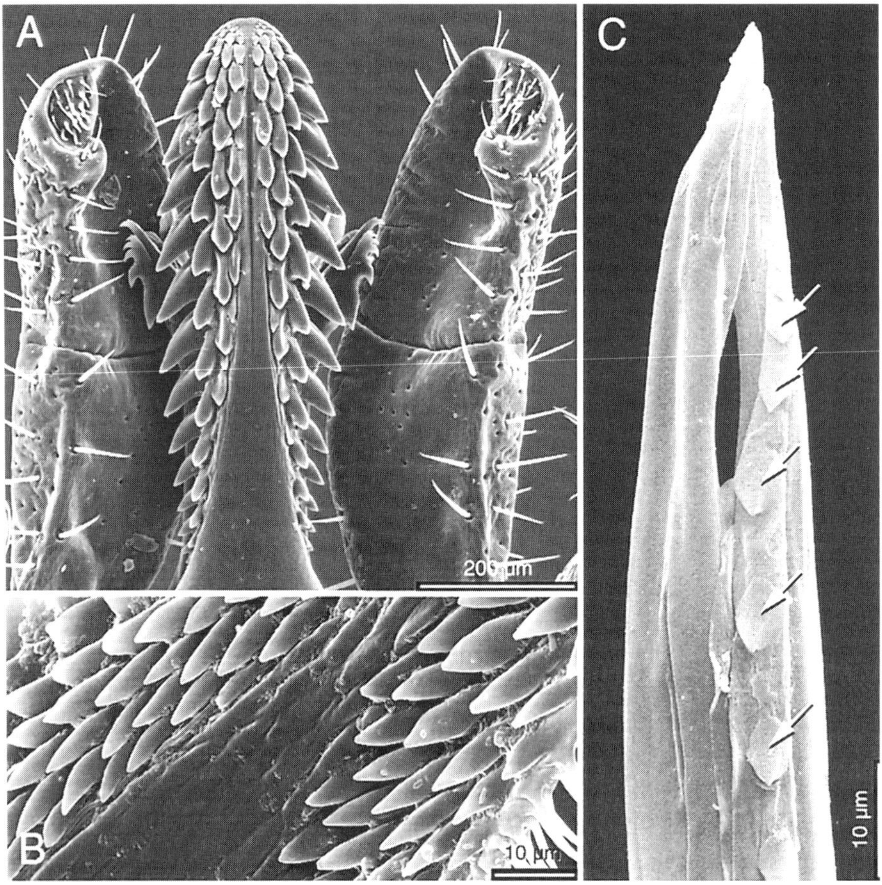

Fig. 4.4. Hook-like structures of the mouth-parts of different arthropods. A, Mite *Ixodes persulcatus* (Acari, Ixodidae), ventral aspect of the gnathosoma; B, *Tabanus bovinus* (Diptera, Tabanidae); C, *Aedes egypti* (Diptera, Culicidae); anchoring hooks are indicated by arrows. A, C, courtesy of Jürgen Berger, MPI of Developmental Biology, Tübingen.

Hook-like devices widely exist in parasitic copepod crustaceans. Intensely modified hooks, which resemble closed locks, are present in some representatives of Copepoda parasitica. These devices enclose suitable host structures from all sides, and therefore guarantee optimised attachment.

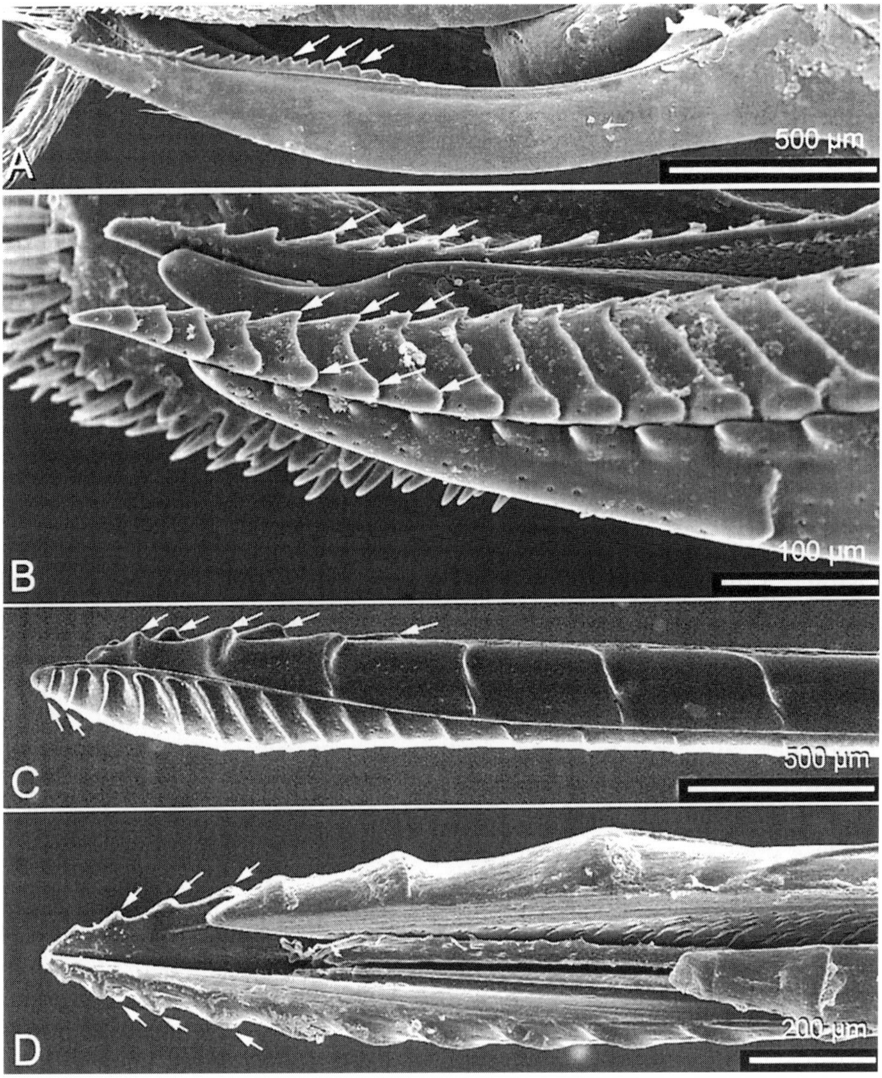

Fig. 4.5. Hook-like structures (arrows) of the insect ovipositor. A, B. Damselfly *Calopteryx splendens* (Zygoptera, Calopterygidae). C, D. Sawfly *Urocerus gigas* (Hymenoptera, Siricidae).

For example, *Shiinoa occlusa* uses such hook-like interlocking to attach itself to the gill bow of fish (Kabata, 1968). This parasite can freely move about the bow axis, whereas the hook-like interlocking of *Schistobranchia ramosa* contains two additional protuberances, which result in limited mobility of the parasite (Kabata and Cousens, 1972).

1.3 Hooking within biological tissues

Parasitic representatives of different groups of arthropods, such as copepod crustaceans, ixodid ticks, or dipteran insects have developed similar hook-like structures to fix their mouth-parts into the host tissue (Fig. 4.4). Ixodid gnathocoxae were transformed into a medially fused, immobile structure, called hypostom (Bregetova et al., 1955). The hypostom is armoured with sloped, teeth-like protuberances (Fig. 4.4 A). In hydracarid mites, the chelae are also armoured with proximally oriented, sharp teeth. Their function is to attach and hold a mite in the host tissues (Mitchell, 1962).

To attach themselves into the host tissues, the parasitic copepod *Hatschekia pseudohippoglossi* uses terminal hooks from the second antennae. In another copepod species, *Trebius clidodermi*, analogous structures originate from the maxillipeds. The terminal hooks of the second antennae in *Achtheinus oblongatus,* are additionally armoured with two rows of spines. This results in a harpoon-like design of the terminal hooks (Titar, unpublished data).

The insect ovipositor is often designed as a penetrating apparatus. However, penetrating is usually followed by hooking into the substratum or often into biological tissues. Many representatives of Odonata and cicadas, which deposit eggs into living plant tissues, or rotten wood, bear on their ovipositor valvulae, hook-like outgrowths to interlock with the plant material (Müller, 1941; Gorb, 1994) (Fig. 4.5).

The sting apparatus in females of Aculeata (Hymenoptera) is derived from an endophytic ovipositor (Snodgrass, 1935). The structure of the sting apparatus is also adapted, not only to penetrate living tissue, but often to attach to it as well. In this respect, several hook-like structures of bees' and wasps' stinging devices are analogous to penetrating-sucking mouth-parts.

A structural study of the sting apparatus in 246 species of 120 genera of sphecid wasps, has revealed that the sting shape correlates with the mobility of the most common species of prey (Radovic, 1985). The quicker the usual prey of the species is, the more curved the sting is. An intensely curved sting is present in species specialised in very agile prey, whereas the sting is more or less straight in species hunting slow moving animals, such as caterpillars or certain beetle species. Female sawflies (Hymenoptera, «Symphyta») drill into wood by using a highly-specialised ovipositor with hooked outgrowths in the apical region (Smith, 1972) (Fig. 4.5 C, D).

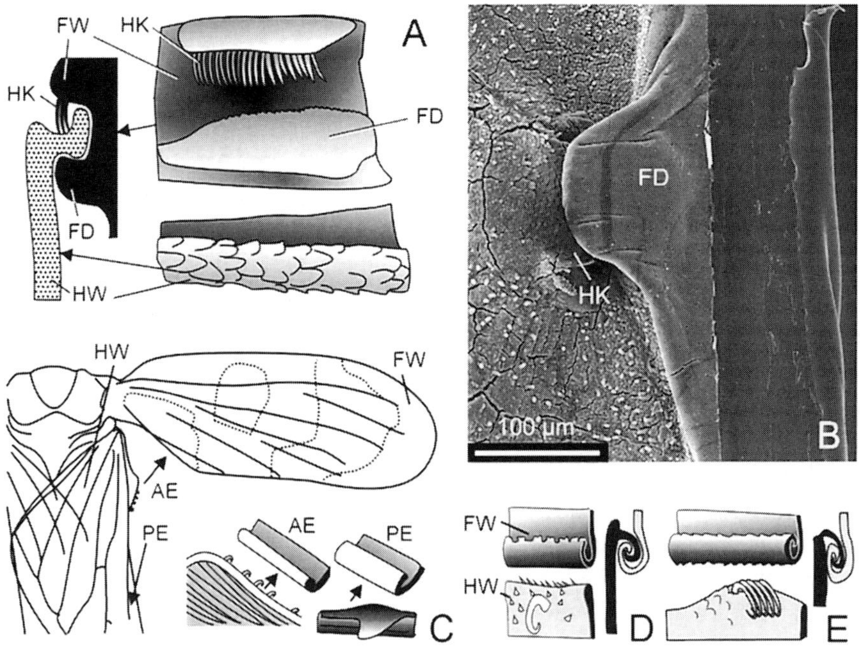

Fig. 4.6. Wing-interlocking devices in Heteroptera and Auchenorrhyncha. A. Pentatomid bug, *Graphosoma italicum* (Heteroptera). B. Forewing of the nepid bug, *Ranatra linearis* (Heteroptera). C. Anterior and posterior fastening-elements in the cicada *Cercopis vulnerata* (Auchenorrhyncha, Cercopidae). D. *Psylla buxi* (Auchenorrhyncha, Psyllidae). E. *Drepanosiphum platanoides* (Auchenorrhyncha, Aphididae). AE, anterior fastening-element; FW, forewing; HW, hindwing; HK, hooks; PE, posterior fastening-element.

1.4 Attachment during copulation

Organs involved in the attachment of males and females during copulation are sexually dimorphic. The attachment structures in males of the bug *Harpocera thoracica* (Heteroptera, Miridae) are curved, hook-like setae located on the surface of the antenna (Stork, 1981). The setae are surrounded by numerous tiny microtrichia, presumably increasing friction when in contact with the female surface. The female prothorax bears scattered setae and a fine reticular surface sculpture.

1.5 Interlocking of body parts

Insect wings can be locked onto the thorax and/or abdomen (*wing-locking devices*). Such systems have been reported for Dermaptera, Coleoptera, Heteroptera, and Hymenoptera. They are usually based on the principles of snap, and/or friction (see also Chapter 8). Only the resting wings of thysanopterans are hooked by rows of hook-like setae, located at the abdominal margin (Shvanvich, 1949).

Insects with functional diptery, when wings on one side of the body are interlocked in flight, have additional devices to attach hindwings to forewings. However, this attachment is dynamic because it allows some mobility in the contact area. These systems will later be referred to as *wing-interlocking devices*.

The evolutionary path of the insect flight apparatus evolved several times, convergently to a functionally dipterous system. Simultaneously, corresponding wing-interlocking devices have appeared to keep the hindwings coupled to the forewings during flight. Wing-interlocking devices occur in representatives of several insect orders: Heteroptera, Auchenorrhyncha, Hymenoptera, Thysanoptera, Trichoptera, and Lepidoptera. These systems are usually a combination of several design principles. However, most of them contain hook-like cuticle outgrowths, or folded wing areas, which interlock with counterparts of the body, according to the hook principle.

A detailed study on the morphology and evolution of wing-interlocking devices has been done on cicadas (Auchenorrhyncha, Cicadina) (Schneider and Schill, 1978). There are two parts of the wing arrester in Cicadina, the primary and the secondary. The primary one consists of the forewing fold and its corresponding fold in the anterior part of the hindwing, where veins C and ScP are fused. The secondary one consists of a single hook, or a series of hook-like spines, located proximad to the main lobe of the anterior hindwing margin. These hooks interlock with the posterior margin of the clavus, or sometimes with other parts of it (Fig. 4.6 C) (Weber, 1930). A similar mechanism has been described in bugs (Heteroptera) (Emeljanov, 1987). However, it differs to some extent from that occurring in cicadas. In the forewing, there is an additional comb consisting of a dense row of short, posteriorly directed setae (Fig. 4.6 A, B).

The anterior margin of the hindwing of *Psylla buxi* bears single, curved setae, which interlock with the S-shaped fold of the posterior margin of the forewing. Additionally, the costal margin of the hindwing is covered with a row of tiny acanthae, functionally corresponding to the surface profile of the S-shaped fold (Fig. 4.6 D). Presumably, such a frictional mechanism prevents wing-sliding in the interlocked position (Weber, 1930).

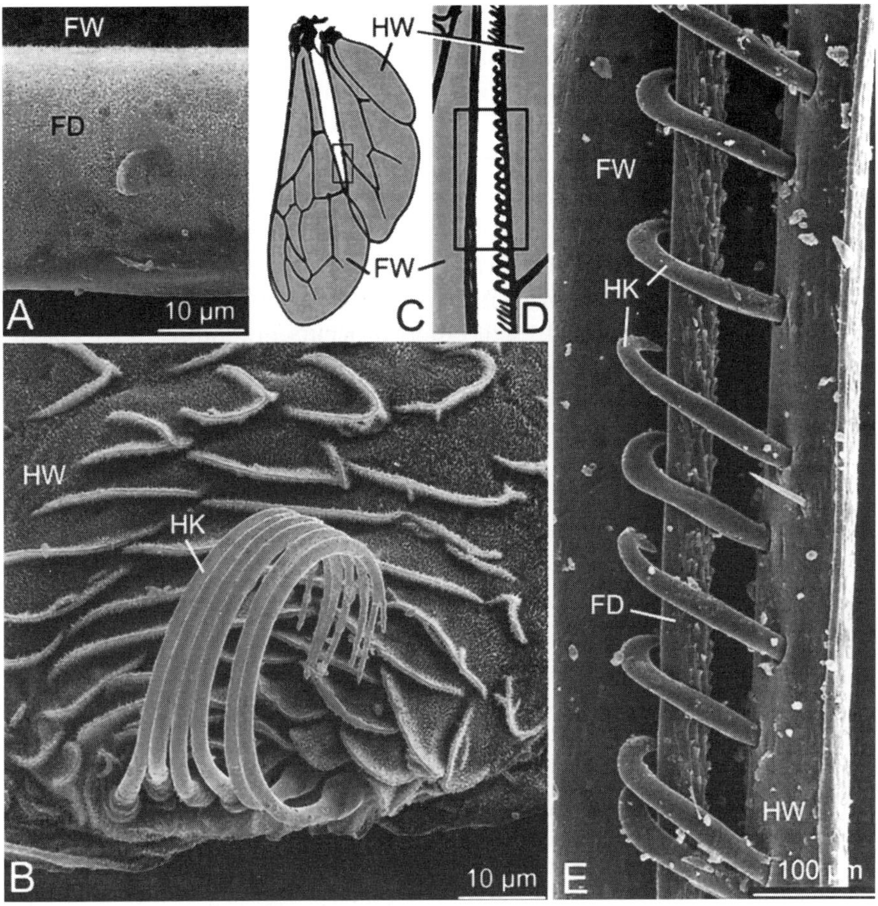

Fig. 4.7. Hook-like wing inter-locking devices in Aphididae (Auchenorrhyncha) (A, B) and Cimbicidae (Hymenoptera) (C-E). A. *Aphis* sp., forewing fold corresponding to hooks of the hind wing. B. same, the hindwing. C-D. Diagram of the wing interlock in Hymenoptera. E. Wing interlock of the sawfly *Cimbex femoratus*. FD, fold; FW, forewing; HK, hooks; HW, hindwing.

Interlocks between the fore and hindwings in aphids (Auchenorrhyncha) are completely different from those existing in closely related Cicadina, or Heteroptera. They are similar to those in Hymenoptera. The antero-dorsal margin of the hindwings bears 2-6 intensely-curved hooks, whereas the posterior margin of the forewings bears an S-shaped fold (Schneider, 1981) (Figs. 4.6 E; 4.7 A, B). If both structures are interlocked, both wings retain their ability to be shifted along each other, and rotated at some angle about an axis, oriented parallel to the costal margin of the hindwing.

In representatives of Hymenoptera, the wing-interlocking mechanism consists of a row of curved hooks on the hindwing and an S-shaped fold of the forewing (Fig. 4.7 C-E). It is likely that stability of attachment between forewings and hindwings has an effect on the flight performance. It has been shown that foraging flight range in the alfalfa leafcutting bee, *Megachile femorata* (Megachilidae), correlates with the number of wing hooks (Abrol, 1986). There is an additional mechanism in parasitic Hymenoptera to interlock both forewings with each other. It occurs in representatives of the families Encyrtidae and Aphelinidae, which parasitise on aphids. A presumed function of such a mechanism is to prevent wing contamination by aphid honey-dew (Hennesey, 1981).

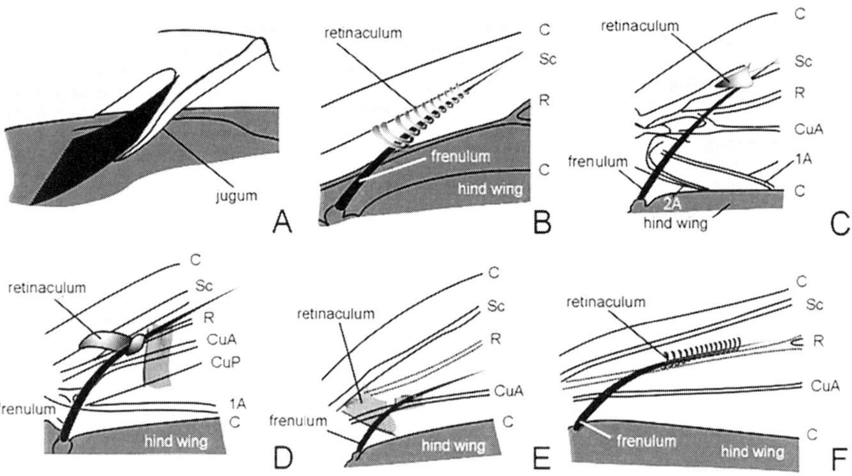

Fig. 4.8. Wing locking devices in Lepidoptera. Ventral aspect showing retinaculum and frenulum. A. *Hepialus sp.* (Hepialidae), jugate type. B-F. Frenate type. B. *Pectinivalva*, male, Nepticulidae. C. *Barea*, male, Oecophoridae. D. *Uresiphita*, male, Pyralidae. E. *Barea*, female, Oecophoridae. F. *Phthorimaea*, female, Gelechiidae.

A variety of hook-like, wing-interlocking mechanisms has been described in representatives of Lepidoptera. There are at least four different types of these systems. The two main ones are *frenate* and *jugate* (Shvanvich, 1949; Bey-Bienko, 1966), with two modifications called *jugo-frenate* and *amplexifom* (Obenberger, 1964; Westheide and Rieger, 1996) (Fig. 4.8 A).

The jugate type exists in all taxa of Hepialoidea, which are often called Jugatae, or Homoneura. The jugate mechanism consists of an outgrowth from the posterior margin of the forewing, called jugum. The jugum hooks to the costal vein of the anterior margin of the hindwing, and in this way connects both wings. The jugo-frenate type is more specialised. The anterior margin of the hindwing bears a group of strong setae. These setae are interlocked between the jugum and the ventral surface of the forewing. This mechanism occurs in basal lineages of Lepidoptera, such as Micropterygidae and Eriocraniida (jugatae group).

The term *frenate* is derived from *frenulum*, a group of 20 thin, non-innervated, strong setae, located at the base of the costal vein of the hind wing (Nielsen and Common, 1991) (Fig. 4.8 B-F). Single setae in this group are not fused to each other. The frenulum exists in both sexes. The corresponding structure is a multicellular seta, which has probably evolved from a group of single setae (Richards, 1981). In some cases, frenulum attaches to so called *retinaculum*, a group of setae located at the ventral surface of the forewing. Both frenulum and retinaculum are sexually dimorphic. The male retinaculum is usually called *retinaculum subcostale*, and the frenulum - *frenulum simplex*. The female retinaculum is called *retinaculum subdorsale,* whereas the frenulum - *frenulum multiplicatum*. An additional hooking structure in the forewing is called *cteniculum,* a group of setae on the posterior margin of the forewing. In the species of the Bombycoidea, the cteniculum is reduced (Obenberger, 1964). In the pyralid moth *Pycnarmon caberalis*, an unusual modification of the frenulum has been described. The male frenulum is extremely long with a clavate apical extension on its tip, whereas females have an unmodified frenulum structure. Interestingly, such a frenulum does not exist in other species of the genus *Pycnarmon,* or other Lepidoptera species (Mathew, 1984).

The amplexiform type of wing-interlocking mechanism is known from some species of the Lasiocampidae, Bombycidae, and Saturniidae. This type lacks jugum, frenulum, and retinaculum, and both wings are mechanically synchronised in flight. This synchronisation is due to a widened costal margin of the hindwing, which is pressed against the posterior margin of the forewing. This contact seems to be essential for wing synchronisation. In fact, this type of wing-interlock is not really a hooking mechanism. All representatives of Rhopalocera have a similar morphological basis of functional diptery. Usually, the amplexiform type is generally accepted to be a kind of frenate type, where both the frenulum and retinaculum have been lost. Males of the tropical genus *Euschemon* have a typical frenate wing-interlocking mechanism, whereas the females posses the amplexiform type.

It is likely that a specialised wing-interlocking device is absent in the groundplan of Lepidoptera. It has been hypothesised that the functional quadruptery was typical for ancestral groups of Lepidoptera, and that functional diptery first appeared in Glossata (Kozlov, 1988). However, since wing-interlocking devices in Trichoptera, which is the sistergroup of Lepidoptera, are similar to the jugate type (Obenberger, 1964), it is likely to assume that jugate type is a groundplan condition for Trichoptera + Lepidoptera.

Morphology of the wing-interlocking mechanism of the fossil *Parasabatinica aftimacrai,* is similar to that of extant micropterygids. Presumably, this mechanism did not provide a mechanical connection of the wings during wing movements, also, frenulum setae initially served as sensory organs.

The hook mechanism is usually comprised of two complementary surfaces. These surfaces are not necessarily mirrored copies of each other, but some dependence on the corresponding surface does exist. If both surfaces bear hooks (wing-interlock), their dimensions are usually predefined in order to optimise attachment, and the probability of attachment as well. When only one surface bears real hooks, they could only attach efficiently, to a particular range of textures (tarsal claws, hooks of phoretic and parasitising animals). The hook design can range from unicellular acanthae and multicellular setae, to spines and cuticular folds (Richards and Richards, 1979).

2. SNAP OR LOCK-AND-KEY PRINCIPLE

This attachment principle includes systems with co-opted surface profiles: outgrowth and depression. In addition, both surfaces can be covered with tiny cuticle protuberances or depressions. In other words, the corresponding surfaces in these snap systems, correspond to each other as a lock-and-key.

2.1 Attachment during copulation

The snap principle is widely represented in organ structures related to copulation. Males of Ephemeroptera (McCafferty and Bloodgood, 1989) and Odonata attach to females with extremely specialised lock-and-key devices. In Odonata, male anal appendages correspond to depressions and cuticular folds on the rear surface of the female head (Anisoptera), and the mesostigmal plate of the female thorax (Zygoptera). Size and shape of the appendages and co-opted structures of the females, are species-specific (Fig. 4.9). This design of attachment structures dealing with copulation, provides not only safe fixation of the male to the female during copulation, but also serves as a species isolation mechanism. The correspondence of the shape and dimensions of male anal appendages to the female structures has been clearly shown for *Epigomphus quadriceps* (Anisoptera, Gomphidae) (Calvert, 1920), for species of the genera *Enallagma* (Robertson, 1982) and *Nehalennia* (Hilton, 1983) (both Zygoptera, Coenagrionidae), and for some other odonate species (Watson, 1966; Knitter, 1998).

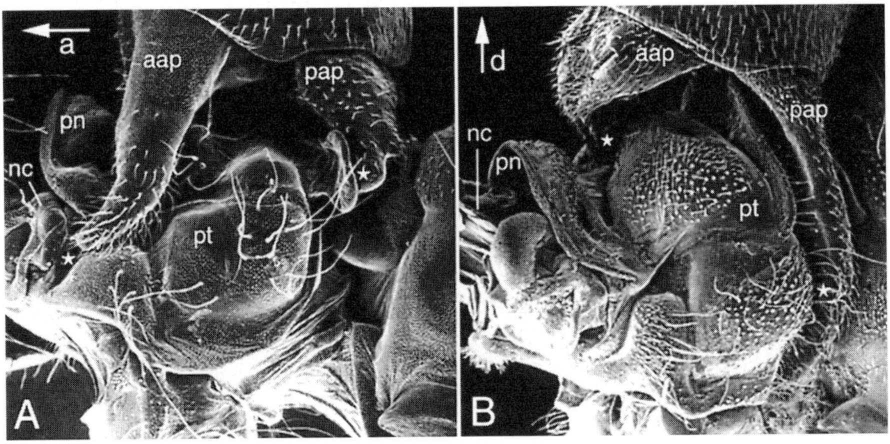

Fig. 4.9. Shock-frozen preparations of the damselfly lock in *Platycnemis latipes* (Platycnemididae) (A), and in *Lestes barbarus* (Lestidae). *Asterisks*, contact regions between female prothorax and male anal appendages. a, anterior direction; aap, anterior (ventral) anal appendages; d, dorsal direction; nc, neck region; pap, posterior (dorsal) anal appendages; pn, pronotum; pt, prothorax. Courtesy of Heiko Knitter, University of Bremen.

Not only abdominal structures can be designed according to the lock-and-key principle. Male forelegs and antennae of the meloid beetle *Cerocoma schaefferi,* are strongly modified to hold the female during copulation (Fig. 4.10) (Matthes, 1970).

The diversity of lock-and-key devices, even among dragonflies, is huge. These snap systems are often supplemented with additional structures designed according to hook, clamp, and friction principles. The lock-and-key principle in the design of insect copulation apparatus, and its biological significance has been reviewed in detail elsewhere (Shapiro and Porter, 1989).

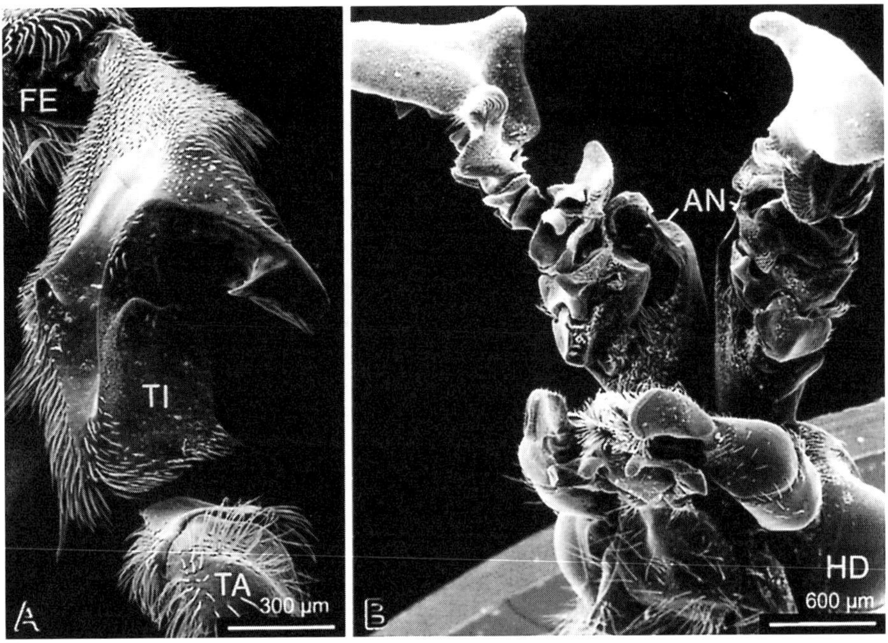

Fig. 4.10. Male beetle, *Cerocoma schaefferi* (Meloidae); details of structures of the foreleg (A) and antennae (B) involved in attachment to female during copulation. AN, antennae; FE, femur; HD, head; TA, tarsus; TI, tibia.

2.2 Interlocking of body parts

To prevent damage and to stabilise the head during the grinding of hard substrata, termite workers arrest the head capsule by the maxillary snap mechanism (Ilychev, 1987). The mandibles begin to work when apical teeth of the maxillae are pressed against the substratum.

During mandible adduction, the head shifts in the posterior direction. The maxillae immobilise the head in this position. The termite is not usually able to cut a plant stem at once, so a lot of repetitive mandible movements are needed. The head, fixed by the maxillary lock-and-key mechanism, enables the termite to apply a higher force to the same spot of the substrate, thus assuring an effective cutting of the stem.

Insects with non-direct flight musculature have so-called bifunctional thoracic muscles. These muscles originate from tergites and attach to the coxa, and/or to the trochanter of the leg. During flight, they are responsible for the wing beat, and when walking they move the legs. In some dipterans and coleopterans, coxae are firmly connected to the thoracic skeleton so that these muscles are exclusively used for flight. Insects with bifunctional flight muscles and mobile coxae have to stabilise their wings during walking. The wing-locking devices are used for this purpose.

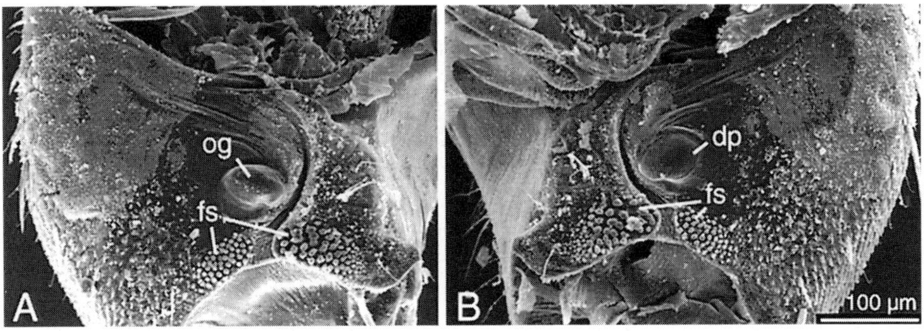

Fig. 4.11. Snap of the hind coxae of the cicada *Euscelis plebeus* (Cicadellidae). A, left. B, right. dp, depression; fs, frictional surfaces; og, outgrowth.

During flight, they have to stabilise their legs to enable the muscles to work as a wing motor. In representatives of Cicadina, these muscles move the hindlegs to the longitudinal body axis and transfer resting wings into flight position (Pringle, 1963). Additionally, cicadas have to lock their coxae in order to synchronise jumping movements of the hindlegs. Different lineages of cicadas have developed several principles to immobilise their coxae (Emeljanov, 1987). In species of Cicadellidae, it is a typical snap, containing outgrowths and depressions on the medial surfaces of the coxae (Fig. 4.11), whereas medial areas of coxae covered with fields of microtrichia, working as frictional surfaces, are present in cercopid cicadas (Chapter 5).

In heteropteran hemielytra and coleopteran elytra, special snap-like devices prevent wing mobility in the resting position. In Coleoptera, medial margins of the elytra are asymmetrical. A fold and a corresponding outgrowth ensure attachment of both elytra (Shvanvich, 1949; Nachtigall, 1974; Gorb, 1998a). Some other structures of elytra fit well into corresponding structures of the thorax and abdomen (Baehr, 1980; Gorb, 1999b). The costal margin of the hemielytron in nepid water bugs, such as *Nepa* and *Ranatra*, bear a groove with a depression in its anterior section (Weber, 1930) (Fig. 4.12). The groove corresponds to the abdominal margin, whereas the depression corresponds to the snap-like outgrowth of the mesothoracic epimeron. All these surfaces are supplemented by frictional surfaces.

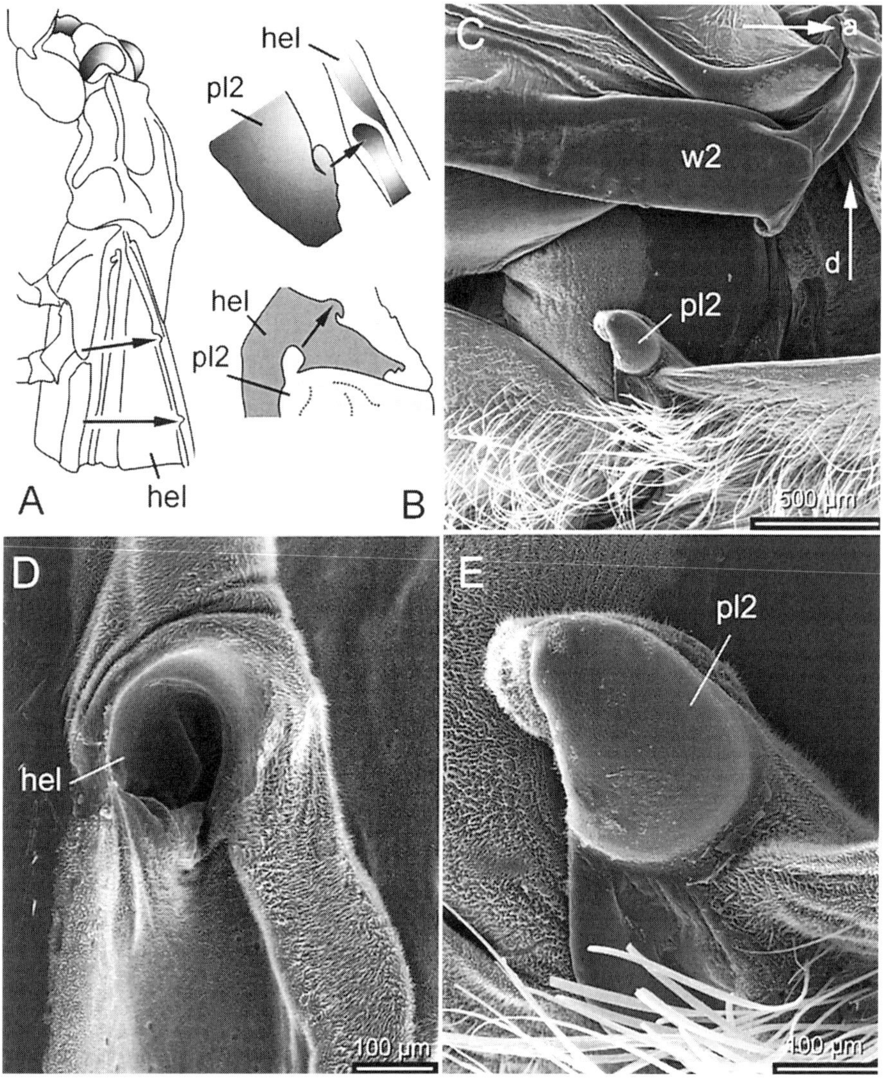

Fig. 4.12. Heteropteran wing-snap in the *Nepa cinerea* (Nepidae) (A, B) and *Notonecta glauca* (Notonectidae) (C-E). a, anterior direction; d, dorsal direction; hel, hemielytra; pl2, pleuron of the mesothorax; w2, hindwing.

Snap-like mechanisms usually provide a long-term fixation and always need two precisely adapted structures. To attach two parts of the snap to each other, or to detach them, muscular force has to be applied.

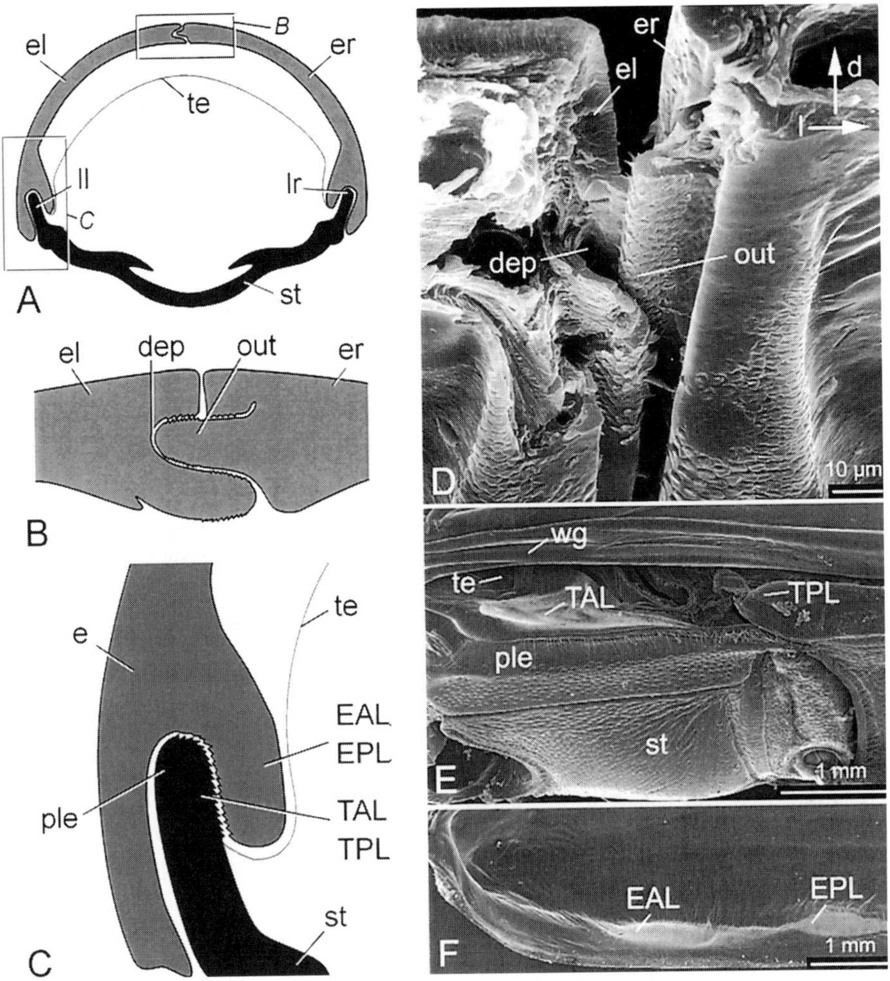

Fig. 4.13. Elytra-locking devices in Coleoptera. A-C. Diagram of the cross section in the mesothoracic region of *Peltodytes caesus* (Haliplidae), with double tongue-and-groove joint between right and left elytron (B), and glossula type lock between lateral margin of elytra and abdominal pleurites (C). D. Double tongue-and-groove joint between the right and left elytra in *Tenebrio molitor* (Tenebrionidae). E, F. Lateral elytra locking devices in *Tribolium castaneum* (Tenebrionidae). d, dorsal direction; dep, depression; EAL, elytra anterior lateral field; e, elytra; el, left elytron; EPL, elytra posterior lateral field; er, right elytra; l, lateral direction; ll, left lock; lr, right lock; out, outgrowth; ple, pleurite; st, sternite; te, tergite; wg, wing.

3. CLAMP

The clamp principle is usually found in complex mechanical systems, which can affix to or hold diverse structures, or substrata, by the use of muscular force. Usually, clamp arms have a particular curvature, and are often supplemented by a variety of coverage. However, they are not adapted to one particular surface. Often, clamp-like systems can also hold non-specialised structures.

3.1 Animal associations

3.1.1 Phoresy and parasitism

Crustacean chelae are the most frequently occurring examples of structures based on the clamp principle. A wide spectrum of chelae, from those specialised for fixation to predatory extremities, is known. It has been suggested for diverse groups of Arthropoda, including insects, that such predatory extremities originated from the same type of extremity previously adapted for fixation (Ass, 1963). On the other hand, initial attachment devices have often developed into even more complex attachment structures in parasitic animals. Chelae at the apex of locomotory limbs are probably a groundplan feature of Euarthropoda (Limulus, Eurypterida, Trilobita). The primary function of these chelae is to obtain suitable food particles and to transfer them into the ventral food rim (Beutel, personal communication). For example, the terminal leg-segments of shrimp of the genus *Atya* (Decapoda), from the seas of South-Eastern Asia, are transformed into structures that function as algae-collecting devices.

The chela of parasitic insects, such as representatives of Anoplura and Mallophaga, is composed of several tarsus segments (Ass, 1963). Therefore, it is actually not a true chela, but a pseudochela. Its immobile part, usually called index, originated from a cluster of setae. The distal part, with the claw, is the mobile part. The working-diameter of this attachment device correlates with the hair-diameter of the host.

The design of the extremities of lice of the genus *Trichodectes* (Echinophthiriidae) is similar, but its mobile parts include the two tarsal segments and the claw, compared with that of the human lice of the genus *Pediculus* (Fig. 4.14 A). *Trichodectes*-species use their forelegs mainly to move about in the hair coverage of their host, whereas *Pediculus* normally use them to attach to the host hairs. In representatives of the genus *Echinothirius*, it is difficult to divide leg structures into the tibia and tarsus. The tibia, and the first and second tarsal segments are fused into a joint segment called tibiotarsus, which contains a mobile claw. The tibial part of the tibiotarsus is transformed into a triangular protuberance, resembling similar structures of other anoplurans. The protuberance bears, in its distal part, four clavate setae. The setae are short, thick, and slightly compressed, resembling little heads on short necks. Three of them are located close to each other, whereas the fourth one is located at some distance from the most proximal one. The tip of the closing claw fits well into the space between this fourth seta and the group of three setae.

Females of parasitic hymenopterans of the Driininae (Bethylidae), such as *Cheloginus carinatus* and *Dicondylus bicolor*, attach themselves to their host insects (Cicadina) through the use of fascinating structures of the terminal-leg segments. The terminal tarsomeres are transformed into a kind of tweezers, with two mobile parts. One tarsal segment is shortened, whereas the next segment is quite large and bears a large, toothed outgrowth. This segment is actually an immobile segment in relation to the next tarsal segment, but it is mobile in relation to the previous segment (Fig. 4.14 B, C). The distally located claw is joined, which is another clamp of these two-levered scissors (Ass, 1963). These complicated transformations of tarsal structures result in a fixatory system with two mobile elements.

The diverse structures of parasitic mites have been modified into clamp-like devices. For example, mites parasitising on rodents have a row of adaptations to attach to the host-hairs. Dispersal stages of the hypopuses of mites from the family Glycyphagidae (*Myacarus hypudaei*), are dispersed by attaching themselves to hairs with the aid of a pair of mobile lobes with two pairs of levers, and a rough surface. This apparatus seems to be ideal for interlocking the host hair between the lobes and the levers.

In mites of the superfamily Listrophoroidea, maxillae are modified in the attaching levers. Margins of maxillae (basis of the hypostom) are transformed into ventrally-curved, wide, thin plates working as clamp-like attachment devices. Segments of the first and second leg pairs in the atopomeline and chirodiscine (genus *Schizocarpus*) mites are also adapted for attachment, whereas the third and fourth leg pairs are involved in attachment in the myocoptine species. Foreleg tarsi of myobiids (Trombidiformes) are very wide and short. The anterior part of the body has a groove-like depression, covered with a plate-like lobe, originating from both the dorsal protuberance of the tarsal margin and the modified claw. The host-hair can be fixed in this canal-like groove (Bregetova et al., 1955).

Many bird-mites use attachment systems based on the clamp principle. These systems often exist in the species parasitising on the softer part of feathers. Some anal and coxal setae (and rarely genital setae as in species of the genus *Compressalges*) transform into special clamp-like attachment structures, which can be actively moved through muscular fibres inserted at their setal base. One pair of such transformed setae exists in the Labidophorinae (Dubinin, 1951). The tibiae of the second pair of legs in the males of *Dogielacarus uncitibia* are hook-like. They can be folded like a jack knife to the inner side of the femur. Two large, triangular teeth are present in this structure. They serve as a stop for the folded tibia. The soft feather is pressed between the tibia and the teeth.

A similar adaptation has been described for the mite *Ingrasiella listrophoroides,* parasitising on ducks of the genus *Dendrocygna.* Clavate tibiae of the fourth pair of legs of the male become more thickened. The distal part of this thickening is divided into two lobes, by a groove. The first lobe (external one) is the tibia itself; the second lobe (internal one) is the ventral protuberance of the tibia. The inside of the internal lobe bears three teeth with their tips directed towards the groove. The hook-like tarsus is at an angle of 90° to the distal part of the tibia. It functions as an accessory fastener of the tibial attachment system. The triangular protuberance is located at the base of the tarsus. When the tarsus is turned, the groove opens, and single barbules of the feather can easily reach into the groove and between tooth-like structures. The mite is safely attached with its hindlegs on the feather, when the tarsus is closed again, thus locking the barbules. Long setae located on the distal part of the tibia, work as gliding structures for the barbules.

In mites of the genus *Analges*, the anterior side of the femur bears a large, ventrally-directed, hook-like outgrowth. It is separated from the surface of the femur by a groove. The internal surfaces of both the tibia and tarsus bear one triangular tooth, each. When the mite attaches itself to the feather, it moves the legs to insert barbules into the groove between the external surface of the tibia and its outgrowths. When the mite locks the barbules in the groove, using its tarsus as a clamp, the parasite attaches to the bird feather.

The clamp-like systems are even more elaborate in parasites living on aquatic birds, because an eminently safe system is required to resist water flow (Dubinin, 1951).

3.1.2 Predation

Devices adapted for prey-capture are often designed as clamp-like structures (Fig. 4.14 D-G). Often they are derivatives of chelae. In fact, fixation is always part of the capturing process. The majority of recent species use their chelae as an organ adapted for both capturing and processing prey. Chelae, and predatory legs in general, have convergently evolved in representatives of many groups of Arthropoda, such as *Squilla* (Crustacea, Stomatopoda), *Mantis* (Insecta, Mantidae), and *Mantispa*

(Insecta, Planipennia) (Obenberger, 1964; Beier, 1968; Frantsevich, 1998). The clamp segment (tibia) flexes during capturing movement, to the basal segment (femur) (Loxton and Nicholls, 1979). With this construction the index is absent, and the clamp just fits into the groove of the basal segment. The groove margin usually bears rows of spines, or setae, to hook captured prey. A similar design has been described in predatory water bugs of Nepidae (Gorb, 1995a).

Among crustaceans, the second maxillipeds in *Squilla mantis* (Stomatopoda) belong to the femoro-tibial class of clamp-like prey-capturing devices (Fig. 4.14 J). In general, this construction resembles the mantid one. However, both constructions differ on some important details. In the mantid system, the tibia moves downward. In *Squilla,* the dactylus moves in the opposite direction (Bremond, 1974). Similar ventro-dorsal movements of predatory limbs have been reported in certain Amphipoda (males of *Cyamus mystiaceti, Caprella equilibra, Phromina sedentaria*) (Ass, 1963).

Aquatic bugs of genera *Belostoma* (Belostomatidae) and *Notonecta* (Notonectidae), capture their prey by latero-medial movements of contralateral legs (Ass, 1963; Gorb, 1995a). Many imaginal dragonflies capture their prey during flight, with highly-specialised legs (Shvanvich, 1949). A modified spatial orientation of coxal axes, and the specialised apparatus of spines on the femur and tibia, are characteristics of this unique capturing device.

The mask in odonate larvae is a specialised type of predatory mouthparts, which is also designed according to the clamp principle. Its most important property is that it retracts by both the set muscles, and an hydraulic mechanism (Olesen, 1972; 1979; Tanaka and Hisada, 1980; Parry, 1983).

Chelicerae of spiders and amblypygids consist of a proximal segment and a second segment, which is immobile in relation to the mobile third segment, often referred to as a fang. (Fig. 4.15 C, D). In the resting position, the fang is enclosed inside a groove in the second segment. When the spider captures its prey, the fangs move out of the grooves and penetrate the prey. Orthognath spiders move their fangs into the sagittal plane, whereas labidognath spiders into the horizontal plane. Presumably, the opposed chelicerae of Labidognatha have the advantage of a larger span, enabling even smaller species to overpower relatively large prey (Foelix, 1982).

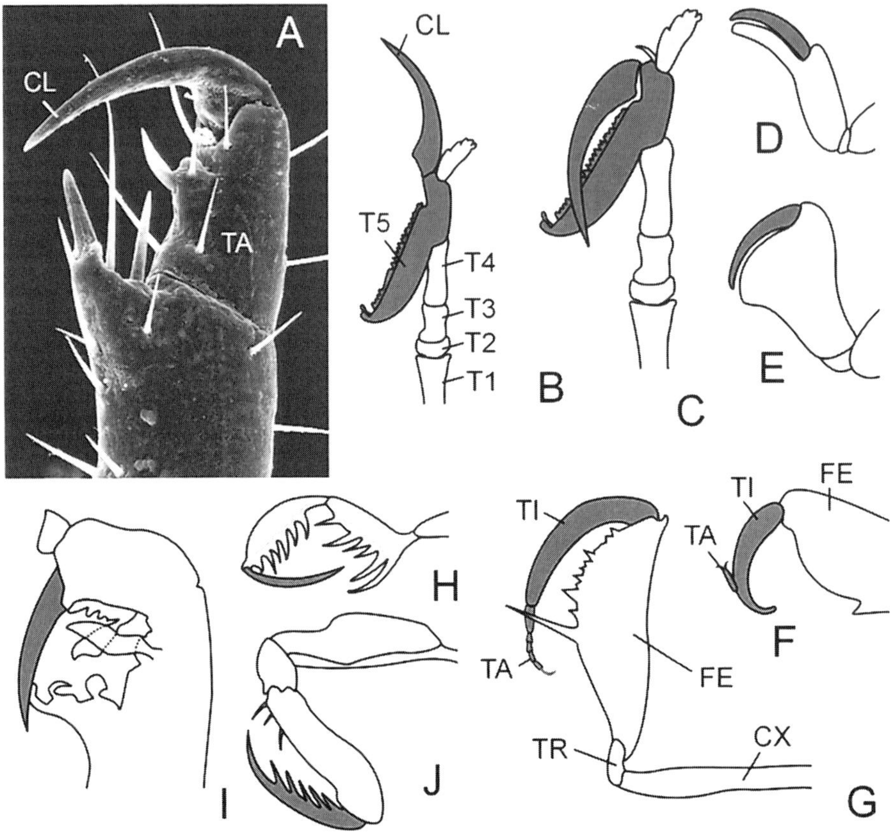

Fig. 4.14. Clamp devices of arthropods. A. Leg of the lice *Pediculus humanus* (Anoplura). B, C. Anterior tarsus of the female of dryinid wasps in an opened (B) and closed (C) condition. D-F. Forelegs of phymatid bugs (Heteroptera, Phymatidae): *Phymata crassipes* (D), *Macrocephalus leucographus* (E), *Carcinocoris binghami* (F). G. Foreleg of *Mantispa sp.* (Planipennia, Mantispidae). H. Terminal claws on the pedipalps of *Obidosus amplochelis* (Opiliones, Gonyleptidae). I. Clasping antenna of *Sminthurus apicalis* male (Collembola). J. Forelegs of *Squilla mantis* (Crustacea, Stomatopoda). CL, claw; FE, femur; T1-T5, tarsomeres; TA, tarsus; TI, tibia; TR, trochanter. A. With permission of J. Berger (MPI of Developmental Biology, Tübingen).

3.2 Attachment during copulation

Clamp-like mechanisms widely exist in functional systems adapted for copulation. Usually, they are species-specific, as that of male mosquitos from the genus *Phlebotomus* (Diptera, Psychodidae) (Ass, 1963), and sexually dimorphic. For example, male chelae of the crustacean *Asellus aquaticus* (Isopoda, Asellidae) are at least twice as large as those of the female. A similar relationship occurs in male forelegs of the lice *Pediculus* (Anoplura, Pediculidae).

Different structures were transformed into «sexual clamps». Males of several species of spiders, such as *Walckenaeria acuminata* (Araneae, Linyphiidae), bear large outgrowths at their prosomata (Foelix, 1982). Females use these outgrowths to support themselves, with their chelicerae, during copulation (Müller and Heimer, 1989). Hindlegs of trichopteran males of the genus *Dipseudopsis* (Polycentropidae) bear species-specific protuberances used to hold the female, during copulation (Obenberger, 1964). Remarkable clamp-like structures related to copulation are known in solitary wasps of the family Sphecidae, such as *Anacrabro fritzi* (Cooper, 1987). It is a modification of the ventral parts of the mesopleurites, which are transformed into clavate protuberances lengthened antero-ventrally. Their shape and curvature are adapted to those of the first segment of the female abdomen. Presumably, these structures aid in male fixation between the metathorax and abdomen of the female. The «sexual clamps» can also derive from antennae. In the males of Collembola, antennae are adapted to grasp female antennae, in order to hold the female during mating (Fig. 4. 14 I).

There are also several examples of «sexual clamps» in the posterior part of the body: (a) appendages of males of copepod crustaceans of Centropagidae (*Centropages typicus*), (b) hindlegs of some chilopods, (c) dermapteran cerci, which are mainly used by males for fighting other males (Briceno and Eberhard, 1988). Males of the water strider *Gerris odontogaster* (Heteroptera, Gerridae) bear, at their abdomina, clamp-like structures derived from two ventral outgrowths of the abdomen (Arnquist, 1988).

Representatives of Acari use clamp-like structures during copulation to hold males and females together. The teleonymphs of bird mites of Pterolichidae (*Cynthiura microdiscus, Xoloptes* spp., *Pseudalloptes* spp.) attach to the male, using a surface depression at the posterior end of the male opisthosoma. The teleonymph fits well into the depression, so that it just uses its large transformed hindlegs to hold itself in the depression. The teleonymphs of the mites from genera *Zachvatkinia, Sokolowiana,* and

Michaelicus hold themselves on the male body with the fourth pair of legs, which bear clamp-like setae.

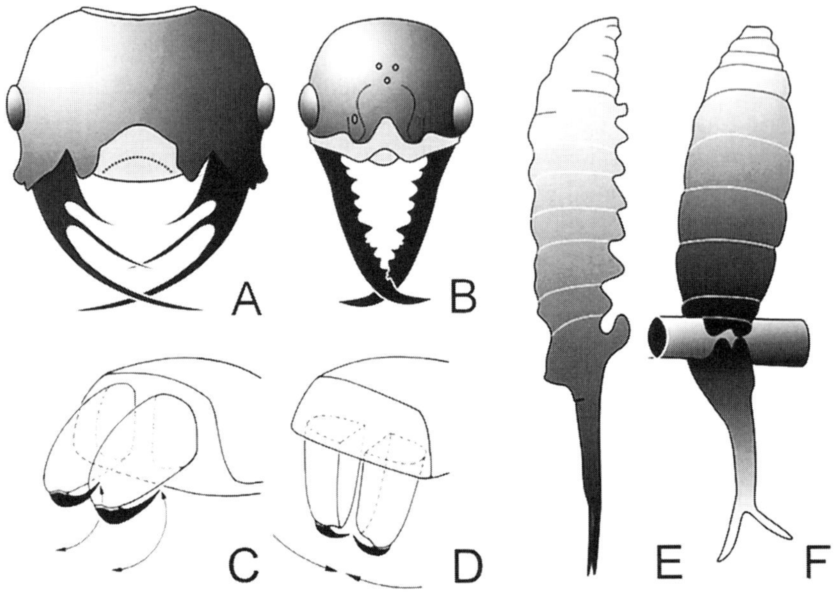

Fig. 4.15. A, B. Mandibles of ants *Myrmecia gulosa* (A) and *Thaumatomyrmex ferox* (B) (Hymenoptera, Formicidae). C, D. Two principal chelicerae designs in Araneae: the orthognath design (C) and the labidognath design (D). E, F. Clamps of ephydrid dipterans. E. Larva of *Ephydra subopacta* (Diptera, Ephydridae). F. Pupa of *E. riparia* (Diptera, Ephydridae). C, D. From Foelix (1982) with permission of Georg Thieme Verlag.

3.3 Holding onto the substratum and holding the substratum

Biting mouth-parts of insects can work as attachment organs designed as a clamp (Fig. 4.15 A, B). For example, cranio-mandibular systems of Hymenoptera can be divided into nine fundamental classes (Dlussky and Fedoseeva, 1988). In at least three of them (sphecoid-type, leptanilloid-type and capturing-type), the attachment function is of primary importance. The sphecoid and leptanilloid are holding types of mouth-parts. The sphecoid-type (Sphecidae, Scolioidea, Andreniidae, Mellitidae, some representatives of Dorylinae) functions as a short-term prey holding-device, and is used prior to stinging. It can also be used for carrying items. The leptanilloid-type (Megachilidae, workers of Leptanillidae, many Dorylinae, some Ponerinae and Myrmicinae) is characterised by an additional 3-5 large needles, which are adapted to hold prey for a longer time, and to manipulate with.

The capturing-type evolved independently in two ant tribes, Dacetini (Myrmicinae) and Odontomachini (Ponerinae). This type is the most complex one. It is characterised by extremely long mandibles, the length of which can surpass the head length. The mandibles bear a few, long, sharp thorns directed almost perpendicularly to the surface, and are controlled by long, powerful muscles with the locking mechanism at their base. Usually, they are widely opened. After the contraction of mandibular muscles, the mandibles remain open at first. The lock can be actively opened by a small trigger-muscle (Gronenberg, 1996a). Lock opening releases a rapid mandible strike. The force created by releasing the lock is the basis for a very unusual jumping mechanism: the ant presses against the substratum and is launched over a distance of several centimetres, when the mandibles close abruptly. (Gronenberg, 1996b; Hölldobler and Wilson, 1990).

Dipteran larvae of the family Ephydridae use split-sleeve clamps to attach themselves to the stems of aquatic plants. A round depression with inwardly directed barrier-hairs along the edge, is present on the ventral side, near the posterior end of the body (Fig. 4.15 E, F). This clamp device is controlled by a muscle. Before pupation the larva clamps a cylindrical plant stem of corresponding diameter (Nachtigall, 1974).

Clamp-like devices are usually supplemented by hooks or rough surfaces. Crustacean chelae bear thorns or spines; mite levers are often covered with hooks; mandibles of some hymenopterans have two pairs of thorns not only to hold the prey, but also to prevent its rotation on the mandibles. There are clamp-like systems which are adapted to one corresponding structure, or to a variety of structures. The linear dimensions of the corresponding structure should not surpass the clamp span, in the opened state. Both geometry and the mechanical properties of the substratum, and its microsculpture, do not play a crucial role. Chelae of insects and crustaceans, and mandibles of many predatory insects, can be included in this group. On the other hand, there are clamp-like attachment systems that are precisely adapted to particular co-opted structures. Usually, these structures are parts of copulatory organs, or serve attachment to particular hosts of parasitic arthropods.

The clamp principle always requires muscular force for attachment. In some crustacean chelae, the muscles belong to the slow type (Costello and Govind, 1984). They contract slowly but produce a large force, and can remain longer in the contracted condition. It can be presumed, however, that predatory chelae adapted for the capture of fast moving prey should be controlled by muscles composed of fast fibres.

4. SPACER

This type of attachment device rarely occurs in biological systems. Such systems are quite similar to the clamp-like mechanisms described above, because muscular force or hydraulics are involved in both principles. In the case of spacers, however, internal forces are used to transfer the system from the free condition into the working condition, and back. Usually, attachment itself does not require any muscular energy.

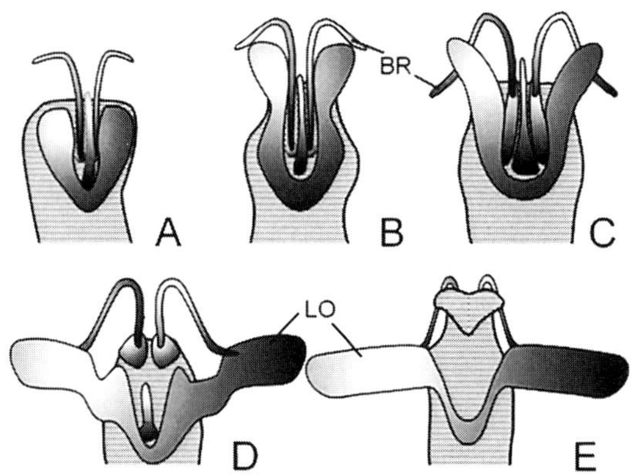

Fig. 4.16. Spacer of the penis of the male *Sympetrum* (Odonata, Libellulidae). BR, bristles; LO, lobes.

The bird mite *Freyanella plataleae,* uses spacer-like structures of specialised shoulder setae to attach itself between the feather barbules (Dubinin, 1951). There is no data about the presence of muscle fibres controlling these setae. The penis of the male libellulid dragonfly is composed of three parts, which can be folded and introduced into the vagina. Increasing pressure unfolds these structures. Lateral lobes and bristles in the unfolded condition anchor the penis firmly inside the vagina (Fig. 4. 16).

In many cases, such systems are supplemented with hooked or rough surfaces to interlock with the supporting substratum, or to increase friction with its surface. The dragonfly head-arresting system is mechanically designed as two spacers (postcervical sclerites) stabilising the head (for detailed description see Chapter 6). In this particular case, the whole system is designed according to several fundamental principles: the spacer is supplemented by hooks, friction surfaces, and/or secretions delivered from epidermal cells in the contact area.

5. SUCKER

In literature, the term sucker is used for a variety of attachment devices, such as pretarsal pads of dipterans and coleopterans. Of course, this is not correct, and also confusing. A sucker, in the proper sense, uses the difference between atmospheric pressure and pressure under the sucker cup. Important properties of the cup are its flexibility, smooth, soft - often damp - surface at its edge, and, in some cases, occurrence of muscles or muscular fibres that are responsible for the generation of low pressure. The sucker principle widely exists in soft-bodied animals, such as plathelminths and molluscs (Nachtigall, 1974). However, numerous examples have also been found among Arthropoda.

5.1 Animal associations: phoresy, parasitism, predation

Some parasitic mites possess intensely developed suction-discs, or lobes, originating from empodium. Additionally, tarsi of the fourth pair of legs in the males of *Analges corvinus*, *Pteronyssus parvinus,* species of the genus *Megninia*, representatives of the families Tyroglyphidae, and Sarcoglyphidae, rodent parasites of the superfamilies Listrophoroidea, Sarcoptoidea, and Gamasoidea bear small suckers representing modified setae (Dubinin, 1951; Bregetova et al., 1955). Tarsal suckers are usually located on stalks, and are therefore called stalked suckers. Mite nymphs of the family Tyroglyphidae, possess an attachment apparatus designed as a round suction-disc, containing 4-5 well developed suckers. There are several additional suckers located closely to the copulatory organs, and often on the first and third coxal fields. Genital suckers are also well known in water mites (Fig. 4.17 B-D).

A suction-disc is also present in phoretic deutonymphs of *Sancassania mycophagus* (Acaridae). It is composed of paired anterior, mesal, and lateral suckers, two pairs of cone-shaped setae, and three pairs of cuticle apodemes (Baker et al., 1987). The lateral suckers originated from the cuticular fold. The mesal and anterior ones are derived from transformed setae. Suckers of heterozerkonid mites resemble pocket-like structures, located lateral to the anus and below the genital lobe. At the base of the sucker, there are cuticular folds with tiny spines. Sucker activity is rendered by a pair of small muscles.

Walking-legs of the mite *Varroa jacobsoni,* are adapted for attachment onto its host, the honey-bee (Akimov et al., 1988; Jelinski, 1988). The leg-suckers are quite large. The tarsus consists of two parts: the basal cuticular stem and the mobile membranous pad. The retracted pad looks like a bilobed sucker. When the pad is protracted, the stem, with its hidden sucker, resembles a cone whose narrow part is connected to the distal end of the pretarsus. The basal stem consists of two lateral plates and one mesal plate. The proximal end of each lateral plate bears sclerotised claw-like structures, the function of which is to support the sucker. On the mesal plate, a long, narrow seam connects the basal stem with the pad. Presumably, this seam is involved in the control of contraction and relaxation of the pad. To prevent mite attachment to the bee surface, beekeepers use dust that contaminates the sucker surface and disables its function (Ramires, 1989).

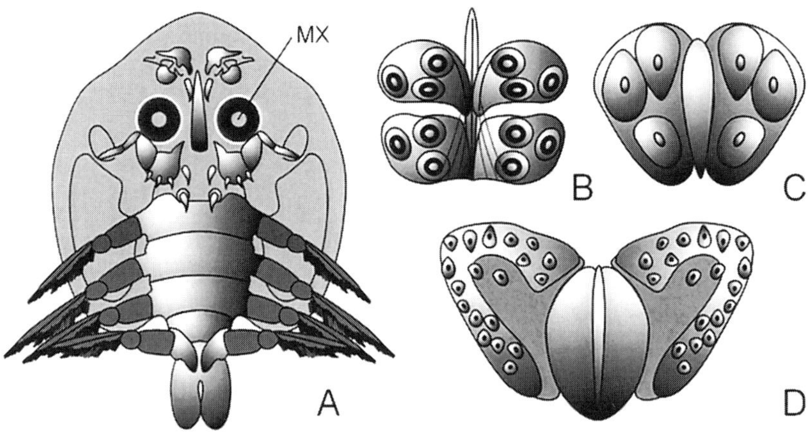

Fig. 4.17. Suckers of parasitic arthropods. A. Crustacean *Argulus foliaceus* (Branchiura). A-C. Genital suckers of water mites (Hydracarina): *Atax grassipes* (B), *Hygrobates longipalpis* (C), *Piona rotunda* (D). MX, maxillae.

The mite species *Poecilochirus necrophori* (Parasitidae), *Macrocheles nataliae* (Macrochelidae), *Iphidosoma fimentarium* (Eviphididae), and *Spinanoetus sp.* (Anoetidae) use beetles of the genus *Nicrophorus* for phoresy. They can safely walk on the insect surface without falling off. All four pairs of legs bear suckers responsible for dynamic attachment-detachment performance during locomotion. Hypopuses of acarid mites of the genus *Spinanoetus,* often concentrate in the subelytral chamber of these beetles, where they attach themselves to the undersurface of the elytra using their suckers (Andreev, 1988).

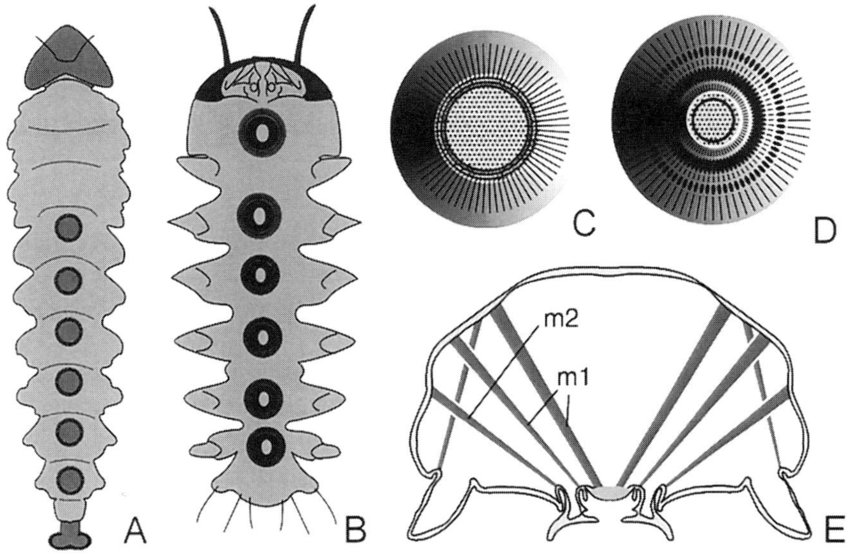

Fig. 4.18. Suckers in larval Diptera. A, C. Larva from genus *Hydropedeticus* (Nematocera, Psychodidae). B, D. Larva of *Blepharocera fasciata* (Nematocera, Blepharoceridae). E. Musculature of the blepharocerid sucker A-D.

Remarkable, bifunctional suckers have been described in water mites. They function as attachment organs, and at the same time are used to suck haemolymph out of the host. A great number of these suckers are usually arranged in several rows around the mouth-parts. Their size greatly varies in most cases. Additionally, they can transform to some extent to improve contact to the host and to prevent water protrusion into the inner part of the sucker, and between mite mouth-parts. Previous studies on the material of the sucker revealed the presence of course fibres in nymphs of some mite species. Presumably, these fibres are responsible for sucker contraction. During such a contraction, the sucker becomes depressed, and its centre protracts. This action results in the decrease of pressure under the cup and attachment, when the sucker is in contact with the surface. The most complex are larval suckers of hydracarid mites of the genus *Eylais*. Each sucker resembles a cup with relatively course walls and a mobile center, which is protracted prior to contact with the substratum. After contact with the host surface, the mobile part of the sucker retracts, providing safe attachment (Tuzovskij, 1987).

Parasitic crustacean *Argulus foliaceus* (Branchiura) uses transformed maxillae on the ventral side of its body to attach to the fish body (Fig. 4.17 A). Fish parasites, belonging to the copepod genus *Trebius,* attach by means of the prothoracic margin, which is transformed into a thin, foil-like structure, so that the whole prothorax functions as a sucker (Titar, unpublished data).

5.2 Attachment during copulation

Probably the most thoroughly studied insect suckers are those located on the palettes of the ventral side of fore tarsi, in males of dytiscine beetles (Fig. 4.19). The palette is formed by the lateral extensions of the second, third, and fourth tarsomeres. They are equipped with modified adhesive setae (Törne, 1910; Blunck, 1912).

The significant correlation between width of the palette and total length of the beetle has been confirmed (Aiken and Khan, 1992). The setae are of the articulo-seta type (Stork, 1980a), and consist of a disc shaped like a shallow bowl, with a shaft attached to its convex surface. Deep folds between the shaft and the disc allow the disc to tilt about any axis, perpendicular to the shaft (Aiken and Khan, 1992). The whole structure is composed of one largest primary, one secondary, and numerous (about 150) small tertiary suckers. The discs of the primary and secondary suckers are reinforced by flexible rods. Such a design suggests that they have originated from a bundle of fused setae (Shvanvich, 1949). Tertiary suckers are actually modified single setae. In *Dytiscus alaskanus,* the diameters of the primary and secondary seta are 0.94 mm and 0.65 mm, respectively. The tertiary setae have a maximum diameter of about 0.05 mm (Aiken and Khan, 1992).

The male attaches with suckers, to the dorsal surface of the female when beginning to mate. The male usually hits the female with his sucker-equipped palettes on the pronotum, or the anterior part of the elytra (Aiken and Khan, 1992). A single palette in males of *D. alaskanus* can generate an attachment force of about 60 mN (primary sucker 22 mN, secondary sucker 15 mN, tertiary suckers altogether 23 mN) (Aiken and Khan, 1992). Actually, palettes of male *Dytiscus* can lift more weight than is required for holding the female's weight (15 mN). However, drag generated by the female can surpass its weight. The male sucker-system is, therefore, designed to overcome these forces.

Males of bird mites of the family Listrophoridae, attach to the females using a pair of large, anal suckers located on both sides of the anus (Dubinin, 1951).

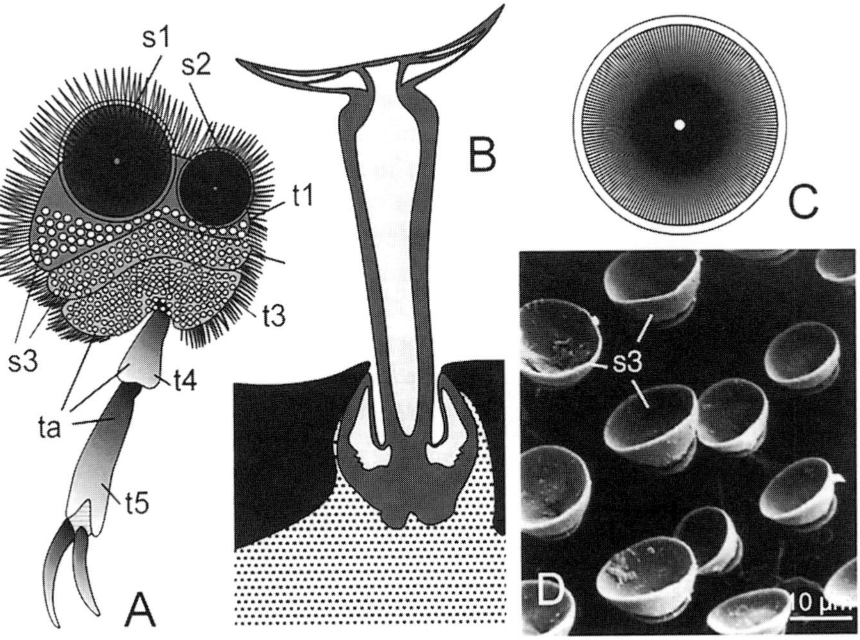

Fig. 4.19. Sucker-bearing tarsus palette of dytiscine males. A. Ventral aspect of the palette. B. Longitudinal section through the primary sucker; C. Setose pattern of the primary sucker; D. Tertiary suckers. s1, primary sucker; s2, secondary sucker; s3, tertiary suckers; ta, t1, t2, t3, t4, t5, tarsomeres.

5.3 Attachment to the substratum

Only representatives of a few insect taxa have true suckers. Larval Diptera of the family Blepharoceridae (Brachycera) live under rocks in river beds. Their suction-discs are located on the ventral side of the body segments (Hofender, 1927; Rietschel, 1961; Hermann et al., 1975) (Fig. 4.18 B, D, E). The disk consists of a membranous pad, a fibrous area, tactile setae, and a so called supporting area. Attachment in water is supplemented by secretion of the sternal gland. A cone-shaped structure on the central area of the sucker undersurface, which is an insertion of strong muscular fibres running at an angle from one another. Fibre contraction shifts the cone structure dorsally. This action causes lower pressure under the sucker. It seems that it is impossible to find any homologous structure to blepharocerid suckers, among Diptera (Shvanvich, 1949).

Larval representatives of the dipteran families Deuterophlebiidae (Brodsky, 1930; Courtney, 1990) and Psychodidae (Nachtigall, 1974) live in similar habitats, and bear suckers on the ventral side (Fig. 4.18 A, C). In deuterophlebiids, the extremities are absent on the three thoracic segments. Seven abdominal segments are widened, and each bears a pair of thick lateral lobes. Each lobe is terminated by a semispherical pad, capable of retraction. The pad is supplemented by several rows of hooks (Shvanvich, 1949).

Prolegs of lepidopteran caterpillars are also true suckers. For locomotion on rough substrata, the rows of hooks surrounding each sucker are ventrally oriented. In such a position, they interlock with the substratum. When the caterpillar walks on a smooth surface, the hooks are turned dorsally, and the suckers attach (Barbier, 1985). One retractor muscle is responsible for the vacuum effect under the sucker (Cox, 1989). The protraction is passively induced by the internal haemolymph pressure (Shvanvich, 1949).

Firefly larvae of the genus *Photuris* (Coleoptera, Lampyridae) bear sucker-like pygopodia (segments) at the end of the abdomen. While resting, they are protracted. During locomotion they can work as small suckers. When the pygopodia are retracted and pressed against the substratum, the larva is attached to the underlying surface. The mobility of the pygopodia is provided by the muscle system and the elastic integument. Histological studies of the pygopodia cuticle revealed the presence of parallel fibrils embedded in the soft matrix material (Domagala and Chiradella, 1984).

Anal lobes of eryophyid mites are transformed into a remarkable sucker, which is not only used for attachment to the substrate, but also for jumping locomotion. The anal lobes are two pocket-like structures at the posterior end of the body, lateral to the anus. Their walls are elastic, thus enabling retraction and protraction of the lobes. Both lobes are mobile and connected to each other (Shevchenko, 1970). The mites use them for their peculiar locomotion mechanism. When the posterior part of the body touches the substratum, both lobes are pressed together and form a sucker. By the contraction of muscle fibres inserted into the lobes, an area of lower pressure is created under the sucker, which guarantees safe attachment of the posterior body-part to the substratum. However, the anterior body-part can still be actively moved. Abrupt oscillations of the anterior body-part can result in a jump, when the sucker suddenly detaches from the substratum (Shevchenko, 1970).

The ventral tube of Collembola, a derivative of abdominal extremities, is a multifunctional organ. One of its functions is attachment to the substratum. Its distal part is smooth and soft, and normally inverted, inside the tube. It can be everted by haemolymph pressure, and pressed against the substratum (Westheide and Rieger, 1996).

True suckers are adapted for attachment to smooth substrata (stones, plant leaves, surfaces of other animals). Suckers are found in different arthropod lineages, such as arachnids, crustaceans, and insects. In different functional systems, they can serve both, for short-term and long-term attachment. The sucker requires a muscle, deforming the cup and specialised types of elastic cuticle. Diverse secretions (Blepharoceridae, Deuterophlebiidae) and excretions (Eryophyidae) are used to provide better contact of the sucker margin to the substratum. This is essential for maintaining lower pressure under the cup.

6. EXPANSION ANCHOR

Expansion anchors are an additional class of attachment devices. A part of the parasitic animal expands within the body of the victim. These devices are often used for food intake, and are therefore called *root organs* (Nachtigall, 1974). Larval aquatic mites of the genus *Hydrarachna,* are parasitic on aquatic beetles and bugs. They anchor themselves to the membraneous areas of the host body with a network of «roots», consisting of cuticular canals produced by evaginated hypodermal cells (Fig. 20 A, B).

Button-shaped, and branching structures of the anterior body region of some parasitic crustaceans (mainly copepods and cirripeds) aid in anchoring into tissues of fish, where the parasite remains for the rest of its life (Kabata, 1968; Kabata and Cousens, 1972; Kabata, 1979) (Fig. 20 C-E).

7. ADHESIVE SECRETIONS

Secretions involved in attachment have different origins, chemistry, exploited forces, and attachment abilities. Adhesive substances usually work as additional agents of attachment, providing better contact of the devices with the substratum. However, in some systems these secretions are the main agents responsible for attachment. Such examples have been reported in literature for many taxa of Arthropoda. In this section, a short review of these systems is provided. Most of the secretions are produced by epidermal cells of the cuticle, or by specialised cuticular glands, and therefore belong to the cuticular attachment systems.

There are two main types of secretions. The first type is based on glues or cementing secretions, which are initially liquid and harden after contact with air or water (Saroyan et al., 1970; Yan and Tang, 1981; Naldrett, 1993). The function of the second type is based on capillary or chemical adhesion. It widely exists as a supplementary agent in diverse attachment devices such as insect pads (Hasenfuss, 1977b; 1978; Bauchhenss, 1979a; b).

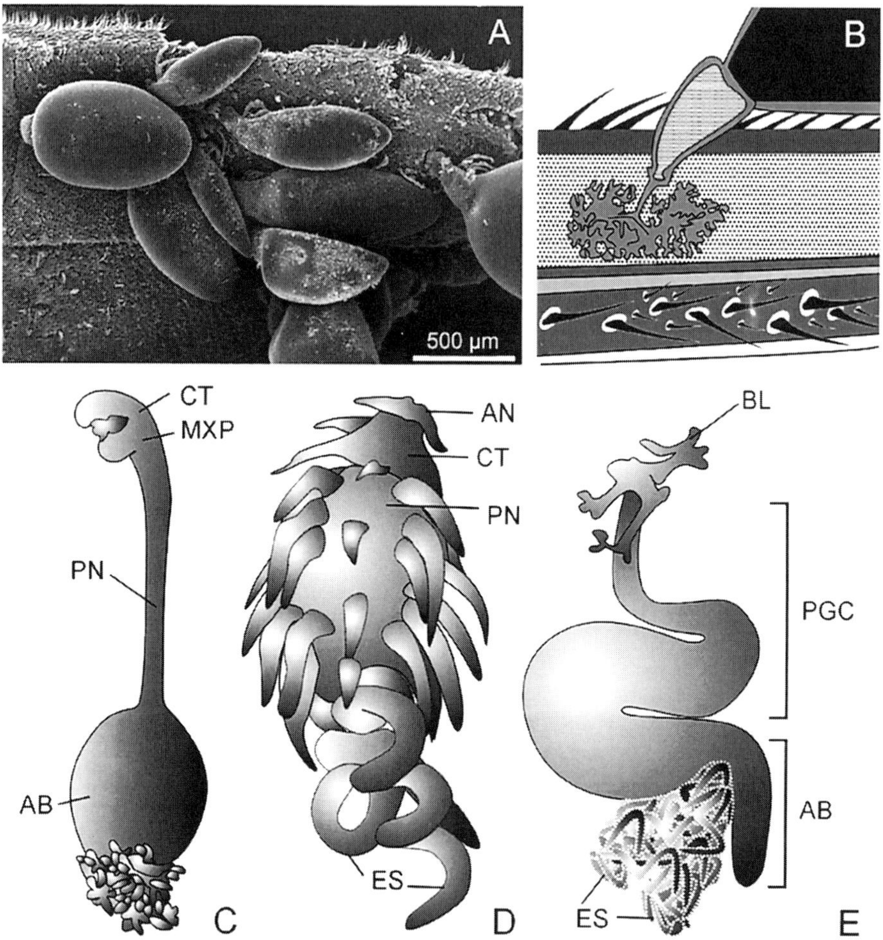

Fig. 4.20. Expansion anchor devices. A, B. *Hydrarachna* sp. anchoring in the cuticle of aquatic insects. C-E. Females of the copepod crustaceans *Sphyrion lumpi* (C) and *Lernaeocera branchialis* (D) (both Siphonostomatoida), and *Chondracanthus neali* (Poecilostomatoida). AB, abdomen; AT, antennae; BL, bulla; CT, cephalothorax; ES, egg sacks; MXP, maxilliped segment; PGC, peraeo-genital complex; PN, peraeon («neck»).

7.1 Associations: phoresy, parasitism, predation

Many parasitic copepod crustaceans use secretions of cementing glands for attachment to fish. In the shark parasite *Achtheinus oblongatus,* they are located within gnathopods (Titar, unpublished data). In other species, such as *Tracheliastes polycolpus,* the glands are maxillary (Kabata and Cousens, 1972). The maxillary cementing secretion of *Salmincola californiensis* is used for attachment, not on the host surface, but within the host tissues. Such modification of the secretory system results in generation of extracellular structures morphologically and functionally analogous to the bulla of parasitic copepods (Fig. 4. 20 E).

Phoretic animals also use glues to attach to mobile animals. The ostracod species *Cyclocypris ovum*, use secretions for phoresy to attach themselves to the skin of amphibians (*Bombina variegata* and *Triturus vulgaris*) (Seidel, 1989). So-called anal stalks in deutonymphs of uropodan mites are actually hardened glue, which is used by mites for attachment to the flea cuticle (Schwan and Corwin, 1987).

Predatory systems containing adhesive secretions are of two main types: active and passive. Active systems contain mobile adhesive parts involved in prey-capture. Such systems have been well studied in larval Coleoptera. Larvae of carabid beetles from the tribe Loricerini are characterised by unusually long and sharp maxillae, with a long galea having a soft and sticky surface of its distal segment. This segment is perforated with tiny pores that terminate in ducts of numerous, large multicellular glands, located in the stipes. Movements of the galea force out secretions from the glands to the surface of the galea (Hintzpeter and Bauer, 1998; Bauer and Kredler, 1988).

The labium of adults of the staphylinid genus *Stenus* is transformed into an extensible prey-capturing, adhesive apparatus. (Betz, 1996; 1998; Kölsch and Betz, 1998). Its mechanism is based on haemolymph pressure. Salticid spiders of the genera *Portia, Brettus,* and *Cyrba* use hairs with adhesive function for capturing prey (Foelix et al., 1984).

The best-known passive predatory systems are silk traps. The silk itself is a kind of hardened secretion. Additionally, traps are often supplemented by adhesive materials. In spider webs, the adhesive properties of threads are not only adapted to the size and mobility of prey, but also to the variety of environmental factors. Similar types of threads used in spider webs originated independently in several groups of Araneae (Strohmenger and Nentwig, 1987).

7.2 Attachment to the substratum, attachment of the substratum

Silk structures described in the previous section are not only used in predatory attachment devices. The most interesting type of secretion is produced by the piriform glands of spiders (Kovoor and Peters, 1988; Peters and Kovoor, 1991). This secretion is responsible for attaching radial threads within the web, and draglines to the substratum (Fig. 4.21) (Foelix, 1982; Gorb and Barth, 1996).

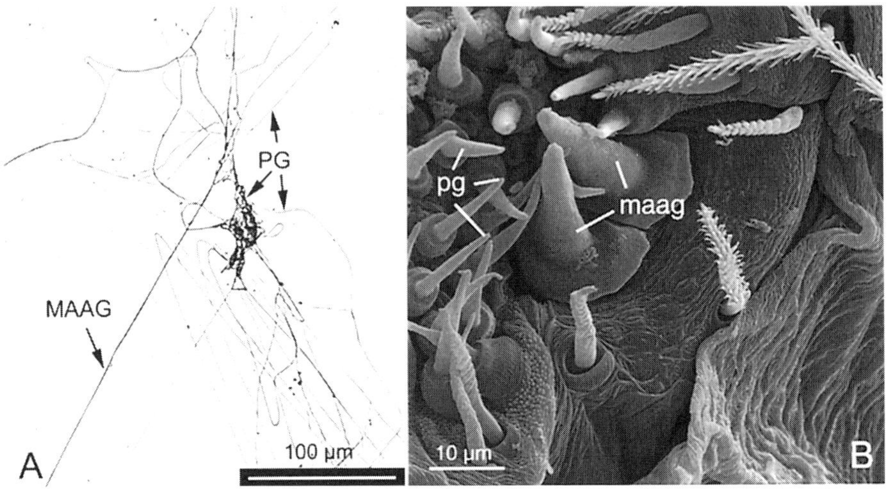

Fig. 4.21. Spider spinning apparatus. A. *Lycosa* sp. (Lycosidae), attachment point produced by secretion of piriform glands (PG) to glue the dragline, which is a hardened secretion of major ampullate glands (MAAG) to the substratum. B. *Philodromus caespitum* (Thomisidae), spigots of piriform (pg) and major ampullate glands (maag) on the anterior spinneret.

Insects also use silk to attach themselves to the substratum. The caterpillars of Lepidoptera, and larval Hymenoptera use salivary glands for silk secretion, not abdominal glands like spiders. Therefore, insect silk is not homologous to spider silk. Chironomid larvae also produce silk from salivary glands, and use their mouth-parts to shape threads. Thin threads are shaped by grooves at the anterior part of the ventro-medial plate, and thick ones by the whole preoral cavity (Kullberg, 1988). Chironomids use silk to build their larval tubes, and to attach themselves in order to resist water current (Tönjes, 1989).

Many herbivorous insects attach their eggs on (or closely to) the food source with cementing secretions of the bursal glands, as do female bugs from the family Reduviidae (Arun Rachel, 1988). Females of *Homarus americanus* (Crustacea, Decapoda) attach their eggs to the pleopod setae using a kind of adhesive secretion produced by the integumental glands (Johnson and Talbot, 1987). Similar behaviour has also been reported for the crab *Paralithodes camtschatica* (Powell and Hurd, 1972). The cuticle of the abdomen in the decapod *Pagurus bernhardus* (Crustacea, Anomura, Paguridae) is not calcified. It contains ducts of cuticular glands, producing substances involved in attachment of the crab to the inner surface of gastropod mollusc shells (Erri Babu and Anger, 1987). Many cirripedian crustaceans, such as representatives of the genus *Balanus,* are attached to the substratum by cementing secretions, the attachment abilities of which surpass many man-made glues (Lacombe and Liguori, 1969; Cook, 1970; Ligouri, 1970; Dougherty, 1990b; Naldrett, 1993).

Safe glues for long-term attachment are a highlight of nature's chemical design. However, it is often necessary to quickly attach and then immediately detach again, for example, during locomotion on sloped substrata. These devices widely exist in the attachment pads of Arthropods. Performance of such dynamic attachment systems is based on the capillary and chemical adhesion, and frictional forces, in the contact area. Usually, these forces are the result of several specialisations: surface texture, material properties, and nature of the secretions occurring in the contact area (Hasenfuss, 1977b; 1978; Bauchhenss, 1979a; b).

7.3 Interlocking of body parts

Wing-locking devices in Coleoptera and Heteroptera, coxal fixators in cercopid cicada, and head-arresting systems in Odonata, are designed according to the friction principle. Their complementary surfaces are covered by acanthae or microtrichia, which increase frictional forces in the contact area. Additionally, these systems contain epidermal secretions delivered to the contact area by pore canals. These secretions presumably have dual functions: they provide soft attachment-detachment by lubrication of contacting surfaces, and increase capillary adhesion in the contact area.

Chapter 5

Frictional systems

The notion that insects are «animals of the surface» reflects the significance of the integument microsculpture in the biology of these animals. The cuticle microsculptures often serve the fixation of the surfaces. In the previous chapter, different attachment systems were grouped into eight fundamental classes based on the form and size of the microsculpture, presence of muscles, cuticle secretions, and according to general design of the system. Friction is the principle, which occurs in most attachment systems.

Two surfaces covered with cuticular micro-outgrowths, such as acanthae or microtrichia, demonstrate high frictional forces when the surfaces come into contact. Since attachment in such systems is based on the use of the surface profile, and mechanical properties of materials resulting in an increase of friction forces in the contact area, they can be called frictional systems. There are frictional systems that have only one specialised surface. The most unusual are «smooth» surfaces with a peculiar material design (Chapter 9). In both of the last cases, the surface may be adapted to a variety of substrata. Parameters of frictional systems, such as surface sculpture and material design, are responsible for optimising contact to one particular co-opted surface, or to many different surfaces. Attachment generated by these systems is fast, precise, and reversible. Another reason for focusing more attention to these devices is that they occur most widely among Arthropoda.

In this chapter, we give an overview of frictional systems of insects and other arthropods. Such surfaces have been described from head arresting systems (Chapter 6), unguitractor plate and other intersegmental fixators of antennal and leg joints (Chapter 7), ovipositor valvulae, tarsal attachment pads (Chapter 9), and wing-attachment devices (Chapter 8) (Fig. 5.1).

Further chapters (6-9) will provide detailed information about design and function of several selected insect frictional systems.

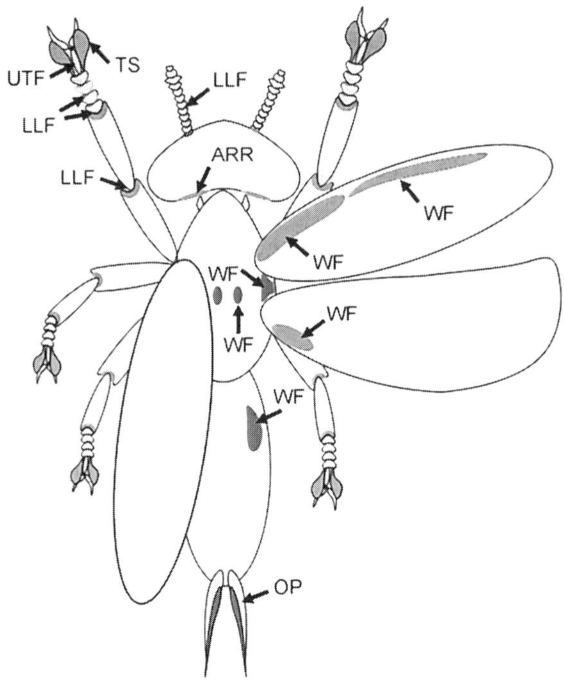

Fig. 5.1. Diagram showing possible locations of the frictional systems (Gorb, 1998c). Grey-coloured areas are usually covered with cuticular protuberances. ARR, head arresting system; LLF, rough surfaces in antennal and leg joints; OP, ovipositor valves; TS, tarsal attachment pads; UTF, unguitractor plate; WF, wing-attachment devices.

1. ANIMAL ASSOCIATIONS: PARASITISM, PREDATION

In the previous chapter, we described different attachment devices, such as hooks, clamps, and spacers, occurring in bird mites. However, the main attachment strategy in this animal group is to increase contact area with the bird feather. In order to achieve better contact with the host surface, diverse membranous, or compressed cuticular outgrowths, have been independently developed in different groups of bird mites. Histological studies show that these structures have an extracellular, cuticular nature. They can be divided into three main groups (Dubinin, 1951): a) membranous structures of the propodosoma and hysterosoma, such as those in species from genera *Freyana* and *Microspalax*; b) membranous lobes occurring at different leg segments in a variety of species; c) elongated leaf-shaped outgrowths on the posterior part of the body (species from genera *Proctophyllodes, Alloptes, Trouessartia, Avenozoaria*). In the mite *Freyanopsis decidilatatusetae,* lateral membranes are enlarged with simultaneous, strong development of the five pairs of caudal setae. The setae are modified into one row of widened, leaf-like plates.

Shark skin has an interesting microsculpture, the function of which is drag reduction during swimming (Reif and Dinkelacker, 1982; Nachtigall, 1998). In the crustacean *Achtheinus oblongatus* (Copepoda), parasitising on the shark *Squalus acanthias*, optimal contact with the host surface is provided by modified legs 2-4. The legs are covered by numerous outgrowths (Titar, unpublished data), the size and shape of which correspond, exactly, to the microsculpture of the host skin.

Ventro-lateral surfaces of tarsi in the spider *Portia fimbriata* (Salticidae) are covered by setae, which are usually flattened to the tarsus of the leg, and are used for capturing and holding prey (Rovner, 1980). These setae are mobile, driven by haemolymph pressure within the leg (Rovner, 1978). The medial surface of each seta is covered with tiny triangular, cuticular extensions, achieving maximum contact with the prey. Similar specialised setae have also been found in other salticid species (*Brettus, Cyrba*), which also use forelegs for prey-capturing (Foelix et al., 1984).

2. ATTACHMENT DURING COPULATION

Most beetles from families Tenebrionidae, Chrysomelidae, Cantharidae, and Curculionidae have the tarsi of their forelegs, and at times their midlegs as well, widened and covered with setae, which build a kind of sole (Chapter 9). In both sexes, such surfaces are adapted to attach to a variety of other surfaces. However, males bear an additional type of setae to hold the female, or to attach onto the female surface during copulation (Bey-Bienko, 1964). The dorsal surface of females has a corresponding microsculpture which, together with male devices, is optimised for male attachment. Specialised male setae are adapted for attachment to a particular surface profile of the female body. For example, tarsal setae in males of the colorado potato beetle *Leptinotarsa decemlineata*, can even disturb male locomotion on a smooth surface, because of their excellent attachment ability on the smooth surface of the female's elytra (Pelletier and Smilowitz, 1987). Such setae differ from female setae adapted for locomotion (Fig. 5.2).

Copulatory organs, or accessory appendages, of many insects often contain fields of cuticular outgrowths (setae, acanthae or microtrichia) to prevent sliding on corresponding surfaces, while contacting. Males of the fly *Dolichopus ungulatus* (Diptera, Dolichopodidae) bear anal appendages covered with setae in their medial surface (Fig. 5.3). Such a coverage provides a firm, but soft hold on the female. Anal appendages of damselfly males from the genus *Enallagma* are densely covered with rounded microtrichia, preventing sliding of the appendages on the female's mesostigmal plate. This list will probably never be completed, because of the extreme variety of copulatory organs among insects.

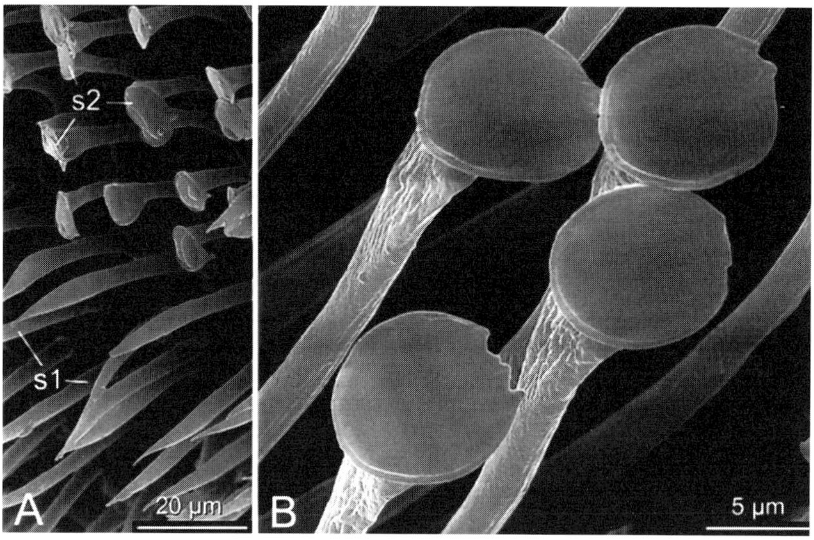

Fig. 5.2. Attachment pads of the male colorado potato beetle *Leptinotarsa decemlineata* (Coleoptera, Chrysomelidae). A. Two types of setae are specialised for different functions: locomotion (s1) and attachment to the dorsal surface of the female (s2). B. Setae of the s2 type.

3. ATTACHMENT TO THE SUBSTRATUM

Pupae of many species of butterflies (Papilionoidea) have frictional surfaces connected with the prominent pupal projections, preventing falling of the suspended pupa (Urquhart, 1970). The projections, which lie on the ventral side of the last abdominal segment, are covered with numerous, minute spines, presumably acanthae (Starnecker, 1998).

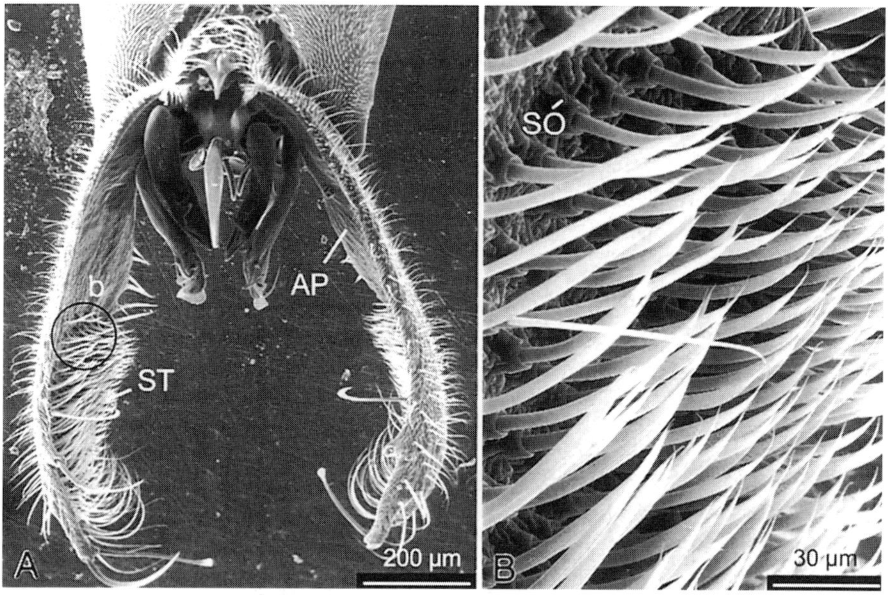

Fig. 5.3. *Dolichopus ungulatus* (Diptera, Dolichopodidae), male copulative organs. A. Dorsal aspect of the posterior part of the abdomen. B. Selected area of the setae field. AP, appendages; b, area shown in B; SO, socket; ST, field of setae.

Among vertebrates, the most prominent, dynamic attachment systems exploited in locomotion, are based on the use of frictional and adhesive forces. Three primary arboreal groups of lizards, the family Gekkonidae, the anoline section of the family Iguanidae, and several Scincidae, have specialised subdigital pads. These animals are able to climb vertical glass surfaces, and even to hang upside-down on a ceiling. Most researchers have reported the presence of microscopic bristles on the underside of the feet (Braun, 1878; Dellit, 1934; Ruibal and Ernst, 1965; Hiller and Blaschke, 1967; Hiller, 1968; Roll, 1995), which are keratinised epidermal derivatives (Maderson, 1964). Similar structures are also known from the ventral side of the tail in geckos, of the family Diplodactilidae (Bauer, 1998). Since no secretory glands have been revealed in these areas, it is presumed that the van der Waals interaction is mainly responsible for the resulting attractive forces (Autumn et al., 2000).

Most arthropods attach themselves to a surface during locomotion, with the aid of a modified pretarsus and/or tarsus, which is often covered by different micro-outgrowths, or composed of specialised cuticle with particular mechanical properties. These structures, often called attachment (or adhesive) pads, are sometimes well adapted to one preferable substratum. Morphology, ultrastructure, and performance of the arthropod pads have been previously studied in representatives of Diptera (Bauchhenss and Renner, 1977; Bauchhenss, 1979a; b; Röder, 1984a; b; Wigglesworth, 1987; Gorb, 1998d); Coleoptera (Stork, 1983a; b; 1987; 1988); Hymenoptera (Freeland et al., 1982); Thysanoptera (Heming, 1970; 1972; 1973); Heteroptera (Ghasi-Bayat, 1979; Ghasi-Bayat and Hasenfuss, 1980a; b; c); Mecoptera (Röder, 1984a; b); and eriophyid mites (Shevchenko, 1970). Diversity and functional adaptations of insect attachment pads will be observed in detail in Chapters 9, 14, and their evolution in Chapter 13.

An interesting attachment function of surface structures has been reported in Hymenoptera. Pollen grains attach to the surface of insect-pollinators with the aid of specialised setae, with a grooved pattern on their surfaces, or covered with tiny microtrichia, that are able to interlock with a grain microsculpture (wave-like, groove-like, needle-shaped) (Hesse, 1981). Thus, such a co-evolution between flowering plants and their pollinators has resulted in specialised cuticular microstructures of insect surface, and in surface adaptations of pollen grains. It has been previously reported that setae substructures in bees from the Megachilidae (Hymenoptera, Apoidea), correspond exactly to both the size and shape of pollen grains of many plants that the bees feed on (Pasteels and Pasteels, 1972).

4. INTERLOCKING OF BODY PARTS

Attachment systems based on the friction principle may be adapted for long-term fixation (wing-lock), and for short-term fixation as well (tarsal pads). In these systems, epidermal secretion often occurs. The lipid nature of this secretion has been shown for several systems. The function of this secretion is not quite clear, especially for systems bearing two corresponding surfaces. There are two alternative hypotheses: (1) the secretion is an additional attachment-agent, responsible for an increase of the capillary force in the contact area; (2) the secretion is a lubricant, enabling soft attachment-detachment. Most of the friction systems, dealing with the interlocking of body parts, will be shown in separate chapters. In this subsection, we focus mainly on the proboscis of butterflies and moths (Lepidoptera), and the inner coverage of the valvulae of an insect ovipositor.

4.1 Proboscis of Lepidoptera

Imagines of Lepidoptera (group Glossata) possess a long proboscis, adapted for extracting nectar from floral tubes. This structure is especially remarkable due to its length and ability to coil, spirally, while resting (Fig. 5.4 A). The spirally rolled proboscis of Lepidoptera is unique in insects.

The proboscis consists of the two greatly elongated galeae, which enclose the central food canal (Fig. 5.4 C). The two parts are interlocked on their dorsal and ventral sides by rows of specialised, cuticular structures. A fluid-tight linkage is achieved by two rows of cuticular plates on the dorsal side, and a row of complexly shaped hooks on the ventral side, in higher Lepidoptera (Fig 5.4 B-D) (Krenn, 1997). Near the distal end of the proboscis, the linking structures are modified, and leave slits between each other that enable the insects to take fluid into the food canal (Fig 5.4 E,F) (Paulus and Krenn, 1996).

There are several interpretations of the mechanism responsible for coiling and uncoiling of the proboscis: (1) haemolymph pressure originating from the stipes; (2) elasticity of the galeal wall, and (3) musculature within the galea (Krenn, 1990). The coiling process begins at the tip and continues towards the base of the proboscis. After complete coiling, the proboscis uncoils whereby the coils slide over each other until they are in close contact, and the outermost coil touches the ventral side of the head. From the point-of-view of attachment devices, the most important question is about the mechanism that keeps the proboscis tightly coiled while in the resting position.

Generally, it is accepted that the elasticity of the cuticle, and muscles, coil the proboscis. However, both these mechanisms are not sufficient for maintaining the tightly coiled resting position of the proboscis. As has previously been shown, the cuticular processes occur on the galeae of most species, at least in areas which touch each other, in the coiled position (Krenn, 1990). These processes are interlocked between single coils sticking together, like the bristles in two brushes, in the resting position of the proboscis. Due to the elasticity of the galeal structure, the coiled proboscis tends to uncoil in a loosely coiled position (Hepburn, 1971). As a result, such a combination of spring loaded proboscis can maintain the resting position, causing the cuticular protuberances to interlock, thus preventing further uncoiling of the mouth-parts without muscle action.

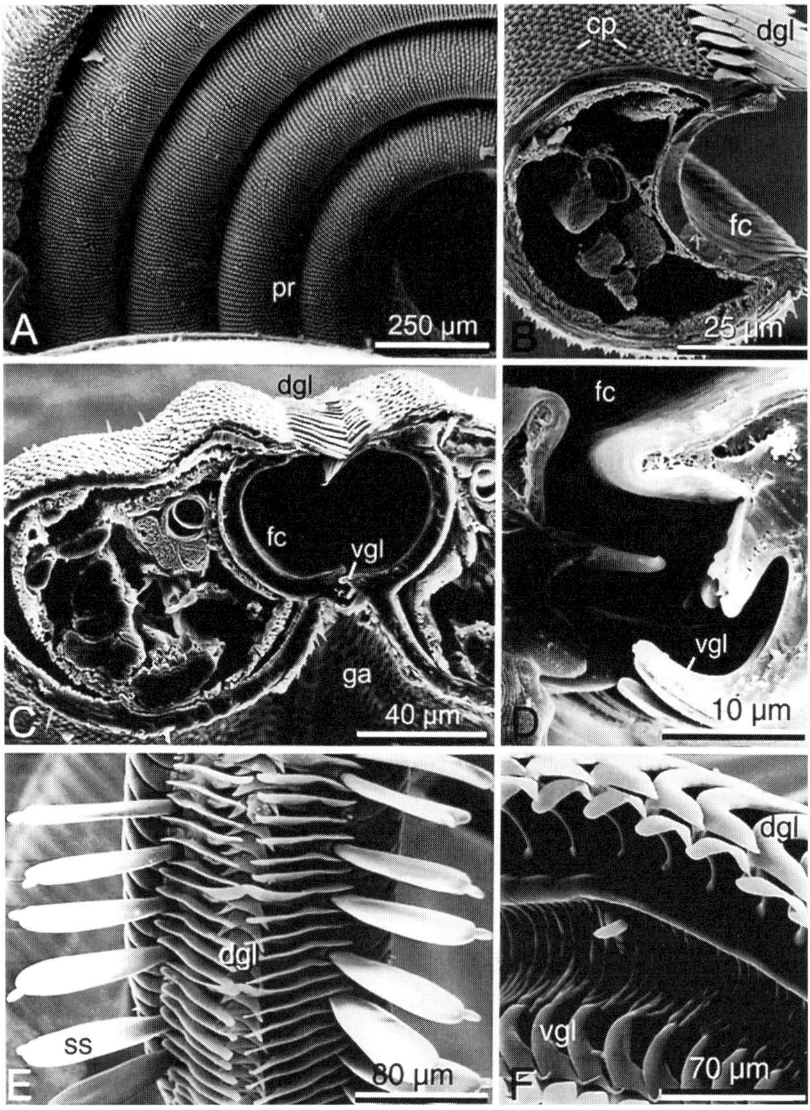

Fig. 5.4. Proboscis of Lepidoptera. A. Coiled proboscis, lateral aspect. B. Galea with dorsal linkage. C. Linked galeae. D. Ventral galea linkage. E. Modified linkage structures in the vicinity of the tip of the proboscis. F. Food canal in the tip region. A, B. *Pieris rapae* (Pieridae). C, D. *Vanessa cardui* (Nymphalidae). E. *Dryas julia* (Heliconiidae). F. *Iphiclides podalirius* (Papilionidae). pr, proboscis; dgl, dorsal galea linkage; ss, sensillum styloconicum; fc, food canal; vgl, ventral galea linkage; ga, galea; cp, cuticular processes. Courtesy of Harald W. Krenn, University of Vienna.

The lateral walls of the proboscis of some butterfly species, also bear little cuticular processes. Presumably, they can provide additional support of the coiled proboscis by interlocking with the setae of labial palpi, which embrace the coiled proboscis on either side. One would expect that the number and density of the cuticular processes correlates with proboscis length, in different species. However, this notion has not been supported by current investigations (Krenn, 1990). Interestingly, in blood-sucking moths lateral processes are modified into barbs and hooks, which are important during the piercing process (Büttiker et al., 1996).

An additional mechanism seems to play an important role in holding the proboscis in the resting position. The ventral side of the head is also covered by microtrichia, such as those occurring in the intersegmental membranes of Diptera (Chapter 7). It has been hypothesised that, in Lepidoptera, the membrane fixes the coiled proboscis in the U-shaped labium.

4.2 Coverage of the ovipositor valvulae

The anterior (lower) and median (upper) processes (valvulae) of the odonate ovipositor are bilateral, narrow, lanceolate appendages, which serve for egg deposition on the plant surface, in the soil, or in rotten wood. The inner surface of the ovipositor valves is covered by microtrichia, which are directed postero-distally (Figs. 5.5, 5.6). The ovipositor movements are controlled by a set of muscles (Weele van der, 1906; Tillyard, 1917; Quentin, 1962; Pfau, 1985; 1991; Matushkina and Gorb, 1997), which generate opposing movements of the anterior and posterior valvulae.

The opposing movements of both pairs of the valvulae result in egg displacement, in a distal direction only. The structures facilitate egg transport within and outside of the ovipositor (Fig. 5.5). An alternative hypothesis of the egg transporting mechanism is egg sliding on a water film, which may be generated by the inner microsculpture of the ovipositor valvulae. Protuberances of the ovipositor may also be used for removal of the sperm rests, after copulation.

Different odonate taxa have a different morphology and arrangement of the ovipositor microtrichia. The spectrum of the microtrichia types, which are present within a single species, may vary from one (*Anax imperator*, representatives of Coenagrionidae) to four types (*Aeshna cyanea*). Additionally, some species have a specific microsculpture in the proximal and marginal parts of each valvula, and at the area of contact (sliding canals and sliding protuberances).

The spectrum of the modification of the structures on the inner surface of the ovipositor valvulae is broad. Three main types of structures can be observed: (1) single sparse microtrichia, which are present in Aeshnidae and on the upper valvula in many Zygoptera; (2) comb-like crest compositions, which are present mainly on the lower valvula, and rarely on the upper valvula in Zygoptera (Fig. 5.6.); (3) plates of Anisozygoptera. The differences in microtrichia structure, and distribution between the upper and lower valvulae are most strongly expressed in a relic dragonfly, *Epiophlebia superstes* (Anisozygoptera). Single sparse microtrichia seem to be an initial structure developed, later in the crest-like compositions (Chapter 13).

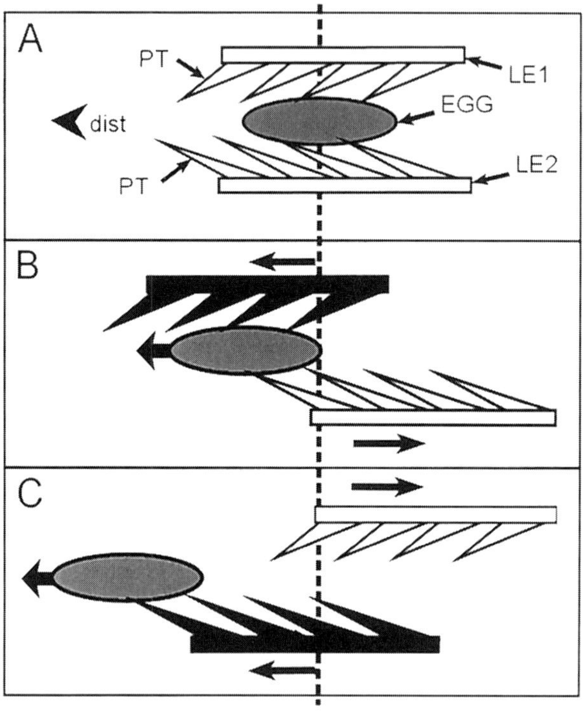

Fig. 5.5. Hypothetical diagram showing how protuberances of the inner surface of the ovipositor valvulae, may generate distally-directed egg movements. The protuberances are arranged in combs and are always angled posteriorly. The distally-moving leaf (darkly shaded) carries an egg out, whereas the leaf moving in the opposite direction (lightly shaded) does not generate egg displacement. dist, distal direction; EGG, egg; LE1, LE2; ovipositor leaves or valvulae; PT, protuberances.

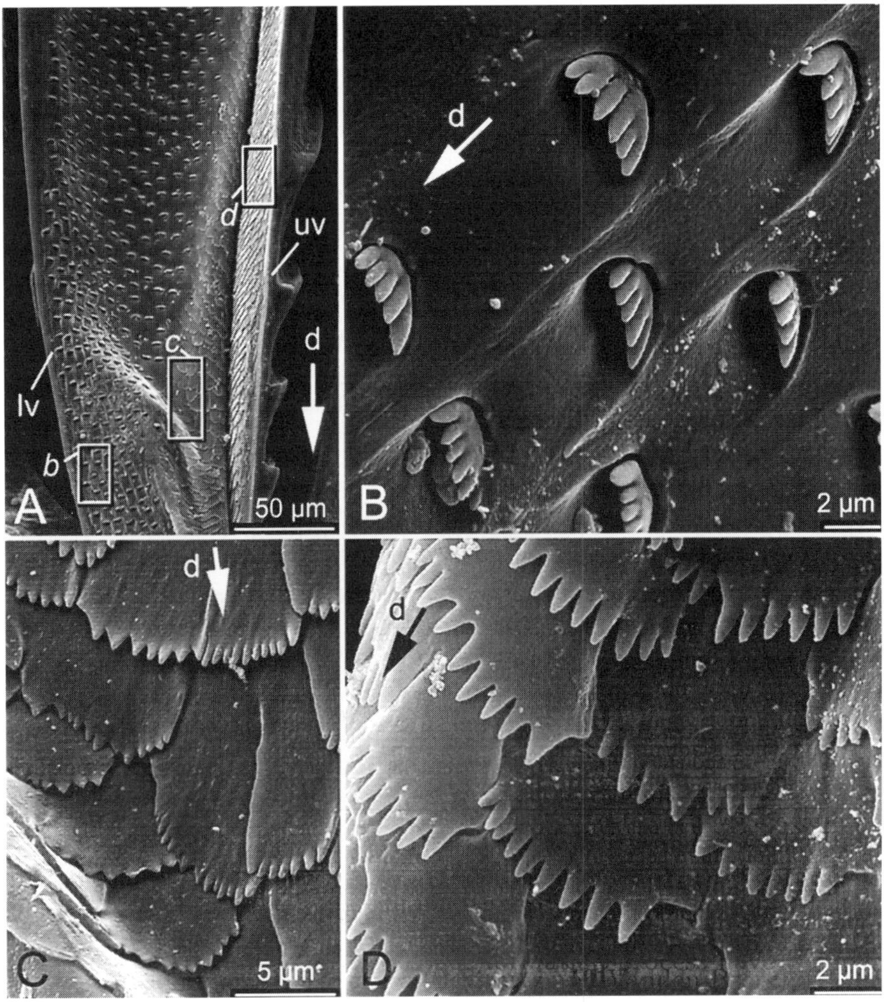

Fig. 5.6. Inner surface of the upper (uv) and lower (lv) ovipositor valvulae in the damselfly *Sympecma annulata* (Lestidae). A. Right valvulae close to the ovipositor tip (areas *b, c, d* correspond to B, C, D). B. Ventral part of the lower valvula. C. Dorsal part of the lower valvula. D. Upper valvula. Arrows (d) indicate distal direction.

4.3 Body coverage

Arthropods living in thin, or liquid substrata with high viscosity, such as dung, living tissues, or mud, bear minute outgrowths (micropapillae, acanthae, microtrichia) on their surface that may provide an ability for directional motion within the substrata. Distally sloped outgrowths result in directionality of the friction during motion in/on the substrate. In such a case, peristaltic body movements generate unidirectional body propulsion (Fig. 5.7).

In eryophyid mites, small cuticle outgrowths are located on the ring-shaped combs of the body (Shevchenko, 1970). These structures are the most developed of living forms connected to the galls (Eryophyidae, Phytoptilidae). Both the dorsal and ventral body sides bear well-developed outgrowths, whereas the posterior part of the body remains smooth, or covered with little papillae. A possible explanation of such a difference is that posterior parts of the body rarely make contact with the substrate.

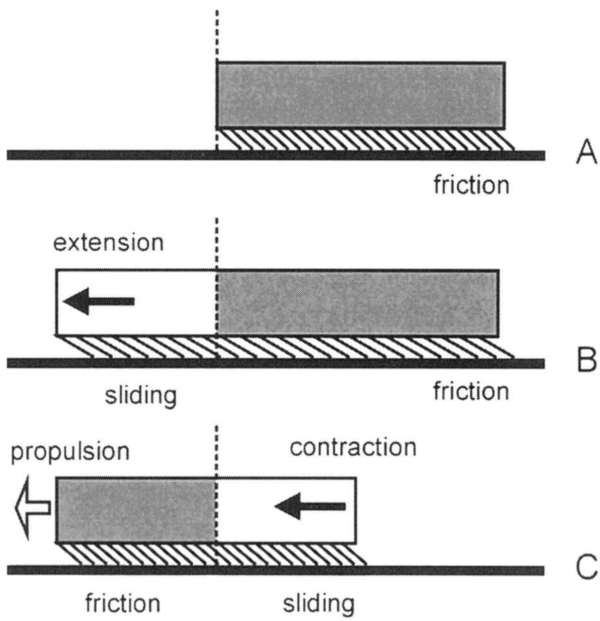

Fig. 5.7. Diagram of the friction mediated locomotion. The body protuberances are always angled posteriorly. Both extensive and contractile changes of the body shape result in propulsive body movements.

Functionally similar structures have been reported in (a) soil inhabiting Oligochaeta (Westheide and Rieger, 1996); (b) dipteran larvae, such as Sciomyzidae (*Sepedon sphegea*) (Gasc et al., 1982) and Cecidomyidae (Spungis, 1990), living in viscous substrata; (c) mites, for example *Nematalicus* (Tydcoidea) and representatives of the family Demodecidae.

Chapter 6

Dragonfly and damselfly head-arresting system

1. THE LARGE HEAD NEEDS ADDITIONAL STABILITY

The design of the head-prothorax transition in adult Odonata has no analogies in other groups of insects. The area of the head articulation with the neck is very small compared to the head dimension; it is virtually a single point providing extreme head mobility on the roll, pitch, and yaw planes (Mittelstaedt, 1950). Such mobility of the head compensates for the absence of a specialised gravity organ. The head is analogous to a statolith, the movements of which are monitored by the head, neck, and prothoracic fields of the hair sensilla.

At the same time, such a design of the head-neck articulation has little mechanical strength. The head, however, must be rigid with the rest of the body, for example during feeding, to prevent disturbances caused by large mandibular muscles, and to stabilise gaze while perching or holding prey while in flight. During the initial stage of tandem formation, the male holds the female by the head, in order to grasp the head or prothorax with its appendages (Gorb, 1996b). This usually happens in flight, and at high speed, and one may expect high loads on the head. Additionally, males hold females by the head (Anisoptera) or prothorax (Zygoptera) in tandem flight, during copulation. In such a position, the female head can be easily damaged if it is not rigidly connected to the thorax (Wildermuth, 1984). The *arrester* is a system that serves to immobilise the head during feeding, or when the dragonfly is in tandem flight (Gorb, 1991a).

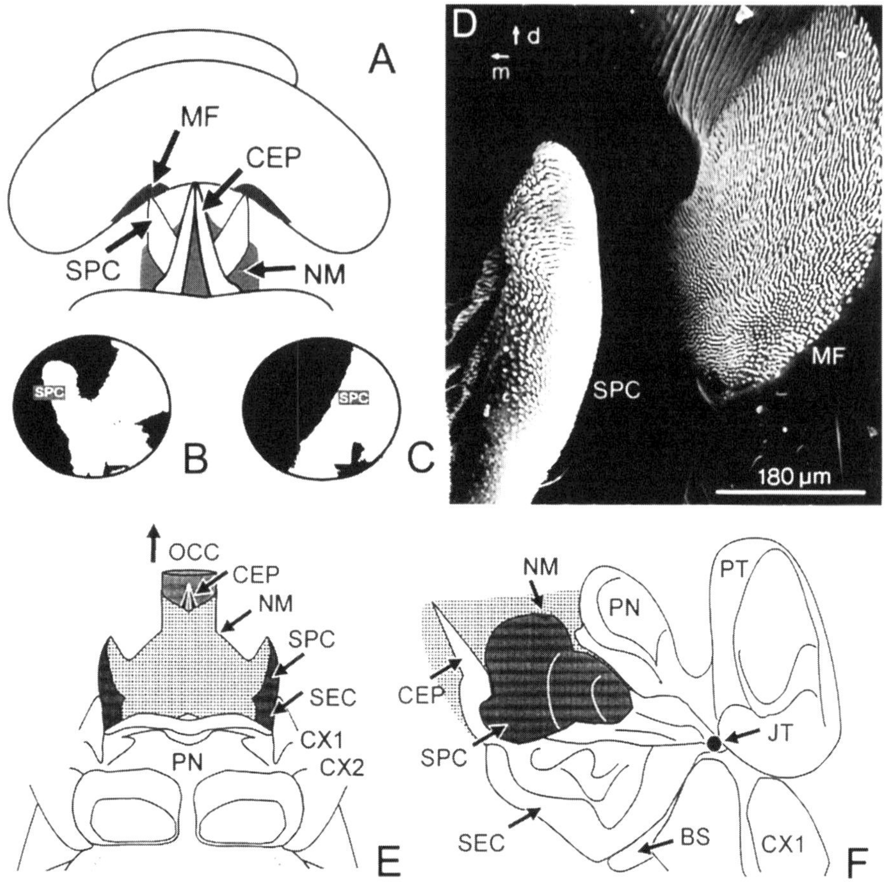

Fig. 6.1. Head arrester in Libellulidae (Odonata) (Gorb, 1998b). A. Sketch of the head-neck articulation, dorsal view. B-C. *Sympetrum vulgatum*, movements of the postcervical sclerite (SPC). B. Lateral position of the SPC. C. medial position of the SPC. D. Relative position of the SPC and microtrichia field on the right side of the head (SEM). E. Neck region of *S. sanguineum*, dorsal view, diagram. F. Neck region of *Orthetrum cancellatum*, lateral view, left side, diagram. BS, basisternum; CEP, cephaliger; CX, coxa; d, dorsal direction; JT, joint between prothorax and eucervical sclerite; m, medial direction; MF, microtrichia field; NM, neck membrane; OCC, anterior direction; PN, pronotum; SEC, eucervical sclerite; SPC, postcervical sclerite.

The arrester system includes adjusting organs of two body segments - the head and neck (Fig. 6.1). It consists of: 1) the skeleton-muscle apparatus that moves the head and neck sclerites; 2) co-opted microsculptures, which are fields of microtrichia on the rear surface of the head (MFH), and on the postcervical sclerites of the neck (SPC) (Figs. 6.1; 6.2); 3) the secretory apparatus, consisting of epidermal cells that produce lipid substances, which pass through pore canals in the cuticle into the region of contact between MFH and SPC; 4) sensory organs monitoring contact between MFH and SPC, and the position of the SPCs relative to the other neck sclerites.

It was initially hypothesised that the arrester was a stridulation organ (Berlese, 1909). Later, this function was placed in doubt (Tillyard, 1917). Finally, H. Mittelstaedt suggested a fixation function for these organs and called this system an «arrester» (Mittelstaedt, 1950). The arrester remained, until recently, a little-known structure. However, a head-arrester is of general interest, because of the microtrichia-covered surfaces providing fixation, due to high friction between interlocked microstructures in the contact area. Such designs seldom exists in biological attachment systems. This chapter briefly summarises studies on the skeleton-muscle organisation of the arrester (Gorb, 1989; 1993), its inner morphology and ultrastructure (Gorb, 1990a; 1997b; d), and the microsculpture of the frictional surfaces (Gorb, 1990b; c). Data on the sensory organs involved in arrester monitoring will be provided in Chapter 12.

2. SKELETON-MUSCLE ORGANISATION OF THE HEAD-ARRESTING SYSTEM

The lateral walls of the imaginal neck consist of a pair of eucervical sclerites, which are symmetrical, relative to the longitudinal body axis (Fig. 6.1). Anteriorly, they are transformed into two sharp processes called cephaligers (CEP). These processes meet at their extreme anterior point, which is the centre of the head mass. At this point, the tentorium, the skeleton of the head, is connected to the neck through a pair of mobile joints. Ventrally, the eucervical sclerites are connected to the basisternum through the neck membrane, which is elastic and has particular folding properties. These properties are reflected in the complex, fine structure of the neck membrane, providing high mobility of the eucervical sclerites. Near their dorsoproximal edges, the eucervical sclerites are connected to the postcervical sclerites (SPC). The latter are isolated from the eucervical sclerites and surrounded by the neck membrane on three sides. Such articulation enables extremely high mobility of postcervical sclerites. Each SPC is differentiated into a sclerotised base and an elastic distal region.

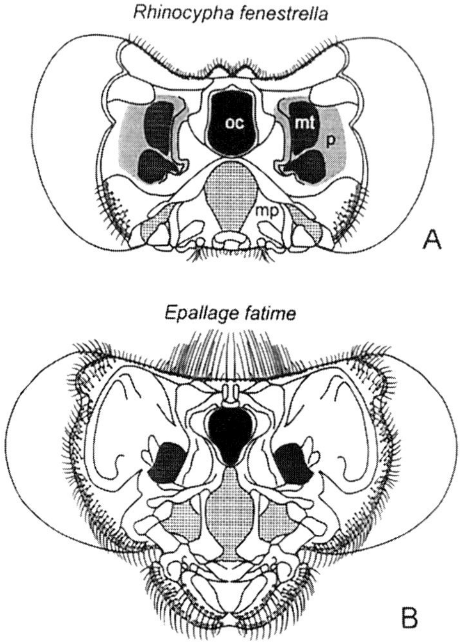

Fig. 6.2. Morphological types of microtrichia fields on the head of Calopterygoidea (Zygoptera) (Gorb, 1998b). A. *Rhynocypha fenestrella* (Chlorocyphidae). B. *Epallage fatime* (Euphaeidae). Dotted areas indicate the membranous cuticle. mp, mouth-parts; mt, microtrichia fields on the rear surface of the head; oc, occiput; p, micropapillae.

The muscle system of the neck-head transition is somewhat similar in many insect groups, and consists of three pairs of muscles responsible for head movement on three planes (roll, pitch, and yaw) (Pringle, 1963; Strausfeld et al., 1987). In Odonata, there are a pair of muscles (M_1) involved in the control of postcervical sclerites (Fig. 6.3). The origin of an abductor muscle is an apodeme of the pronotum. The insertion is located in the proximal region of the SPC, between both the eucervical sclerite and SPC. The location of the insertion is somewhat similar in all odonate taxa studied. However, the apodeme of the pronotum is diversely pronounced in representatives of different taxa. In the species series *Cordulegaster*, *Sympetrum*, *Cordulia*, *Aeshna*, and *Gomphus*, the apodeme tends to elongate. This can result in a different direction of the muscular force depending on the relative positions of the origin and insertion.

When the abductor muscles contract, the SPCs move medial to the longitudinal body axis and lose contact with the microtrichia fields (the head is free). This movement provides for the disabling of the arrester system when the animal needs to move its head. The antagonistic muscle responsible for SPC adduction was not found; thus, the SPC adduction is passive. In the transitional area between the eucervical sclerite and SPC, there are a few regions of cuticle with high elasticity due to peculiarly organised cuticular structures. These are cuticular «springs», which can return the SPC to the lateral position when the abductor muscle relaxes. The elastic cuticle of the distal part of the SPC is adapted to provide soft contact with the MFH.

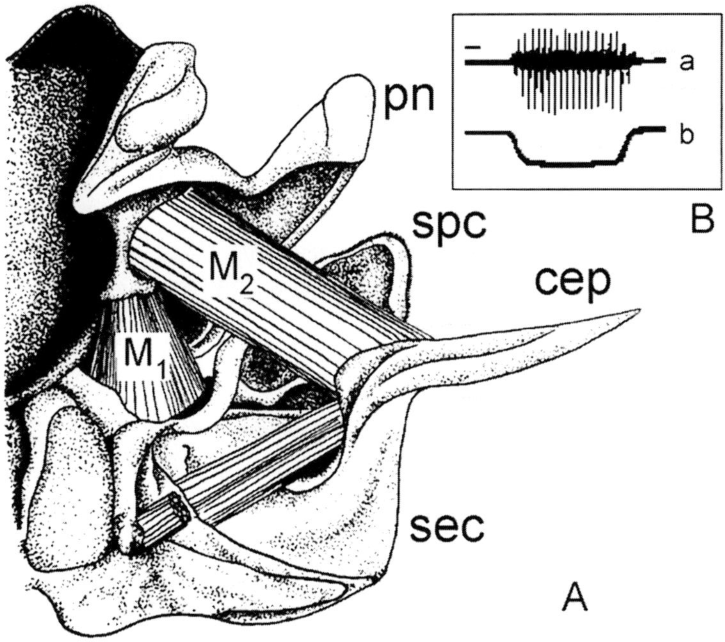

Fig. 6.3. A. Skeleton-muscle organisation of the arrester system in *Orthetrum cancellatum* (Anisoptera, Libellulidae) (Gorb, 1991a; 1998b). CEP, cephaliger, M_1, muscle controlling postcervical sclerite; M_2, muscle involved in head control; PN, pronotum; SEC, eucervical sclerite; SPC, postcervical sclerite. B. Responses of the dragonfly wing-muscle anterior depressor (a) and SPC (b) to a wind pulse, scale bar = 0.1 s.

The MFHs on the rear surface of the head, and the postcervical sclerites covered with microtrichia, are functionally corresponding structures corresponding in the shape, size and density of microtrichia (Figs. 6.4; 6.5).

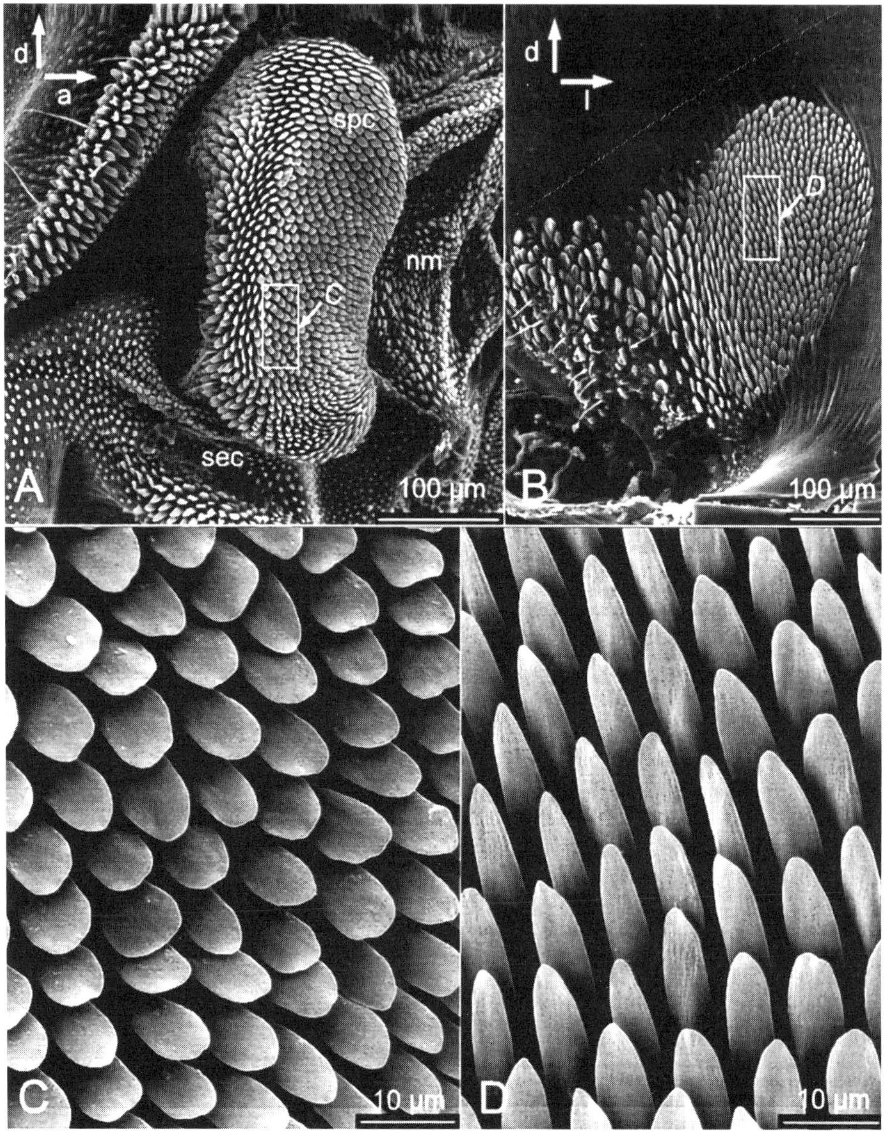

Fig. 6.4. *Perissolestes romulus* (Zygoptera, Perilestidae) (Gorb, 1998b). Corresponding surfaces of the damselfly arrester system. Postcervical sclerite (A, C) and microtrichia field on the back surface of the head (B, D). a, anterior direction; d, dorsal direction; l, lateral direction; nm, neck membrane; sec, eucervical sclerite; spc, postcervical sclerite.

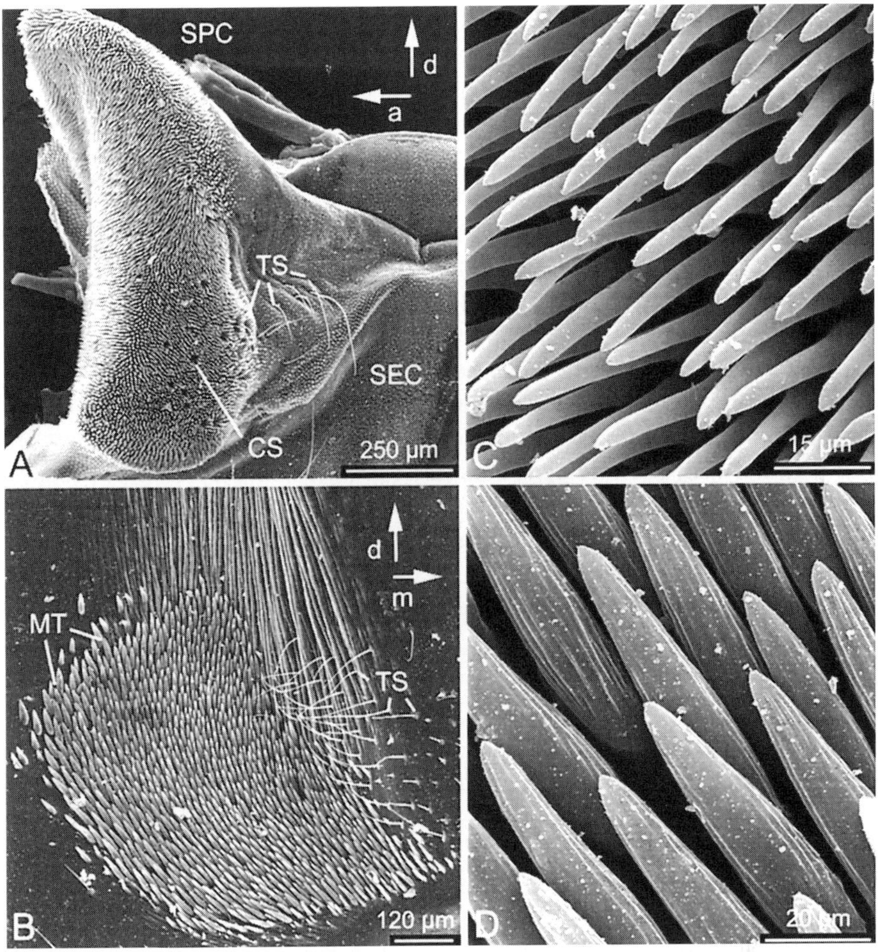

Fig. 6.5. *Aeshna mixta* (Anisoptera, Aeshnidae) (Gorb, 1998b). Corresponding surfaces involved in the dragonfly arrester system. Postcervical sclerite (A, C) and microtrichia field on the back surface of the head (B, D). a, anterior direction; d, dorsal direction; m, medial direction; CS, campaniform sensilla; MT, microtrichia; SEC, eucervical sclerite; SPC, postcervical sclerite; TS, trichoid sensilla.

3. FUNCTIONAL SIGNIFICANCE OF THE ARRESTER

Electrophysiological recordings of flight-muscle activity with simultaneous photodiode registrations of SPC movements, showed that the release of the head was synchronised with the initiation of wing-muscle activity only when flight was elicited by substrate removal, or a wind pulse (Gorb, 1995c) (Fig. 6.3 B). However, the head-arrester system can work independently of the wing apparatus, at least in two behavioural situations. These are: while eating, or flying in tandem. A dragonfly shows wing muscle activity even when the head is arrested (Fig. 6.6).

Dragonfly legs serve as a basket-trap used for prey-capture, or mate-holding during flight. When a dragonfly captures prey, its front legs bring the victim to the mouth. Dragonflies tear apart their prey with mandibles and pull it down with the forelegs. This must require a highly efficient locking mechanism by the neck, so that the dragonfly does not pull off its own head when eating. Complex flight manoeuvres have not been observed in flying dragonflies holding prey. A dragonfly uses gliding flight when feeding on the wing. The reason for this is that the high-speed manoeuvres are only possible when the head is free. It serves as a gravity organ that controls turning during flight. Another foreleg action that would perturb head position in flight is when the insect cleans its eyes. Presumably, the head is also arrested in this situation.

The female head is arrested in tandem to prevent head damage during initial stages of copulation or in tandem flight. Such cases have been described for gomphids (Dunkle, 1991) and libellulids (Wildermuth, 1984). Thus, the arrester system may serve as an adaptation to save the head from violent mechanical disturbances, and to stabilise gaze in a variety of behavioural situations.

An intriguing idea explaining the arrester system function in Anisoptera has recently been proposed (Beckemeyer, 1997). Since the male has a firm hold on the female head capsule, he can control her flight more easily. He controls the orientation of the flight directional sensor - the head - and in essence, forces the female to fly in formation not only by mechanical contact, but by direct manipulation of the flight control, and steering system, of the female. However, the female's head is arrested in tandem, as is indicated above. It is plausible that flight control may also be possible when the female's head is fixed by the SPCs, because the arrester sensilla may be involved in control of female flight, in tandem.

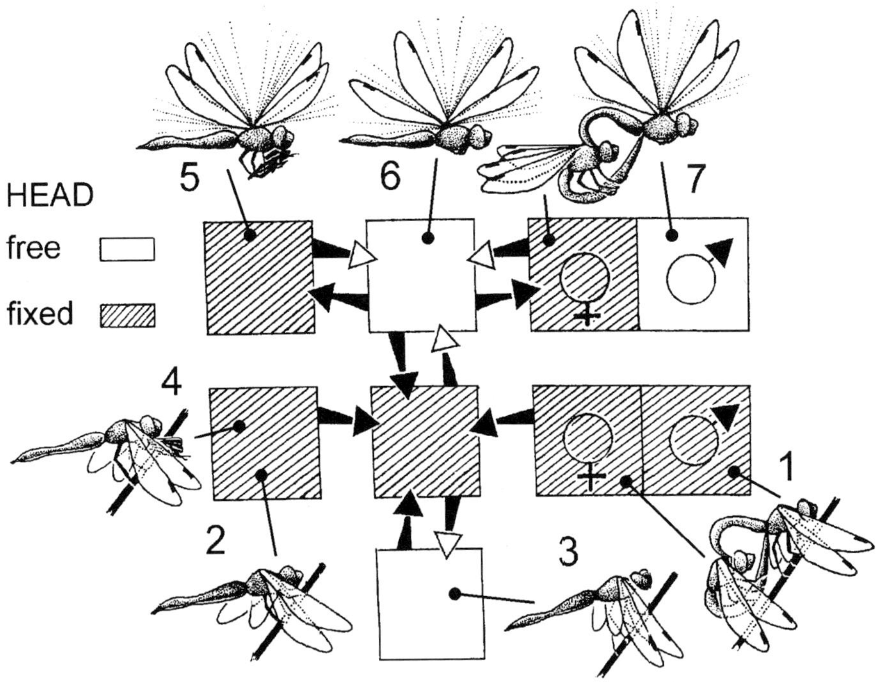

Fig. 6.6. Diagram of diverse behavioural situations, when the head is free or fixed, by the arrester system (Gorb, 1998b). 1, tandem sitting; 2, perching and roosting; 3, saccadic head movements during ambushing; 4, feeding; 5, flying with captured prey; 6, patrolling flight; 7, tandem flight. Shaded squares indicate that the head is fixed by SPCs. Blank squares indicate that the head is free. Arrows indicate fixation (black arrowheads) or release (white arrowheads) of the head, during the transition from one behavioural situation to another.

4. HEAD MOBILITY AND HEAD STABILITY: A COMPARISON WITH OTHER INSECTS

Dragonflies have several characteristics, such as the male secondary copulation organs and larval mask, that are unique among insects (Tillyard, 1917). The relative dimensions and weight of the head in adult dragonflies surpasses those of many other insects. It may be suggested that the ability to perform head movements in every plane in space, is provided by a highly-developed neck area. Indeed, adult Odonata have large cervical (neck) sclerites, when compared with those of other insects or even with odonate larvae (Crampton, 1926; Asahina, 1954; Popham and Bevans, 1979). The larval head has a larger base of articulation to the prothorax when compared with the imaginal one. The head mobility in the larva is lower, because the larval head is firmly connected to the thorax to maintain stability during powerful predatory strikes (Olesen, 1972; Tanaka and Hisada, 1980; Parry, 1983). The larval neck is small, the cervical sclerites are tiny, and the postcervical sclerites are fused with the eucervical sclerites (Crampton, 1926). In the adult, the prey is usually captured with the front legs, and the mouth-parts seem to be no longer involved in prey-capture. The neck sclerites, specialised for head fixation, were found only in adult Odonata. Postcervical sclerites of representatives of the closest group to Odonata, Ephemeroptera, are very small, and are not specialised for head fixation. Clearly-cut fields of micro-outgrowths on the rear surface of the head are absent in odonate larvae, and in representatives of Ephemeroptera and Plecoptera as well.

Many other insects have large, heavy heads, and most flying insects need to have independent head movement (Olberg, 1978; 1981; Olberg and Pinter, 1990). For instance, mantids have exceptionally high mobility of the head (Liske, 1982; 1984; 1989; Liske and Mohren, 1984; Liske and Davis, 1986), but do not posses a specialised head-arresting system as that of Odonata. Being ambushing predators, mantids do not feed on the wing, as dragonflies do, and they do not escape immediately when disturbed by predators. Therefore, mantids may not need such a specialised gaze stabilisation device as dragonflies. Nor do mantids employ odonate-like, tandem formation, during copulation. Males dragonflies hold females by the head; even damselfly males hold the female's head during the initial stage of tandem formation (Gorb, 1996b).

In some representatives of Insecta, head-arresting systems are known, which are completely different from those of odonates. While cutting a plant stem, representatives of Isoptera lock the head to the thorax, using mandibular locking devices (Ilychev, 1987). In dipterans, the head is also very mobile (Sandeman and Markl, 1980). The range of head-turning ability in the blowfly *Calliphora* is ±20°, both horizontally (yaw) and vertically (pitch), and ±90° for rotations about the line of sight (roll), which roughly corresponds with the main body axis (Hengstenberg, 1991). Head stability, in some taxa of flies, is presumably provided with armoured membranes (Gorb, 1997a).

The Odonate arrester is a unique fixation system due to the following features. First, the head-arresting systems, designed as mobile neck frictional devices, have not been previously reported in representatives of other orders of insects. The second feature is the absence of antagonistic muscles, which are normally involved in mobile structures of most arthropods. Such an absence is usually compensated for by different hydraulic adaptations, such as the Odonate larva mask (Olesen, 1972; Tanaka and Hisada, 1980), spider legs (Foelix, 1982), or by elastic springs, as in the cranio-mandibular system of the Dacetini and Odontomachini ants (Dlussky and Fedoseeva, 1988), insect pretarsus (Gorb, 1996a), and fly labellae. In the arrester system, the SPC-SEC transitional area is elastic, and acts as an antagonist to the SPC abductor muscle.

The third, unique feature is a variety of modifications of the fixation type within the Odonata. The fixation type depends on the structure of the microtrichia on the corresponding surfaces of the head and neck. There is a wide variety of microtrichia types existing in the head arresting mechanism (Fig. 6.7).

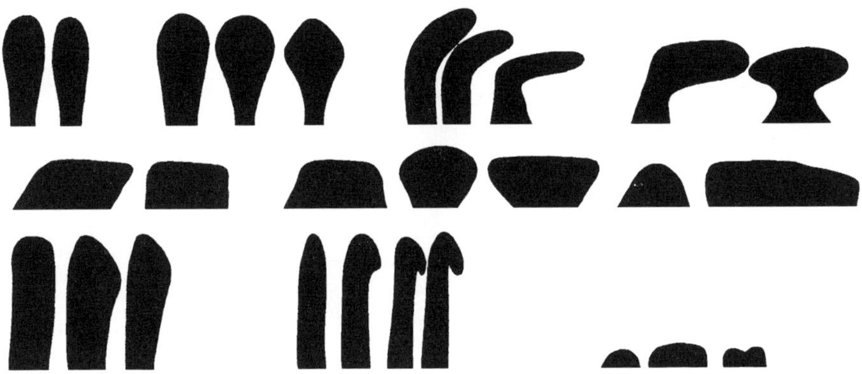

Fig. 6.7. Diversity of microtrichia shapes existing in representatives of Zygoptera (Gorb, 1998b).

However, five main sets of correlating microtrichia (MT) types exist: 1) hook-like, or cone-shaped MT, and mushroom-like MT (*Sympecma*, Libellulidae), 2) cone-shaped MT and thin seta-like MT (Aeshnidae) (Fig. 6.5), 3) mace-like MT on both surfaces, presumably acting like an elastic snap (Lestinae), 4) smooth surface and compressed MT (Gomphidae, Petaluridae), and 5) plate-like microtrichia on both surfaces (Coenagrionoidea) (Fig. 6.4). This may explain the lowered dependence of the MFH size on the form, and size of the head (Gorb, 1998b).

Presumably, attachment stability is provided by different fixation types, and not by an increase of the area of contact. Contact is additionally supplemented by adhesive secretions, delivered through a pore canal system in the contact area (Gorb, 1990a; 1997b) (see also Chapter 10). Detailed information about sensory equipment of the arrester, and the evolution of this system among Odonata, will be given in Chapters 12 and 13, respectively.

Chapter 7

Intersegmental frictional devices

As shown previously, frictional surfaces widely exist in attachment systems adapted to afix two structures, or parts of the body, together. In leg-joints, antennae and large spines, such surfaces serve to prevent segment mobility in certain directions or even immobilise completely, in order to save muscle energy. There are more specialised structures, such as the insect unguitractor mechanism, armoured membranes and the mid-coxal prong in Brachycera (Diptera). The frictional surfaces in joints are also called link-fixing systems. Other systems, such as coxa-locking devices in the hind legs of cercopid cicada, are responsible for synchronising jumping, for instance. In this chapter, several case studies on frictional surfaces in joints, and interlocking mechanisms of insect legs are reported.

1. FRICTIONAL SURFACES WITHIN JOINTS

The external skeleton in arthropods is the main advantage, and constraint, in the evolution of their extremities. The principal design of arthropod extremities is completely different from that of vertebrates, in respect to location of skeletal and muscular structures. In arthropods, the leg is a cuticular tube containing muscles that control parts of the tube, called segments. The material properties of various parts of the tube are completely different. The segments are, more or less, hard and stable (Hepburn and Chandler, 1976). The intermediary membranous area is flexible, and/or elastic (Hackman and Goldberg, 1987). The most fascinating aspect is that the tube is not interrupted between the segments, but consists of a continuum of various material properties. The design of insect joints varies from simple articulations with an endless number of rotation axes, monoaxial joints with a singe rotation axis, to complex structures with several precisely defined rotation axes. Usually, profiles of contacting surfaces of both segments are smooth within the joint. This results in decreasing friction within the joint, during segment movements. However, many insect joints contain structured surfaces that are covered by cuticular protuberances or depressions. Such surfaces exist not only in leg joints, but also in joints between segments of antenna, or at the base of large mobile spines (Fig. 7.1).

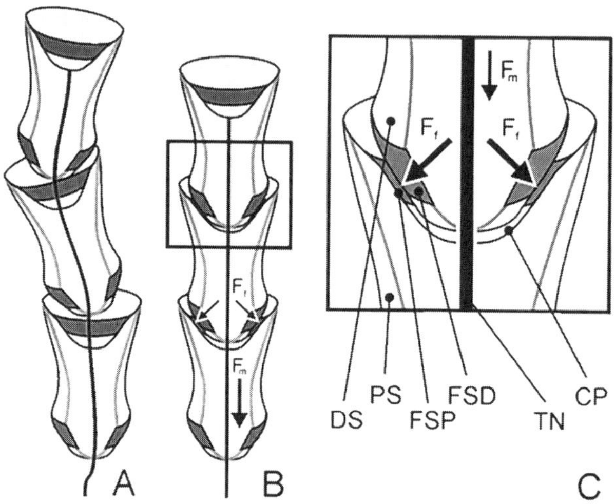

Fig. 7.1. Principal diagram of the telescopic design of the multisegmental chains, such as tarsi or antenna. Membranous areas are not shown. A. Mobile chain-like structure of two joints. B-C. After application of muscular force to the tendon, running through the chain, the structure became stiff due to the interlocking of frictional surfaces at both ends of each segment. CP, joint cup; DS, distal segment; F_f, frictional force; F_m, muscular force; FSD, frictional surfaces at the distal end; FSP, frictional surfaces at the proximal end; TN, tendon.

Tarsal segments between the first and last tarsomeres lack muscles. They are connected to each other through a series of joints that resemble a telescopic structure, where the proximal end of the next segment fits into the cup-like distal part of the previous segment. The whole tarsus is driven by a single muscle, usually located in the femur, and connected with the distal part of the tarsus (pretarsus) through a long tendon. When the muscle contracts, the entire tarsus becomes stiff due to joint-interlocking (Fig. 7.1).

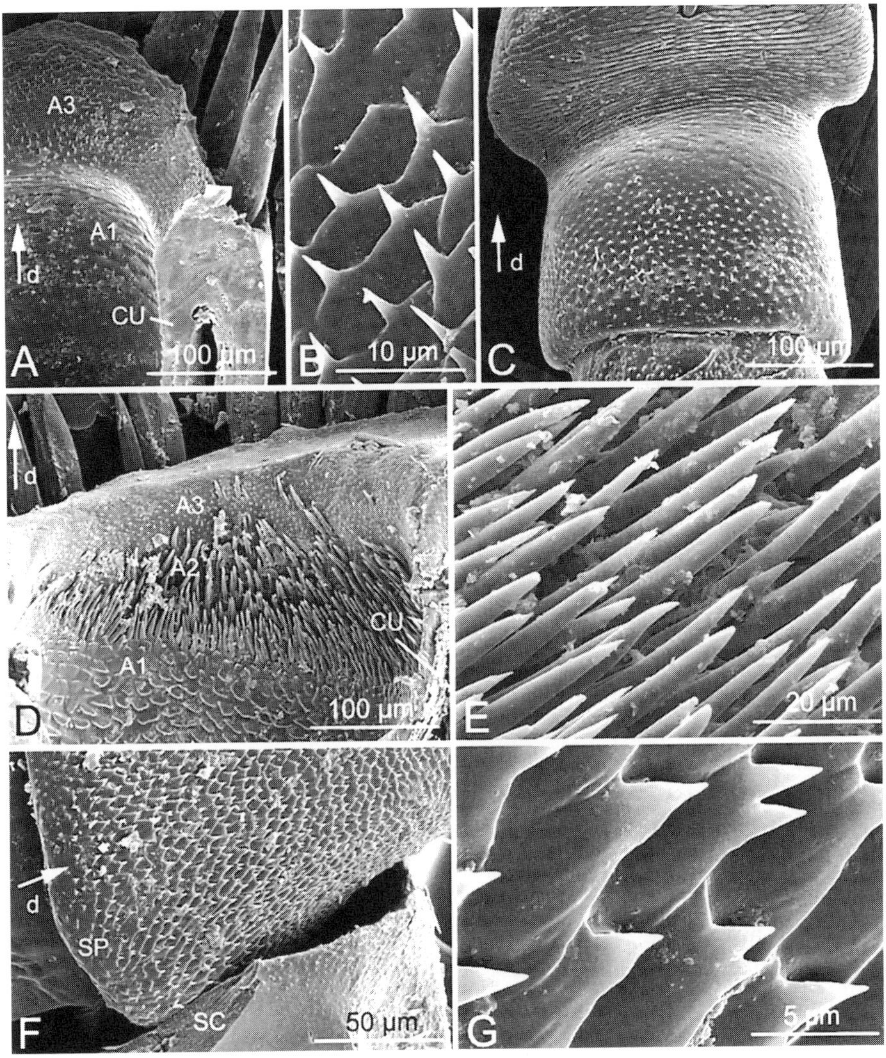

Fig. 7.2. *Melolontha melolontha* (Coleoptera, Scarabaeidae). Examples of link-fixing systems in insect joints. A-B. Distal end of the first tarsomere, joint cup. C. Proximal end of the first tarsomere. D-E. Distal end of the tibia, joint cup. F-G. Basis of the large tibial spine. A1, rough area; A2, acanthae-covered area; A3, microtrichia-covered area; CU, joint cup; d, distal direction; SC, spine socket; SP, spine base.

The structures responsible for this mechanism are fields of microtrichia, located on the cup surface at the distal end of the proximal segment (Figs. 7.2 A-B), and at the proximal end of the next distal segment (Fig. 7.2. C).

The tibio-tarsal joint has an even more complex design due to the presence of short, wide outgrowths located deeply in the joint cup, long acanthae in the center of the cup, and tiny microtrichia in the upper part of the cup (Figs. 7.2 D, E). The outgrowths on the frictional surfaces of the leg joints are usually distally-directed. Large thorns and spines that are connected with the underlying cuticle through mobile sockets, bear distally-directed microtrichia in their basal parts (Figs. 7.2 F, G).

2. DESIGN OF THE INSECT UNGUITRACTOR APPARATUS

The pretarsus of insects originated from crustacean dactylopodite, but is intensely modified in comparison with it. An insect pretarsus only has a flexor muscle; an extensor muscle is absent (Shvanvich, 1949; Walther, 1969). The pretarsus is connected to the terminal segment of the tarsus by a sclerotised plate called unguitractor plate (or simply unguitractor) (Goel, 1972). The unguitractor is an insertion point of the claw-flexor-muscle, which is responsible for claw-bending. The return of the claw to an extended position is passive, due to resilin filled cuticle areas at the base (Gorb, 1996a).

It has been indicated by previous authors, that the cuticle of the unguitractor region varies in the degree of sclerotisation, flexibility, and elasticity (Seifert and Heinzeller, 1989). Scanning electron microscopy of the anterior surface of the unguitractor has revealed the presence of the denticle-like structures, which have been interpreted as a frictional system providing anchorage for the unguitractor (Conde-Boytel et al., 1989). Additionally, diverse unguitractor structures have been reported to have mechanosensory and secretory functions (Seifert and Heinzeller, 1989).

2.1 General morphology of pretarsus

The insect pretarsus includes the last tarsal segment (terminal tarsomere) with two tarsal claws, empodium, and unguitractor plate (Fig. 7.3 A, C). These parts are connected to each other through membranous cuticle. The tarsal claws in their anterio-proximal areas are connected to the unguitractor through two tendon-like regions of a flexible cuticle. Between claws there is often a small cuticular plate (empodium), immovably connected to unguitractor. In the dorsal region, the tarsal claws are connected to the terminal tarsomere wall, which is modified as an apodeme-like protuberance.

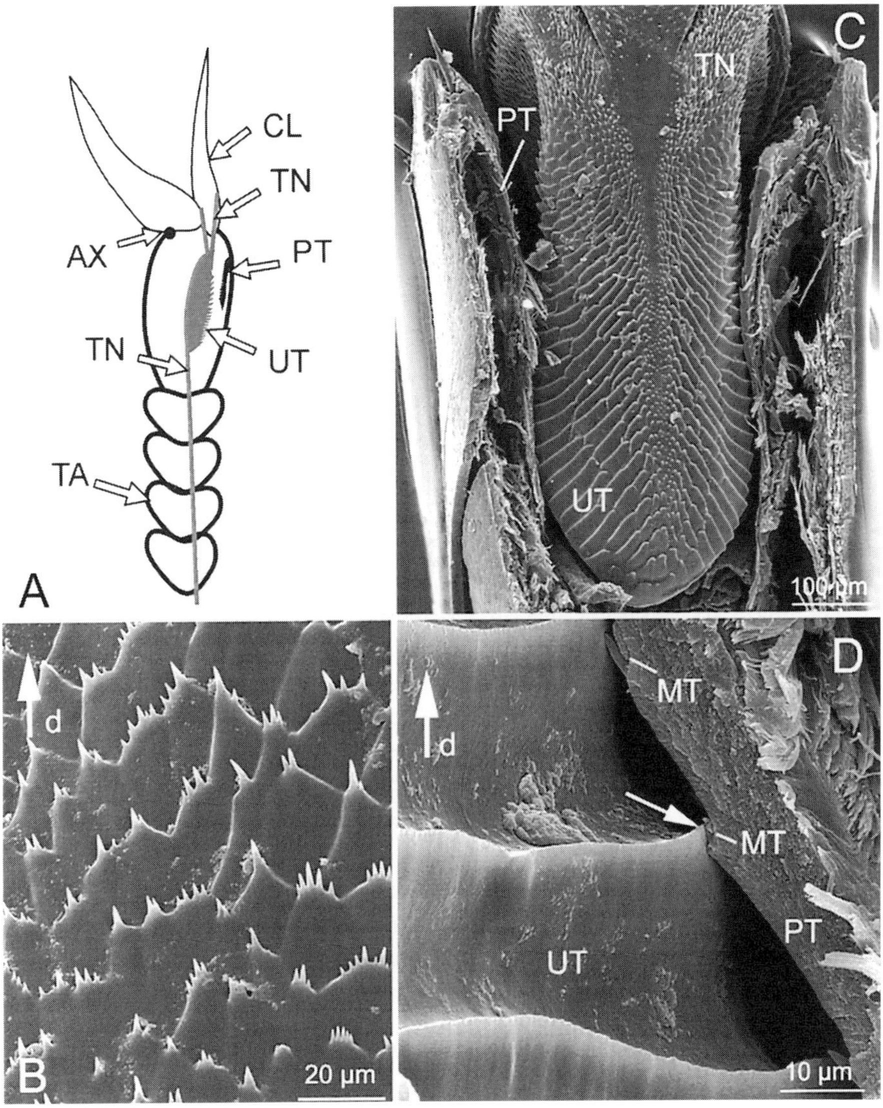

Fig. 7.3. A. General diagram of insect pretarsus. B-D. Unguitactor apparatus of the beetle *Melolontha melolontha* (Scarabaeidae). B. Surface of the anterior plate of the tarsal wall. C. Terminal tarsomere, ventral wall is removed. D. Contact between the two co-opted surfaces. AX, joint between claw and tarsus; CL, claw; d, distal direction; MT, microtrichia of the corresponding surface; PT, anterior plate of the tarsal wall; TA, tarsomere; TN, tendons; UT, unguitractor plate; *white arrow* indicates contact between unguitractor sculpture and microtrichia of the corresponding surface.

The anterior surface of the unguitractor is covered by denticle-like microtrichia, or cuticular plates, (2-7 μm) distally-directed. The anterior part of the terminal tarsomere bears a wall-invagination, that forms an additional plate-like structure facing the unguitractor plate. This structure comprises the unguitractor from the lateral and anterior side. This structure is also covered by microtrichia on its distal part. Its microtrichia are elongated and are usually proximally-directed (Fig 7.3 B, D).

2.2 Ultrastructure of the unguitractor cuticle

The ultrastructure of cuticle, composing different pretarsal structures in the damselfly *Platycnemis pennipes* (Platycnemididae) is varied (Gorb, 1996a). The membrane connects all parts of the pretarsus and divides the unguitractor into external (plate), and inner (tendon) parts. The cuticle of the membrane is composed of the electron-dense, external, exocuticular layer and the electron-lucent, inner, endocuticular one. Intermediary regions contain pieces of external layer plunged into an inner layer.

The unguitractor cuticle has three principal layers. Both external ones are similar in design to the layers of the membrane. The inner, third layer is endocuticle, containing fine-scattered, electron-dense material not clearly separated from the electron-lucent material. The hard cuticle contains electron dense chitin-crystallites, located closely to the surface (Fig. 7.4 A). Platinum-carbon coating revealed that the material properties of these structures differed from the surrounding material (Fig. 7.4 B).

The wall of the terminal tarsomere in the damselfly has a thick exocuticular layer containing sublayers, the orientation of which, presumably, provides material stability in this region. Cuticle surrounding the apodeme of the posterior wall of the terminal tarsomere, contains large pore canals, whereas the cuticle of the apodeme contains small pore canals. Elastic cuticle connects the claws to each other and to the apodeme. These regions are composed of electron-light, unstructured resilin-containing cuticle.

2.3 Diversity of the unguitractor plates in insects

Both the unguitractor and the corresponding plate of the terminal tarsomere in three dragonfly species, revealed a high resemblance in the morphology (Gorb, 1996a). This can be explained by similar leg function in adult dragonflies. The odonate unguitractor is more elongated, compared to that in other insects; its length-to-width ratio surpasses 2 (Fig. 7.5 A). Its protuberances are papillae-like. Elongated crests are located only in the tendon-like transitional area between the unguitractor and claws (Fig. 7.5 B).

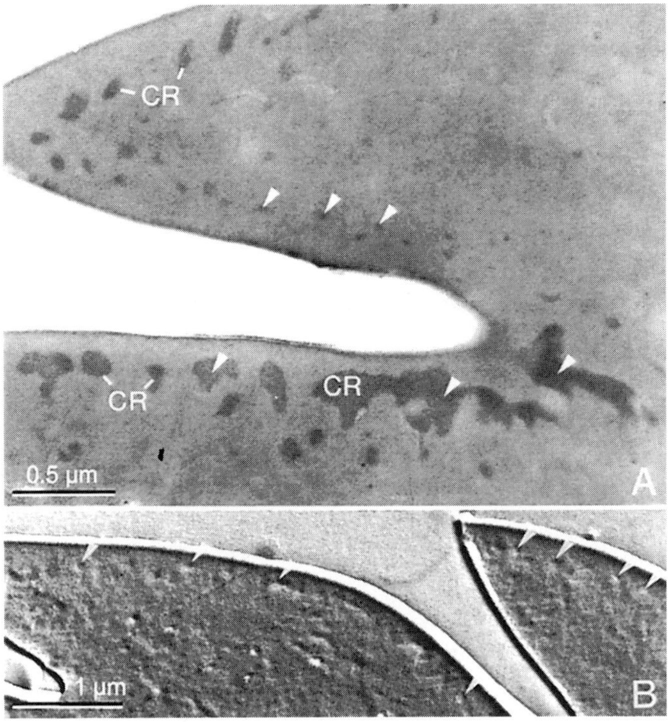

Fig. 7.4. Unguitractor material in the fly *Calliphora vicina.* A. Ultrathin section, TEM. B. Ultrathin section coated with carbon-platinum, TEM. CR and arrowheads show chitin chrystalites, responsible for stiffness and stability in this area of the cuticle.

Coleoptera and Heteroptera exhibit a variety of leg functions. Within these insect groups, morphology of the unguitractor also varies to a larger extent. Differences were found mainly in the microsculpture of the plates. The scarabaeid beetle, *Lethrus apterus,* which can burrow into dense soil, bears relatively large unguitractor covered by diverse microstructures (Gorb, 1996a). The plate-like microtrichia are located on the lateral surfaces. The medial part is covered by small crests composed into comb-like groups (2-5). The distal part is covered by long, sharp microtrichia. The corresponding plate of the terminal tarsomere is densely covered with sharp microtrichia that are distally-directed. In contrast to *L. apterus*, the unguitractor of the beetle *Geotrupes stercorosus*, which burrows in relatively soft soil, has uniform plate-like microsculpture (Fig. 7.5 C).

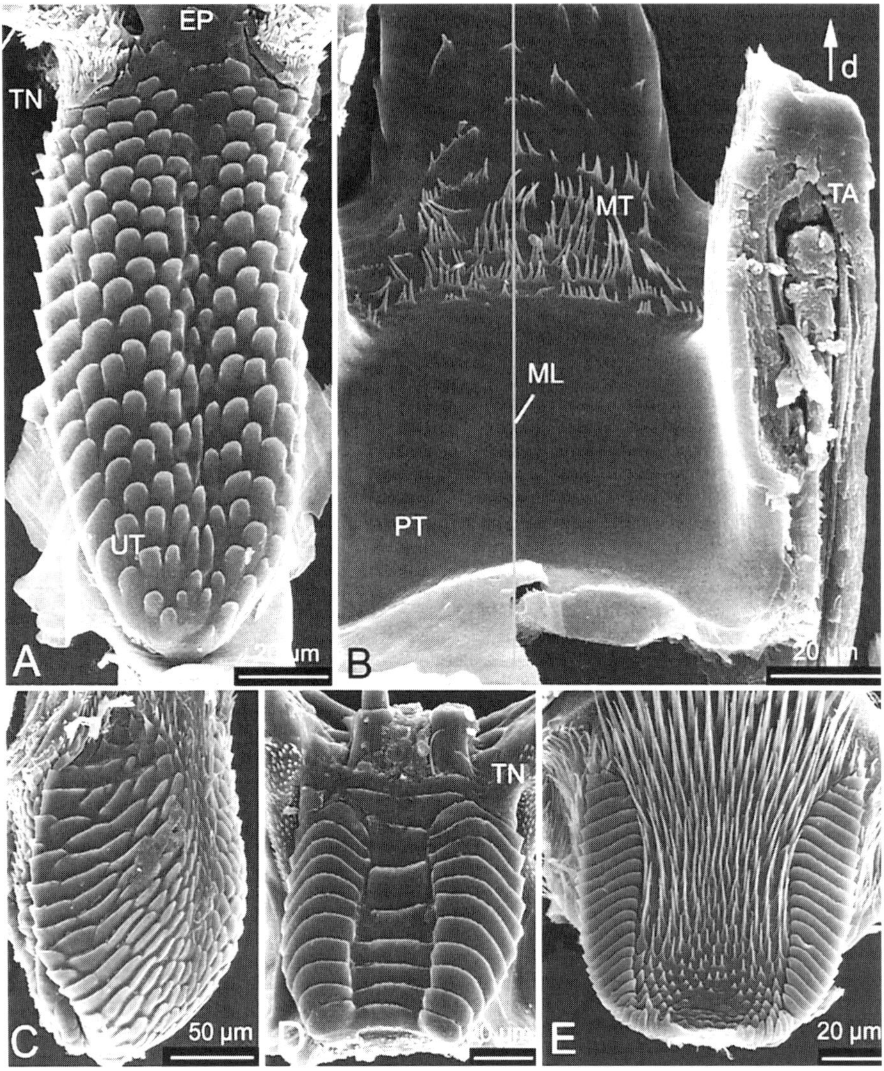

Fig. 7.5. Insect unguitractor. A. *Coenagrion puella* (Odonata, Coenagrionidae), unguitractor, ventral aspect. B, *C. puella*, surface corresponding to the unguitractor, dorsal aspect. C. *Geotrupes stercorosus* (Coleoptera, Scarabaeidae), unguitractor, ventro-lateral aspect. D, *Graphosoma italicum* (Heteroptera, Pentatomidae), unguitractor, ventro-lateral aspect. E, *Calliphora vicina* (Diptera, Calliphoridae), unguitractor, ventro-lateral aspect. EP, empodium; ML, midline; PT, an invagination of the wall of the terminal tarsomere; TA, wall of the terminal tarsomere; TN, tendon; UT, unguitractor plate.

Small insects, which are able to walk quickly along thin rods, such as the Colorado potato beetle *Leptinotarsa decemlineata* (Gorb, 1996a), and the pentatomid bug *Graphosoma italicum* (Fig. 7.5 D), have convergently developed wide microplates oriented crosswise, to the longitudinal axis of the unguitractor. Such microsculpture probably provides for quicker anchoring and releasing of the unguitractor. Similar unguitractor microstructure exists in the calliphorid fly *Calliphora vicina*. They have two lateral rows of microplates setting off a medial part with long microtrichia, which are distally-directed (Fig. 7.5 E). Two microsculpturally different areas on the fly unguitractor may serve for claw anchorage during locomotion, using a different pretarsal structure: claws on a rough surface and pulvilli on a smooth one. The unguitractor of the sphecid wasp *Bembix rostrata,* also consists of three main parts covered by different surface structures: denticle-like microtrichia located laterally, hair-like microtrichia located latero-distally, and setose protuberances located medially (Gorb, 1996a).

In water-bugs, unguitractor microstructures of walking legs (*Ranatra linearis,* Nepidae) and predatory legs (*Notonecta glauca,* Notonectidae) are mainly papillae-like, and similar to those in Odonata. Leg specialisations for predatory (*R. linearis*, legs 1) or rowing functions (*N. glauca*, legs 3) result in reduction, or loss, of claws and unguitractor microstructures.

2.4 Functional interpretations of unguitractor structures

Tendons terminating in hard sclerites are known as tibial flexor-muscle in some Coleoptera, Megaloptera, Planipennia, Hymenoptera, Heteroptera (Furth and Suzuki, 1990), and Heteroptera (Gorb, 1995a). Such an eccentric sclerite presumably serves as an additional lever, providing an increase in the tibial working angle of the femur-tibia articulation. However, the sclerite is usually smooth, to minimise friction in the contact area.

The unguitractor system has been previously interpreted as an energy-saving device, anchoring claws in the grasping position (Heinzeller et al., 1989; Seifert and Heinzeller, 1989). The surface of the unguitractor has a complex microstructure, providing for fixation to a corresponding surface of the anterior wall of the terminal tarsomere. Microtrichia of both the unguitractor and the tarsomere wall are oppositely directed, and provide anchoring of the unguitractor whenever the flexor-claw-muscle contracts.

The insect pretarsus exhibits a wide range of material design. This makes it possible for such a complex mechanical structure to actually be driven by a single muscle. It has been suggested that the unguitractor of *Chironomus riparius* (Diptera) could be released from interlocking by a change in the haemolymph pressure, with resulting changes in the shapes of the surrounding structures (Seifert and Heinzeller, 1989). Recent experimental data supports the hypothesis that highly-elastic materials play an essential role in insect tarsus control (Frazier et al., 1999). The elastic, non-layered cuticle, located between the claws and the posterior wall of the terminal tarsomere, is also involved in a claw-retraction mechanism.

The cuticle surrounding the apodeme of the posterior wall of the terminal tarsomere, contains large pore canals. Apparently, the canals in the solid material concentrate mechanical tensions and, through this, provide an increase in the material stability in the claw-joint area. A similar structural principle occurs in the vertebrate bone (Boyde and Jones, 1996).

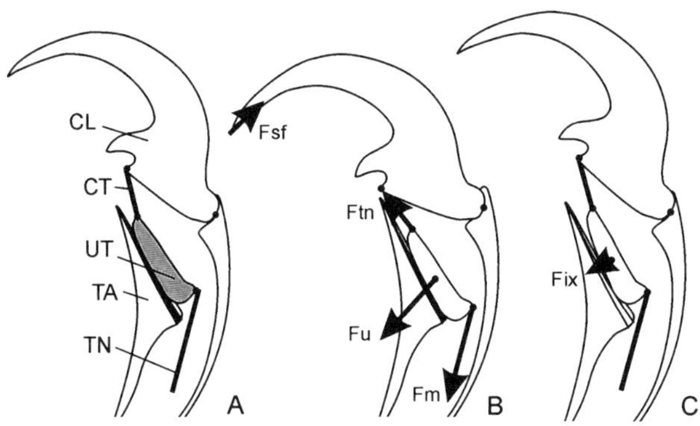

Fig. 7.6. Diagram of a possible fixation mechanism of the unguitractor. CL, claw; CT, tendon connecting claw and unguitractor plate; Fix, point of fixation; Fm, tractive force of the flexor muscle; Fsf, substrate resistance; Ftn, tendon force; Fu, resulting force of shifting unguitractor plate; TA terminal tarsomere; TN, tendon of the claw-flexor-muscle; UT, unguitractor plate.

The analysis of pretarsal structures suggests the functioning of the anchorage mechanism as follows (Fig. 7.6). During the contraction of the flexor-muscle, the claw moves proximally from an extended position. Substrate resistance (Fsf) is trying to return the claws to their initial position. Therefore, the resulting forces of both Ftn and Fm, which are directed along both tendons, CT and TN, are directed to the anterior wall of the segment where the corresponding field of microtrichia is located. The system fixates itself in this area (Fix).

3. COXA-SYNCHRONISING MECHANISM IN CICADA

Mechanisms of fast leg-movements in which the speed can surpass the velocity limitations for muscle contraction, are known for only a few insect groups such as locusts (Bennet-Clark, 1975), fleas (Bennet-Clark and Lucey, 1967; Rothshild et al., 1975), and click beetles (Evans, 1973; Furth et al., 1983). These catapult devices are designed to increase acceleration for short legs, and are provided with a specialised joint design (Gronenberg, 1996a). Joints of such devices usually contain active latch, or power muscles, which produce tractive force and trigger muscles responsible for the release of the saved force.

In Auchenorrhyncha, jumping is achieved by metathoracic muscles which are inserted into the trochanter of the hindleg. Left and right legs cannot move absolutely synchronously, because of possible, slight differences in their morphology and neural control. Even slight asynchrony can result in failure of a jump, when leg movements are very fast. Both the synchronisation of the movements of a pair of jumping legs, and their stabilisation during a jump, seem to be important problems for small jumping insects. Interestingly, different lineages of this insect group have developed different attachment devices for this purpose. Some fulgoroid leaf-hoppers (Auchenorrhyncha) have a remarkable mechanism of leg synchronisation. This mechanism involves cogwheel sectors of the cuticle of the medial surface, of hindleg coxae. Previous authors have presumed that these attachment devices aid in attaching left and right coxa, during jump performance (Sander, 1956; Heilig and Sander, 1986). Other Cicadina, such as the Cicadellidae, interlock their coxae using the snap system in which two asymmetrical structures (outgrowth and depression) on the left and right coxae, are involved (Emeljanov, 1987) (see also Chapter 5). In cercopid cicadas, median surfaces of the hind coxae bear frictional surfaces, which interlock the coxae during a jump (Gorb, 1997c).

In the cicada *Cercopis vulnerata* (Cercopidae), hind coxae are rounded in their anterior and lateral parts, whereas medial parts are planar, and contact each other over a rather large area. Microtrichia fields (270 x 320 μm) are located on the medial surface of each hind coxa (Fig. 7.7). Single microtrichia are a kind of microplate in the central and ventral regions of the field, or comb-shaped crests in the dorsal region. The microplates are 3-4 μm high, 3-10 μm wide, 2-5 μm long, and elongated along the longitudinal body axis.

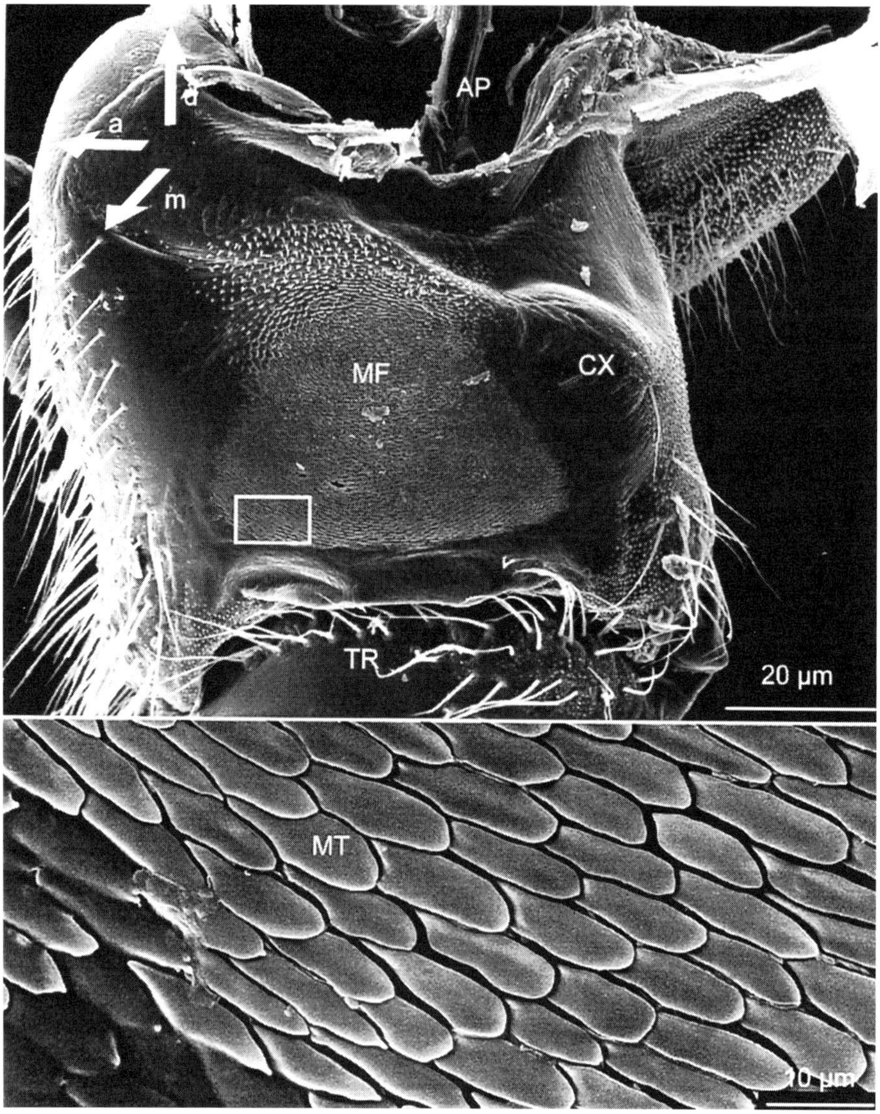

Fig. 7.7. Coxa fixation mechanism in the cicada *Cercopis vulnerata* (Auchenorrhyncha, Cercopidae). This surface is presumably responsible for synchronising jump movements of the hindlegs. A. Median aspect of the hind coxa. B. Microtrichia from the median field. a, anterior direction; AP, apodeme; CX, coxa; d, dorsal direction; m, medial direction; MT, microtrichia; TR, trochanter.

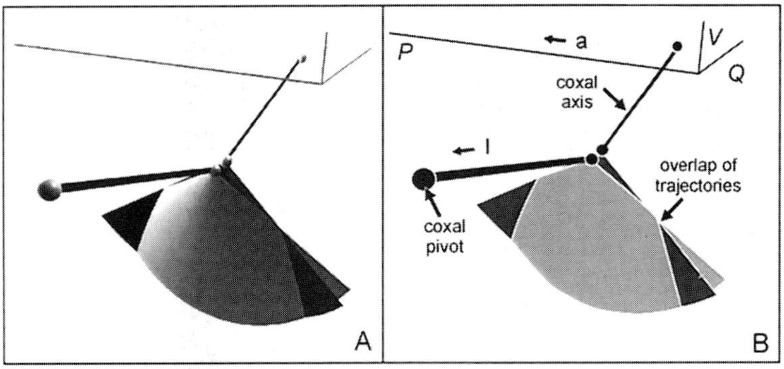

Fig. 7.8. Coxa fixation mechanism in the cicada *Cercopis vulnerata* (Auchenorrhyncha, Cercopidae) (Gorb, 1997c). 3D kinematic model of the hind coxa movements. Medial trajectories of both coxae resemble overlapping cone surfaces. In the overlapping area, microtrichia fields of the left and right side can interlock coxae. A. Cut of the rendered 3D model. B. Diagram explaining the model given in A. a, anterior direction; l, lateral direction; P, longitudinal body axis; Q, cross body axis; V, vertical body axis.

The inverse-kinematic model of propulsive leg movements was used to draw the trajectory of the medial surface of the coxa, during the jump movement. This trajectory is a cone surface, faced with its bulged-in side, medially. Trajectories of the right and left coxae overlap in their most anterior, and most posterior positions (Fig. 7.8). In both extreme positions, coxae are presumably connected to each other by coupled microtrichia fields of medial surfaces of coxae. In other words, in extreme positions, both coxae can be moved together synchronously.

4. SELF-FOLDING ARMOURED MEMBRANES IN DIPTERA

Armoured membranes are newly discovered, friction-active structures, investigated in eight fly species from six families (Tabanidae, Stratiomyidae, Calliphoridae, Syrphidae, Muscidae, Drosophilidae) (Gorb, 1997a). The flexible, membranous cuticle is covered by single or grouped microtrichia, often located on thickened areas of the membrane or directly on it (Fig. 7.9). Such membranes exist in joints between basal segments of legs (pleurocoxal, coxo-trochanteral), and in transitional areas between segments, such as head-proboscis and prothorax-neck-head transitions.

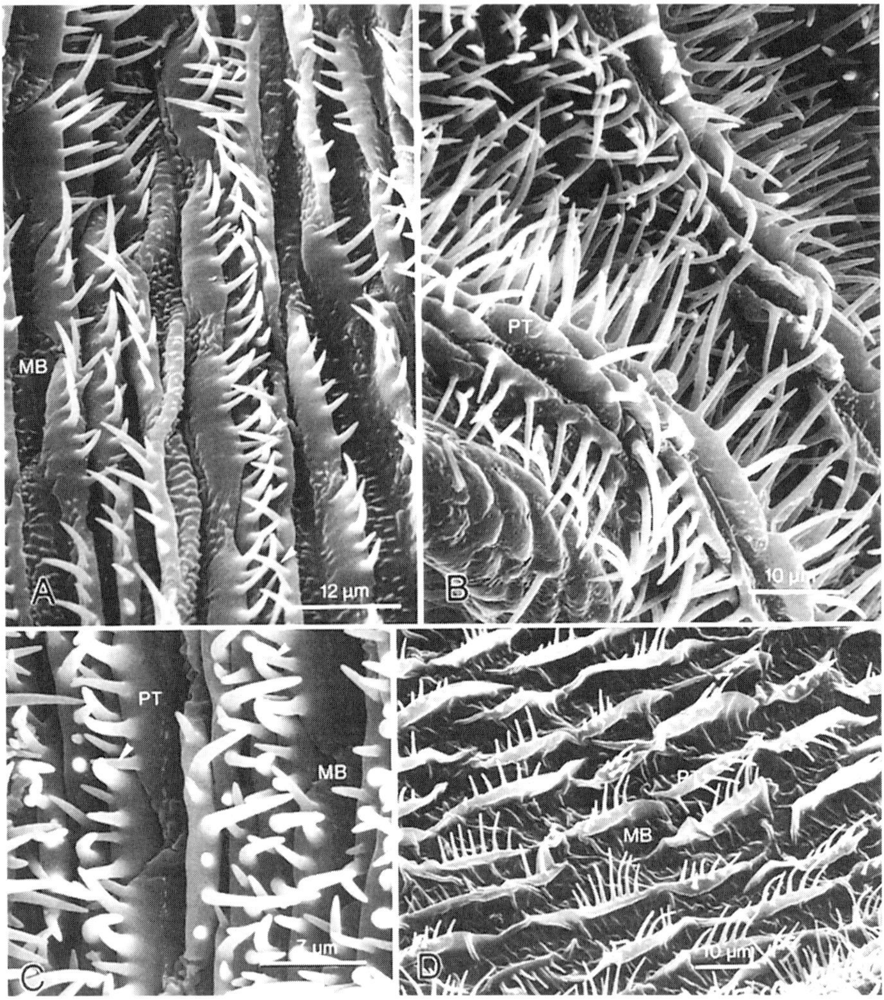

Fig. 7.9. Membranous cuticle of the head-proboscis joint, in the syrphid *Eristalis tenax* (A-C) and the tabanid *Tabanus bovinus* (D) (Gorb, 1997a). It is designed as microplates connected with flexible areas. This construction of the cuticle may result in parallel folding during movement within the articulation. MB, flexible areas; PT, microplates. White arrowheads indicate points of contact between microtrichia.

Presumable mechanical functions of armoured membranes are fixation of the folds, and determination of the direction of folding. In Diptera, five types of membrane microstructure were identified with the aid of scanning electron microscopy.

The first type of membrane is covered by single, short, papilla-like microtrichia arising directly from the flexible membrane (Fig. 7.10 A). This type is located in the prothorax-neck transition of the majority of species studied, and in the leg joints of the drosophilid fly *Drosophila melanogaster*. The density of these papillae is quite high (2-4 μm^{-2}).

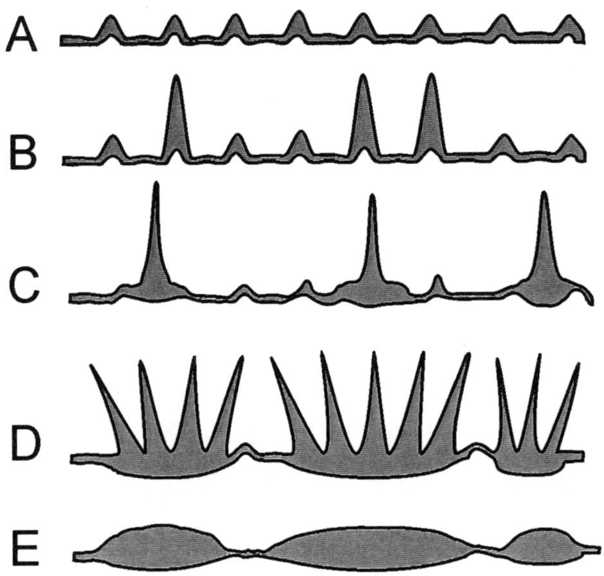

Fig. 7.10. Types of microstructure of the membranous cuticle existing in representatives of Brachycera (Diptera) (Gorb, 1997a). A. Single, short, papilla-like microtrichia directly connected with the flexible membrane. B. Single, elongated microtrichia directly connected with the flexible membrane. C. Single microtrichia located on the microplates. D. Microtrichia groups located on the microplates. E. Membrane composed of microplates without microtrichia.

The second type is characterised by single, elongated microtrichia, such as those from the neck-head transition in the muscid *Musca domestica,* and the syrphid *Myathropa florea* (Fig. 7.10 B). Areas between these elongated microtrichia are often covered by micropapillae.

The third type is a membrane in which single microtrichia are located on small areas of less flexible cuticle, called microplates (Fig. 7.10 C). Such membranes were found in the pleurocoxal joint of the calliphorid *Lucilia caesar,* and in the coxo-trochanteral joint of the syrphid *Eristalis tenax* (Fig. 7.10 D). The flexible, membranous areas between such microplates usually contain small papillae.

Membranes with groups of microtrichia located on the microplates belong to the fourth, and most complex membrane type. (Fig. 7.10 D). They were reported from the head-proboscis transition of the syrphid *E. tenax,* and tabanid *Tabanus bovinus*, in the pleurocoxal joint in *E. tenax*, and in the head-neck transition of *T. bovinus*. Microplates arising from the flexible membrane are usually oval in shape, and are orientated parallel to each other. The fifth membrane type consists of microplates completely lacking microtrichia. Such structures are known from the coxo-trochanteral joint of the stratiomyid *Stratiomys chamaeleon* (Fig. 7.10 E).

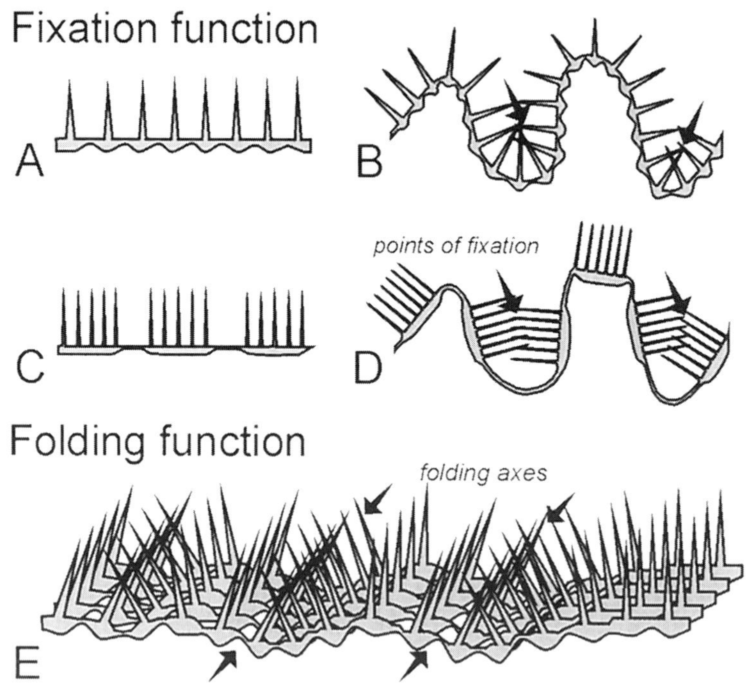

Fig. 7.11. Diagram demonstrating possible functions of armoured membranes in Diptera (Gorb, 1997a). A-D. Fixation. Microtrichia, located between two folds, can readily come into contact with each other, and fix the folds in a stable position until the fly actively changes the relative positions of the joint segments. E. Folding. Membranous cuticle, that is designed as parallel-orientated microplates connected by flexible membrane, may provide folding directionality during joint movement.

The prothorax-neck transition seems to be less specialised in the membrane armature as compared with the head-neck, and especially, the head-proboscis transition. The latter has the most complex membrane armature in terms of the number of single microtrichia per microplate, length of microtrichia, and shape of the microplates. In the pleurocoxal articulation, which is the most basal joint of the legs, the membrane has larger microplates and a higher number of microtrichia per microplate, than in more distal joints.

When the membrane is actively folded by contraction of the muscles moving the segments, fold surfaces covered with microtrichia can contact each other. According to the microsculptural design of the membrane, such mechanical functions as (1) predetermined folding and (2) fold fixation, are suggested (Fig. 7.11). Parallel-orientated microplates, which are connected to the flexible membrane, are responsible for the first function (Fig. 7.11 E). The rows of plates predetermine directionality of folding and unfolding, during movements in the articulation (Fig. 7.9). Directional activity of such armature was observed at the head-proboscis transition in the syrphid *Eristalis tenax,* the tabanid *Tabanus bovinus,* and in the leg joints of the calliphorid *Calliphora vicina*.

If the microtrichia are longer than the radius of the fold, the microtrichia located between two closely-located folds can readily come into contact with each other, and can fix a fold in a stable position, until the fly actively changes the relative positions of the segments at a joint (Fig. 7.11 A-D). Multiple connections between folds may prevent sliding within them. Such microtrichia-fixed folds, were observed in the shock-frozen preparations of the pleurocoxal joint in the calliphorid *Lucilia caesar*, in both prothorax-neck and head-proboscis transitions of the tabanid *Tabanus bovinus*, and in the neck-head transition of the syrphid *Myathropa florea*.

Observations of living *Calliphora* show that roll perturbations of head posture, up to 60°, remain until immediately before the fly starts to walk. At this moment, the fly can actively correct its head posture (Gilbert et al., 1995). It has been hypothesised that the frictional forces in the head-neck region must be quite high to keep the head in the perturbed position. The armoured membranes may be responsible for this mechanism. However, rigorous experiments are still needed to prove the mechanics of the neck-head articulation in flies. It should also be noted that the biomechanics of membranes have an additional complexity because of the presence, under the membrane, of a fluid-filled sinus (Poodry, 1980); this must have a strong effect on the membrane geometry during the displacement of segments at articulation.

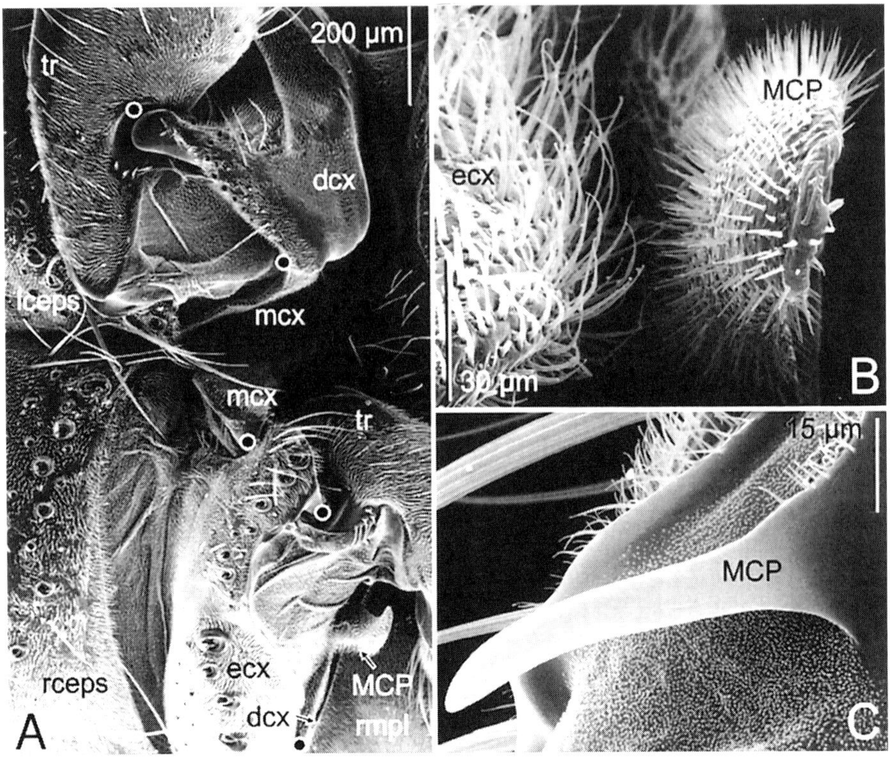

Fig. 7.12. Middle coxae and mid-coxal prongs in flies (Frantsevich and Gorb, 1998). A. Ventral view of the coxae of the calliphorid *Lucilia caesar* in pronating configuration (below), and in supinating configuration (above). Anterior direction points to the left side of the panel. B. The prong in the syrphid *Eristalis tenax*, as seen from the apex; the hairy, anterior face of MCP against the hairy area of the eucoxite. C. The prong in the drosophilid *Drosophila melanogaster*. ecx, eucoxite; dcx, distocoxite; lceps, left episterne; MCP, mid-coxal prong; mcx, mediocoxite; rceps, right catepisterne; rmpl, right meropleurite; tr, trochanter. Articulaion points are marked with white circles filled with black.

5. COXO-TROCHANTERAL LOCKING MECHANISM IN FLIES

The mid-coxal prong is a tiny, dagger-like process protruding from the posterior side of the coxa into the space between the coxa and the trochanter.

It is present in all but a few families of the huge group of Brachycera (Frantsevich and Frantsevich, 1999). Its covert position was among the reasons that entomologists neglected this organ. Its purpose was unknown until recently (Frantsevich and Gorb, 1998). During flight, the prong is clumped between the coxa and trochanter, its anterior, hairy surface faces a hairy field on the anterior coxal sclerite. The flight position of the middle leg is caused by the trochanter rotation (Fig. 7.12 A). The trochanter presses both hairy surfaces together. Morphology and the biomechanics of the middle coxa suggest that the prong is a sort of Velcro fastener, locking the middle leg into its flight position.

The prong is soft and bare in smaller species, such as *Drosophila melanogaster* (Fig. 7.12 C), and sclerotised at its base in large flies. Its hind surface, facing the trochanter, is absolutely smooth. The trochanter, in turn, has a smooth groove on its antero-dorsal face, which the prong glides into during leg movement. In large species, the front side of the prong is covered with microtrichia. Viewed from the top, it looks like a hedgehog on its back, with a bare belly and spines. (Fig. 7.12 B). The same site, which bears the prong in muscomorph flies, has a pyramidal outgrowth like the articulation membrane in horse flies (Tabanidae). The front face of this outgrowth is densely covered with hair.

No sensilla were found on the prong or at its base. No reflexes in the subcoxal muscles were obtained in electrophysiological recordings, upon tactile stimulation of the prong. This process performs, presumably, some mechanical function, which is supported by data of fluorescence microscopy. The prong exhibits an autofluorescence in UV light. This suggests the presence of resilin, a rubber-like protein, in the prong cuticle.

The middle leg contributes to several types of motion, including walking, grooming, grasping, flight start, flight posture, and the landing reflex. In all of the named modes, except the flight posture, the trochanter and the femur (which act together) are more or less depressed, and the prong slacks freely, or slightly touches the trochanter.

No significant differences were revealed in the gait, leg positioning relative to the body, or in parameters of the step (duration of the swing and stance phases, and stride length versus velocity) in flies, before or after prong removal. The leg span increased by 5% after the operation.

The position of the fly's legs during flight, is rather peculiar (Fig. 7.13 A, B). The femora of the front and middle legs are elevated, tibiae are flexed, and the legs are pressed to the thorax. The tarsi point ahead and inward. In such a position, the legs offer only slight aerodynamic resistance. During flight, the prong is clumped between the coxa and the trochanter.

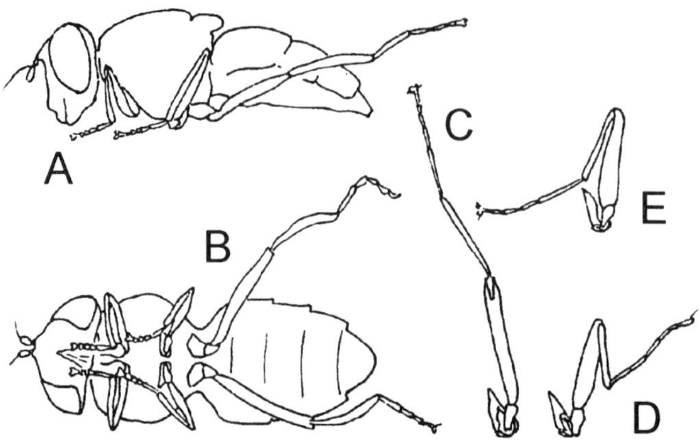

Fig. 7.13. Leg positions during flight in the syrphid fly *Eristalis tenax* (from photographs) (Frantsevich and Gorb, 1998). A. Lateral aspect. B. Ventral aspect, the fly yaws to the left. C-E. Left middleleg configurations seen from below, as in B. C. The leg is extended. D. The leg is elevated, the tibia flexed, the femur pronated. E. The same, but the femur is supinated.

The middle femur points backwards and upwards. If the middle coxa was simply retracted, the femur raised and the tibia flexed, then the tarsus would point outwards instead of inwards (Fig. 7.13 C, D). The leg must also be rotated about the longitudinal axis of the femur at an angle of about 20-60°, in order to attain flight position. Additional rotation is performed by internal mobility of the middle coxa, which consists of three separate sclerites called the eucoxite, the mediocoxite, and the distocoxite (Fig. 7.12 A; 7.14).

The kinematics of the middle coxa are rather complex (Fig. 7.14). The coxa is attached to the thorax at two fixed points: at the pleural ridge and at the evaginated peg on the furcasternum. The pleural ridge is articulated with the eucoxite, and the furcosternal peg is a pivot for the mediocoxite. Rotation of the coxa about this subcoxo-coxal axis results in protraction, and retraction, of the leg. The trochanter articulates anteriorly with the eucoxite and posteriorly with the distocoxite. This pair of articulations defines the coxo-trochanteral axis. Rotation of the trochanter about this axis results in elevation, or depression, of the femur. Movements of the coxal sclerites, relative to each other, bring the distocoxite into the inferior or superior position, thus changing the direction of the coxo-trochanteral axis. One may describe this movement as the rotation of the trochanter about the intracoxal axis. If the femur is depressed, this rotation adds to protraction-retraction. The elevated femur lies almost parallel to the intracoxal axis, hence rotation about this axis, in this configuration, results in supination or pronation of the femur. There is a difference between the two configurations of the coxa: the exposed eucoxite in the pronated right leg, and the exposed distocoxite in the supinated left leg (Fig. 7.13 A). Three coxites and the trochanter form a closed kinematic chain of four segments, with one degree of freedom.

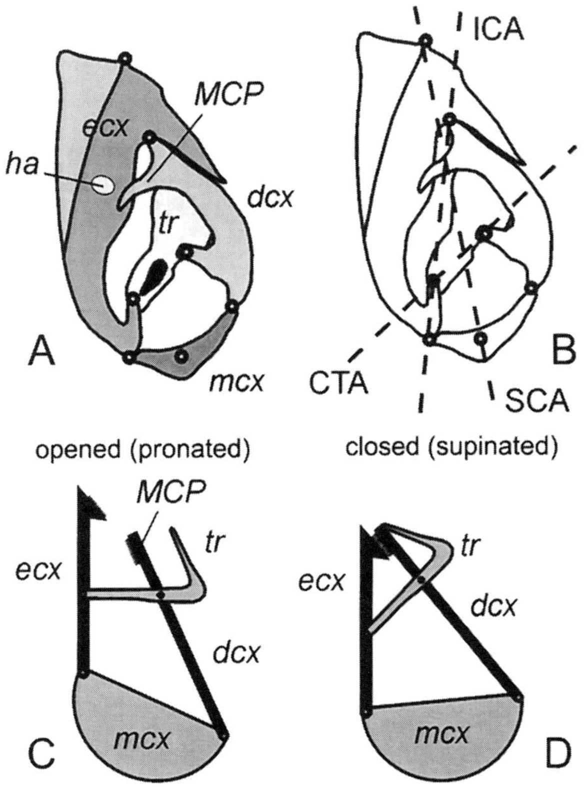

Fig. 7.14. Coxo-trochanteral locking mechanism in flies (Frantsevich and Gorb, 1998). A. Coxal sclerites. B. Equivalent kinematic diagram of the coxa. C, D. Flat model of the locking mechanism. Only the anterodorsal face of the trochanter is shown. C. Free position. D. Locked position. In the supinated configuration, hair brushes on the distocoxite (i.e., on the prong) and on the eucoxite are pressed together by the trochanter. Circles indicate articulation points. CTA, coxo-trochanteral axis; ha, hairy area; ICA, intracoxal axis; SCA, subcoxo-coxal axis; for other abbreviations see Fig. 7.12.

The supinated trochanter bends the prong and presses it against the eucoxite. In large flies, the part of the eucoxite against the hairy face of the prong, is covered with similar microtrichia. Mechanical interaction of both hairy surfaces suggests that in this position, they can interlock. The coxa becomes strapped in the supinating configuration. When the trochanter presses against the prong, the coxa is self-locked (Fig. 7.14 C, D).

This hypothesis explains the shape, and position of the prong, and predicts that an individual without this structure, would be awkward putting its legs in the flight position, and holding them against the opposing air stream. However, no significant difference was observed in flight performance between intact, and altered flies. This result may be explained by certain mechanisms (muscle activity) compensating for the lack of the lock mechanism.

Chapter 8

Wing-locking devices

There are three functionally different groups of wing fixators: (1) the first one serves for interlocking fore and hindwings in flight; (2) the second one serves to affix forewings to each other, in order to prevent them from pasting together; (3) the third one locks the wings, in the resting position, to the body. The fixators of the first type are known among Lepidoptera (Shvanvich, 1949; Obenberger, 1964; Mathew, 1984), Hymenoptera (Schneider and Schill, 1978; Abrol, 1986), Heteroptera (Emeljanov, 1987), and Auchenorrhyncha (Weber, 1930; Schneider, 1981; Emeljanov, 1987). They are usually designed as (1) two co-opted folds of wings, (2) wing fold and hook-like setae, or (3) two sets of hooking setae (Chapter 4). The fixators of the second type, reported only in parasitic wasps from the families Encyrtidae and Aphelinidae (Hymenoptera, Chalcidoidea), are also designed according to the hook-principle.

For Hymenoptera (Schrott, 1986), Heteroptera (Weber, 1930), Coleoptera (Hammond, 1989; Samuelson, 1994; 1996), Dermaptera (Haas, 1995), some Diptera (Rodova, 1968), and some Lepidoptera (Common, 1969), the third type of wing fixator is known. These insects have convergently developed the ability to attach their wings to the body when resting. Such wing-to-body locking devices have very different designs and origin in different lineages of Insecta. They range from snap-like structures in some Heteroptera and Hymenoptera, to typical fields of microtrichia, or acanthae, in Coleoptera and some Diptera, and finally to the thoracic hair brushes consisting of large setae in Dermaptera. Independent of the general design of the system, the largest similarity of these devices is that co-opted fields of cuticular outgrowths are present on two separate parts of the body.

1. COLEOPTERA

In beetles, the system responsible for attachment of the forewings (elytra) to the thorax is composed of a few interlocking structures, located between thorax and abdomen, and between left and right elytra. These structures have been studied in detail in Carabidae (Baehr, 1980). They are designed as co-opted lock-and-key devices. The shape of the inner surface of the elytra repeats the shape of the corresponding structures on the pterothorax (Chapter 4). The macrostructure of co-opted surfaces is optimised to provide mechanical interlocking between elytra and thorax + abdomen. There are three main areas where elytra are locked to the body: 1) medial margins of both left and right elytra; 2) lateral margins of elytra, and pleural area of the pterothorax and anterior abdominal sternites; 3) anterio-medial area of elytra, and corresponding structures of tergites of mesothorax and scutellum. Median margins of the left and right elytra are asymmetrically designed. The right margin has a furrow, whereas the left one has a groove. Such lock-and-key construction locks left and right elytra to each other, and prevents sliding movements along the dorso-ventral axis of the body. The furrow is covered with intensely depressed microtrichia with short, fine tips, oriented in a posterior direction. Scutellum holds the locked elytra together in the anterio-medial region of elytra.

In the majority of beetle taxa that can fly, these structures are supplemented with microtrichia fields on the inner surface of the elytra, and the adjacent parts of the pterothorax. Such fields differ in the density, design, and directionality of the microtrichia (Chapter 11). The configuration of microtrichia fields, as well as the microtrichia structure on elytra and the thorax, are known as an important set of characters used in the systematics of some groups of Coleoptera (Hammond, 1989; Samuelson, 1994; 1996). The ultrastructure of the microtrichia, underlying cuticle, and epidermal layer has been previously studied in detail in tenebrionid beetles (Gorb, 1998a; 1999b).

1.1 Microstructures

In the three species of tenebrionid beetles (*Tribolium castaneum*, *Tenebrio molitor*, and *Zophobas rugipes*), there are five microtrichia fields on the thorax and abdomen, and eight fields on the elytra and hindwings (Fig. 8.1 A-C). Each of these fields prevent deviation from the closed system, in some preferred direction. However, the locations of fields and the directionality of microtrichia, make it possible for the whole system of both elytra, to be stable against shifting in any direction, when elytra are fixed to the body (Chapter 11).

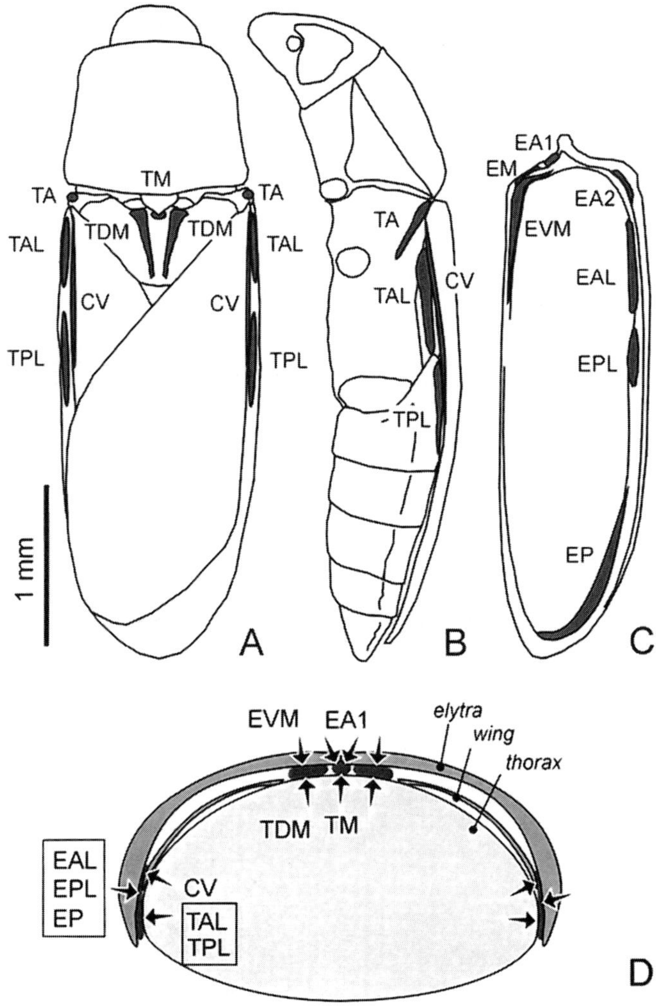

Fig. 8.1. Locations of microtrichia fields in *Tribolium castaneum* (Coleoptera, Tenebrionidae) (Gorb, 1998a). A. dorsal aspect, elytra removed. B. lateral aspect, elytra removed. C. Left elytra, ventral aspect. D. Cross section of the beetle-body in the area of pterothorax; the location of co-opted fields is indicated with arrows. CV, costal vein; EA1, first anterior field of the elytra; EA2, second anterior field of the elytra; EAL, anterio-lateral field of the elytra; EM, medial field of the elytra; EP, posterior field of the elytra; EPL, posterio-lateral field of the elytra; EVM, ventro-medial field of the elytra; TA, anterior field of the thorax; TAL, anterio-lateral field of the thorax; TDM, dorso-medial field of the thorax; TM, medial field of the thorax; TPL, posterio-lateral field of the thorax.

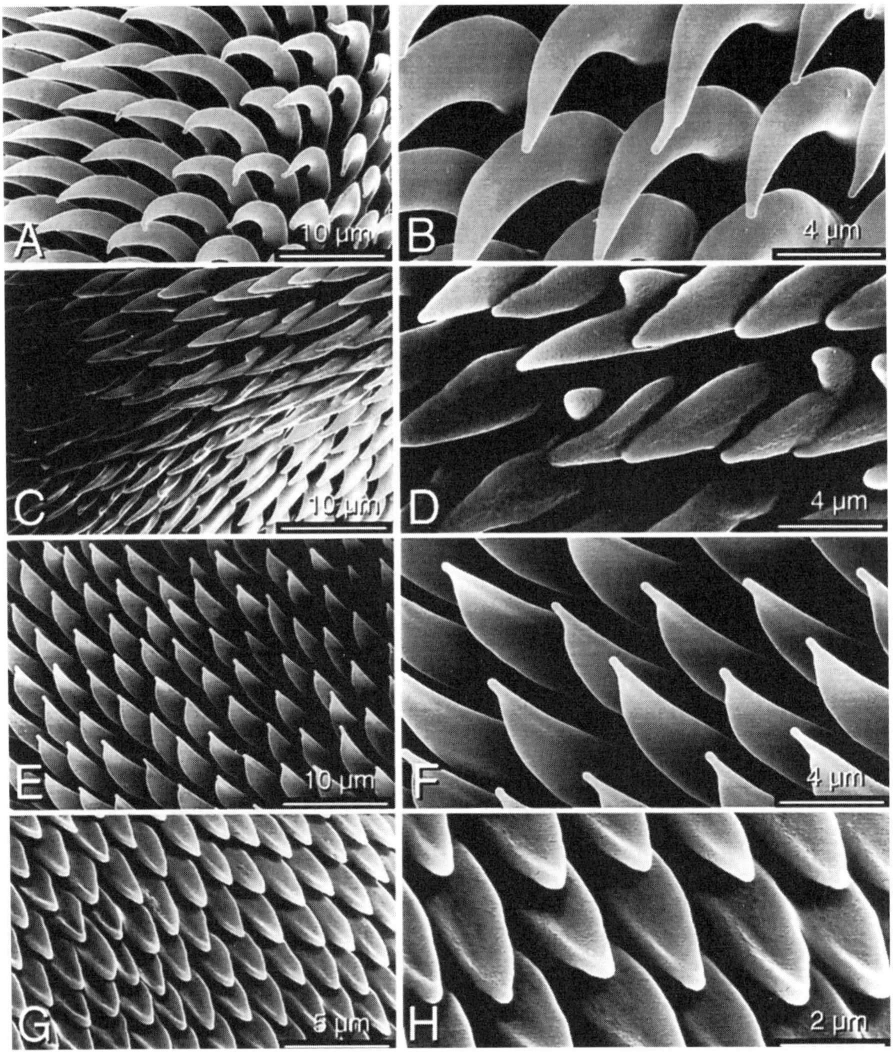

Fig. 8.2. Microtrichia fields in elytra-locking devices in tenebrionid beetles *Tenebrio molitor* (A, B, E, F), and *Tribolium castaneum* (C, D, G, H) (Gorb, 1998a). A-D. The TAL field. E-H. The TPL field.

Hindwings and elytra. Eight fields occur on the hindwings and elytra (Fig. 8.1). The costal vein of the hindwing (CV) is covered by microtrichia. In *T. castaneum*, it contains approximately 4-6 rows of microtrichia. In *T. molitor*, it has 12-16 rows. The microtrichia of the costal vein can contact both anterio-lateral (EAL) and posterio-lateral fields (EPL) of elytra. The fields occur on the lateral wall of the elytra, and are densely covered by slightly curved microtrichia. Microtrichia are distributed quite regularly. In *T. castaneum*, an additional field (EP) occurs at the posterior and posterio-lateral margins of the elytra. Both medial (EM) and ventro-medial fields (EVM) are located at the anterio-medial edge of the elytra. They differ in the design and directionality of microtrichia. The EM field contains cone-shaped crests, whereas the EVM field is covered by triangular or rounded microplates. Close to its posterior border, the latter field is covered with long microtrichia. The first anterior field of the elytra (EA1) is located near the medial side of the elytra-thorax joint. It is oval and consists of 60-90 single, cone-shaped microtrichia, in the smallest species studied. The second anterior field (EA2) is located near the lateral side of the elytra-thorax joint, and runs to the lateral margin of the elytra.

Thorax and abdomen. Five microtrichia fields occur on the thorax and abdomen (later we refer to them as thoracic fields). Both the anterio-lateral (TAL) and posterio-lateral fields (TPL) are binding patches corresponding to those EAL and EPL fields located on the elytra (Fig. 8.1). The TAL field is located on the pleurites of the lateral part of the pterothorax. The TPL field is located on the pleurites of the anterior abdominal segments. Both fields are densely covered by microtrichia, which are curved or slightly compressed (Fig. 8.2). The dorso-medial field of the thorax (TDM) corresponds to the EM and EVM fields of the elytra, and is located on the tergites of the pterothorax (Fig. 8.3). In fact, microtrichia of the TDM field cover the cuticular fold, in a shape which corresponds to the shape of the anterio-medial edge of the elytra. Microtrichia of the TDM field are widened and intensely compressed. The anterior field of the thorax (TA) is located on the lateral side of the mesothorax, and is lengthened in the posterio-ventral direction. This field usually contains cone-shaped microtrichia. The medial field of the thorax (TM) is the smallest of those involved in the elytra-locking system. It contains several cone-shaped microtrichia located just beneath the scutellum. This field functionally corresponds to the EM field of the elytra.

1.2 Functions of the elytral lock

If we compare frictional systems of wing-locking devices among insects, the beetle's system seems to be organised in the most complex way, according to the design of the macro-shape of both elytra and thorax, and to the number of fields of microtrichia involved. Elytra have a variety of additional functions, which are not common for the forewings of other insects.

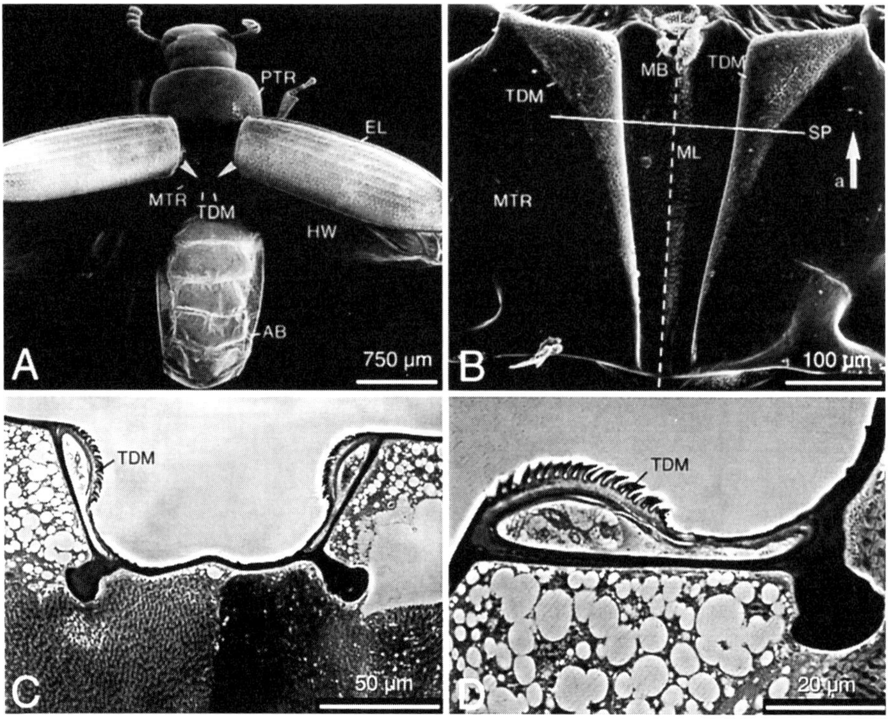

Fig. 8.3. Thoracic dorso-medial field (TDM) in the beetle *Tribolium castaneum* (Tenebrionidae) (Gorb, 1999b). A. Dried specimen, dorsal aspect. B. Spurr-resin-cast of the TDM-field. C, D. Semi-thin sections stained with tolouidine blue. AB, abdomen; HW, hindwing; MB, membrane; ML, midline; MTR, metathorax; SP, section plane of semithin sections shown in C and D; PTR, prothorax.

The elytra provide complete cover for the delicate hindwings and abdomen. When they are interlocked with the thorax and abdomen, the body forms a stable unit. Elytra protect the flight-wings from becoming wet, dirty, or damaged. In many species of Coleoptera, elytral coverage minimises loss of water. As previously suggested (Hammond, 1989), the evolution of body strength and protection necessary for life in the soil and thin substrata, and in arid or wet environments, resulted in successive changes of the beetle's forewing design and wing-locking mechanisms. The arrangement of bifunctional muscles in Coleoptera suggests that the effective attachment of both elytra and hindwings to the body, may be important in enabling the bifunctional muscles to work effectively during walking (Karner, 1997). Such structures are gradually lost, or strongly reduced, in the course of evolution, in species with fused left and right elytra, or species that have lost flight ability (for example *Lethrus apterus*, and also flightless Carabidae and Tenebrionidae) (see also Chapter 13).

It seems that the elytra are multifunctional structures, and that the correlation between different functions, and their relative importance for selective pressure, is different, in different taxa and ecological groups of beetles. All functions mentioned above require tight connection of elytra to the thorax, in a non-flight condition. Microtrichia fields can hold resting elytra, and probably seal contact areas to prevent water loss, and hindwings from becoming dirty. Indirect evidence for this hypothesis is that, in the scanning electron microscope, the microtrichia fields are often viewed while submerged in the amorphous substance, which is secreted by epidermal cells and transported through the system of pore canals, to the surface (Chapter 10). Since microtrichia, within one field, are directed in one preferred direction, they may indicate preferred direction of gliding for elevating elytra.

It has been experimentally shown that some scarabeid beetles (Geotrupini) use cuticular structures of the elytra for sound production (Palestrini et al., 1987). The noise generation, which may be involved in communication or in defence against predators (Palestrini et al., 1987), was hypothesised to be another possible function of the microtrichia fields. However, the stridulation structures are usually arranged as rows of cuticular micro-combs. The sound-generation groups of beetles can be easily recognised due to their typical combs-grooves pattern, existing on the ventral surface of the elytra (Hammond, 1989; Hyder and Oseto, 1989; Schmitt, 1991), and that they are not identical to structures that are responsible for attachment. Since the structures discussed above are very similar in beetle species from a variety of families, such as Tenebrionidae, Scarabaeidae, Haliplidae, Gyrinidae, Dytiscidae, Hydrophilidae, Anobiidae, Dermestidae, Carabidae, Cicindellidae, Staphylinidae, Chrysomelidae, and Elateridae, it is doubtful that such a similar design of corresponding fields could be involved in species communication.

2. DERMAPTERA

Earwigs probably have the most unusual hindwing venation and folding pattern among insects (Haas and Wootton, 1996). Both patterns correlate with unusual wing-folding mechanics, in which folding is achieved from within the wing, and unfolding is done by cerci. Resilin-bearing elastic springs exist dorsally in the radiating veins, but ventrally in the intercalary veins (Haas et al., 2000). This distribution determines the folding direction, since resilin is the major driving mechanism for wing folding.

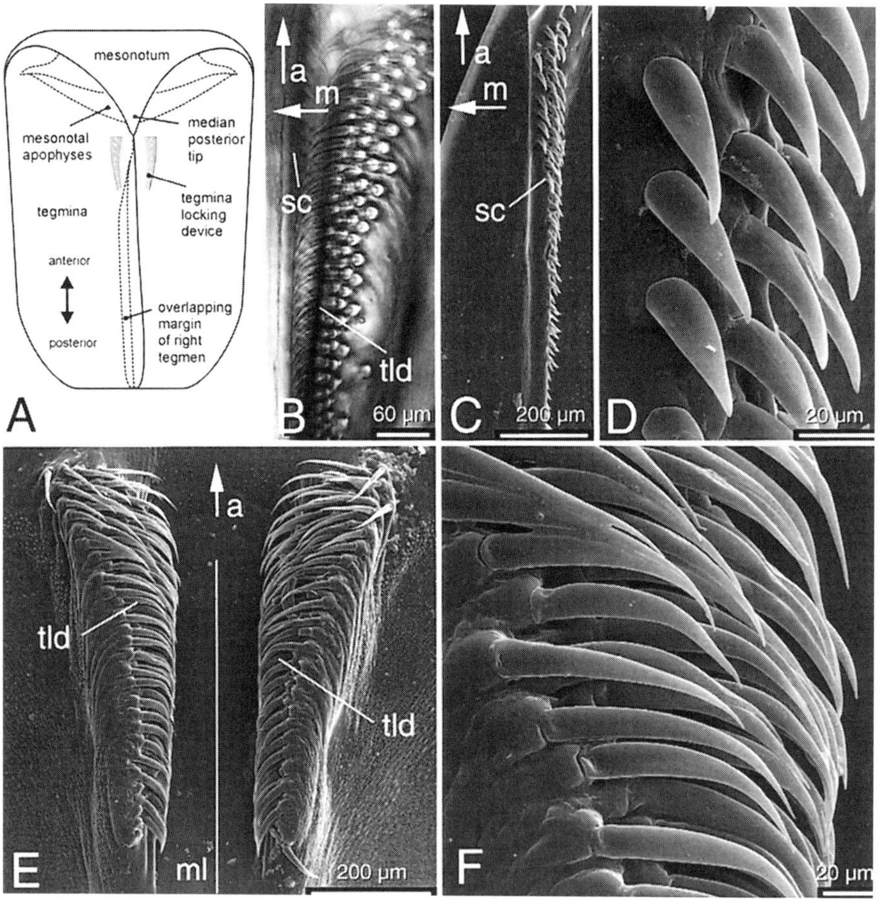

Fig. 8.4. Wing-locking mechanism in *Forficula auricularia* (Dermaptera, Forficulidae). A. Schematic dorsal view of the mesonotum with tegmina of Chelisochidae, Forficulidae, and Spongiphoridae. B. Wholemount of the wing-locking device in the locked condition (light microscopy). C-D. Spiny crest of the tegmen. E-F. Tegmina-locking device of the metanotum. a, anterior direction; sc, spiny crest, m, medial direction; ml, midline, tld, tegmina-locking device. A, B. Courtesy of Fabian Haas (University of Ulm).

The folded wings are covered by tegmina (covering leathery forewings), which can be locked to the metanotum in the resting position. Such a mechanism not only prevents the sliding apart of the tegmina, but also protects the wings, metanotum, and the third thoracic stigmata.

In *Forficula auricularia* (Forficulidae), there are two rows of postero-medially oriented setae, situated beside a median groove of the metanotum (Verhoeff, 1902; Giles, 1963) (Fig. 8.4). These thoracic fields are called the tegmina-locking device. The corresponding structures are setae, located medially on a cuticular ridge of each tegmen (covering wing), and together with the ridge they are called the spiny crest (Haas, 1995). The setae bear sockets at their bases and are quite large structures (90-100 μm long and 10-15 μm wide).

Haplodiplatys orientalis (Diplatyidae) lacks the tegmina-locking device on the metanotum. However, a weak, spiny crest of the tegmen can interfere with the, more or less pronounced, median longitudinal groove, thus preventing the sliding apart of the tegmina, even without a tegmina-locking device. It has been hypothesised that a spiny crest, without tegmina, is a kind of preadaptation in the evolution of a complete locking device bearing both structures, spiny crest and tegmina-locking device (Haas, 1995). The presence of these structures is an autapomorphy of Dermaptera.

3. DIPTERA

Wing-locking devices in Nematocera (Diptera) were originally described for species of Chironomidae, using light microscopy (Rodova, 1968; 1971; 1974; 1980). They consist of a pair of brush-like fields of microtrichia, situated on the thorax at each side of the body, laterally from the scutellum. The corresponding wing-brushes are located on the alulae (postero-proximal wing area) at the wing-bases. Both, the thoracic and wing fields consist of cone-shaped, elongated microtrichia. The protuberances of the thoracic fields are shorter, and more dense than those on the alula. While resting, the wing alula is bent at a sharp angle to the wing plane, and pressed against the thorax. In such wing position, both brushes interlock with each other. Similar structures were found in *Bryophaenocladius illimbatus* (Chironomidae) and *Neosciaria* sp. (Sciaridae), with the use of the SEM technique.

Representatives of the Bibionidae posses dual wing-lock. They can attach not only wing alula to the thorax, as described above, but also two parts of a folded wing. The wing-thoracic lock consists of a *wf1* microtrichia field in the lateral part of the alula, and a dense microtrichia brush at the lateral side of the thorax (Fig. 8.5). The two additional brushes (*wf2, af3*) are situated at the bases of the two principal veins. They can interlock with each other by pulling together and slightly folding the wing into resting position. Such a mechanism has a quite complex kinematics, whose morphological background involves a variety of joints, and several resilin springs.

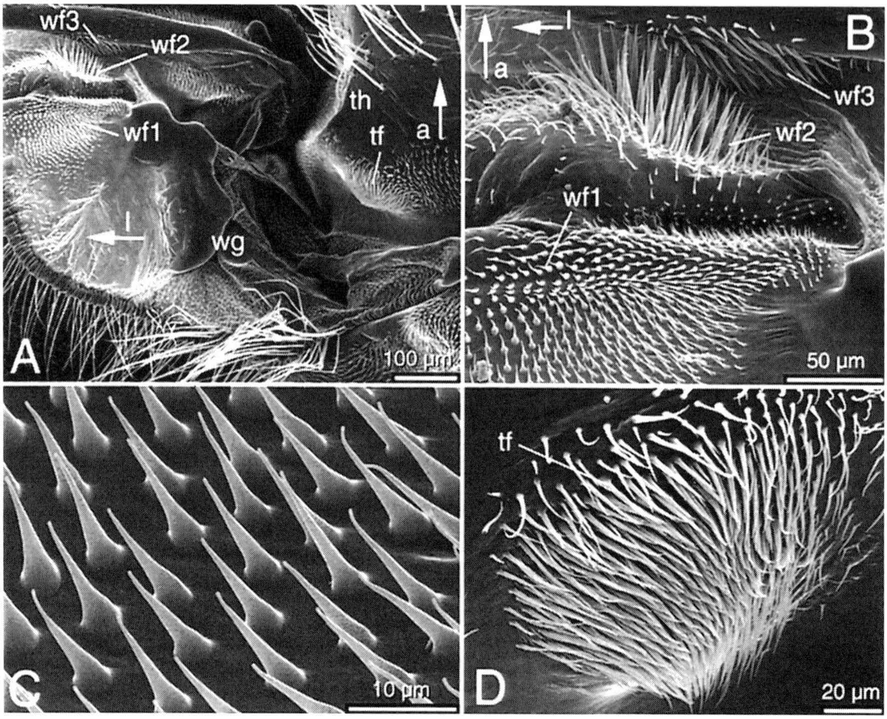

Fig. 8.5. Wing-locking mechanism in *Bibio nigriventris* (Diptera, Bibionidae). A. Base of the left wing. B. Wing fields. C. The wing field (wf1) of the wing-thoracic lock, the left side of the body, dorsal aspect. D. The thoracic field (tf) of the wing-thoracic lock, right side of the body, dorso-lateral aspect. a, anterior direction; l, lateral direction; tf, thoracic field; wf1-wf3, wing fields.

4. HYMENOPTERA

The so called «cenchrus-thorn» apparatus, is an autapomorphy of sawflies («Symphyta»). Representatives of all their families, with the exception of Cephidae, have this wing-locking system. Dorso-laterally, in the area between mesoscutellum and metapostscutellum, two symmetrical balloon-like structures exist that are called cenchri (Mallach, 1936; Schedl,

1991). They are usually light coloured, and have an area of about 0.05-0.40 mm^2 (Fig. 8.6 A).

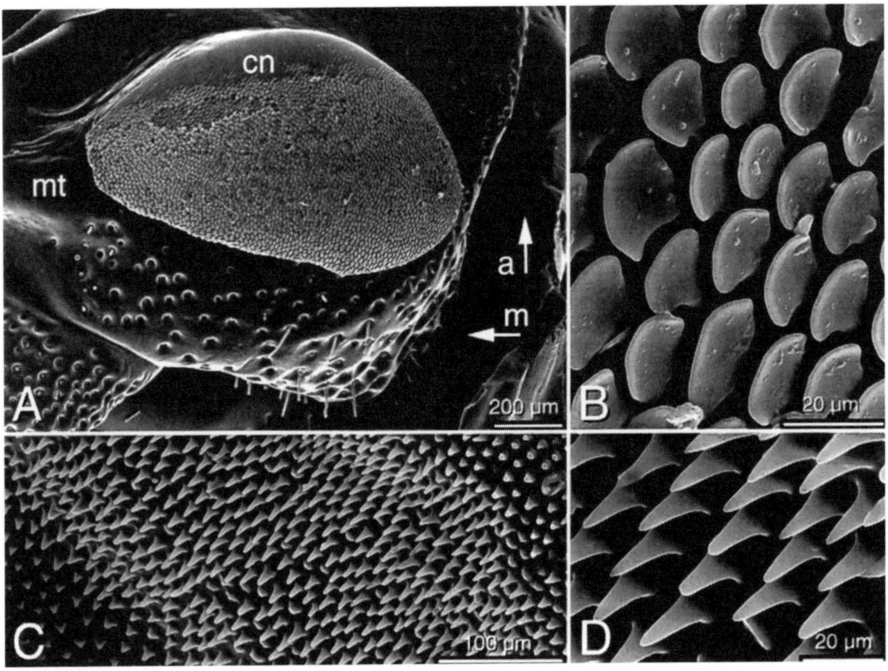

Fig. 8.6. «Cenchrus-thorn» locking mechanism of the forewing in *Urocerus gigas* (Hymenoptera, Siricidae). A-B. Dorsal aspect of the right side of the metathorax. C-D. Ventral surface of the forewing base. a, anterior direction; cn, cenchrus; m, medial direction; mt, metathorax.

The structures are composed of 100-400 tiny cuticular protuberances. While resting, these areas are in contact with the ventral surface of the anal region of the forewing. Both thoracic and wing fields, consist of microtrichia (Fig. 8.6 B-D). On the thorax, the microtrichia often have a quite complex shape, with strongly compressed distal parts, whereas on the wings they are cone-shaped (Schrott, 1986).

In the series of experiments with living «Symphyta», it has been shown that the removal, or damage, of the cenchri caused wing-gliding on the thorax. The ultrastructure of the cuticle of the cenchri is different, if compared to the rest of the thoracic cuticle. In the region of the cenchres, it is thinner, and its layers are clearly cut. Presumably, epidermal secretion is an additional attachment agent, as in a variety of other frictional systems (Chapter 10). In the region of the cenchres, underlying epidermal cells contain well developed, smooth endoplasmic reticulum, and many lipid vesicles. The ducts transporting vesicle content to the cuticle surface have not been found (Schrott, 1986).

5. HETEROPTERA AND OTHER GROUPS

Representatives of Heteroptera have a snap-like wing lock (Chapter 4). The pleural outgrowth fits into the hollow of the costal area of the forewings (Weber, 1930). However, both surfaces are covered with microtrichia, which can increase frictional forces in the contact area, and hold the system in the locked condition. The microtrichia are short, and cover both surfaces quite densely. In the smaller species of the Corixidae, the outgrowths are wide and plate-shaped. Nepidae posses an additional arresting mechanism between the base of the hindwing and the dorsal surface of the metathorax. It consists of frictional surfaces covered by microtrichia.

The wing-locking systems, based on the friction principle, were also found in representatives of Raphidioptera, Megaloptera (*Sialis*), and some cicada. It is also known that representatives of Mecoptera can use their wing-locking devices to hold both partners during copulation (Hlavac, 1974; Kaltenbach, 1978). However, these were not studied in detail.

Chapter 9

Attachment pads

Interest in the mechanism used by the insect for walking on smooth surfaces, has a very long history. The very first reports, which described structures responsible for the adherence, and provided ideas about possible mechanisms of attachment, are known from the 19th century (West, 1862; Dewitz, 1884; Rombouts, 1884). Different hypotheses, ranging from micro-suckers to the action of electrostatic forces, have been proposed (Gillett and Wigglesworth, 1932; Maderson, 1964). Some comparative accounts on hexapod pretarsal structures and a terminology, were given later (De Meijere, 1901; Holway, 1935; Dashman, 1953). To date, both the morphological and ultrastructural bases of the ability of hexapods to walk on vertical surfaces, have been studied in detail only in representatives of selected taxa, including Orthoptera (Slifer, 1950; Kendall, 1970; Henning, 1974), Blattodea (Roth and Willis, 1952; Arnold, 1974), Thysanoptera (Heming, 1970; 1972), Heteroptera (Gillett and Wigglesworth, 1932; Hasenfuss, 1977a; b; 1978; Carver et al., 1991), Auchenorrhyncha (Lees and Hardie, 1988; Dixon et al., 1990), Hymenoptera (Snodgrass, 1956; Federle et al., 2000), Diptera (Bauchhenss and Renner, 1977; Bauchhenss, 1979a; b; Röder, 1984a; b; Walker et al., 1985; Gorb, 1998d), and Coleoptera (Stork, 1980a; b; 1983a; b). A lot of specific information about the general design of insect leg structures, adapted for attachment, is scattered within the vast field of systematic literature. This chapter summarises morphological data on such devices, provides some insight into the ultrastructural architecture of materials involved in their design, shows their mechanical, adhesive, and frictional properties, and gives related information on attachment-detachment performances.

1. HAIRY AND SMOOTH: TWO ALTERNATIVE DESIGNS OF INSECT PADS

In their evolution, insects have developed two distinctly different mechanisms to attach themselves to a variety of substrates: with smooth pads or with setose, or hairy surfaces. To put it simply, due to the flexibility of the material of the attachment structures, both mechanisms can maximise the possible contact area with the substrate, regardless of its microsculpture (Fig. 9.1).

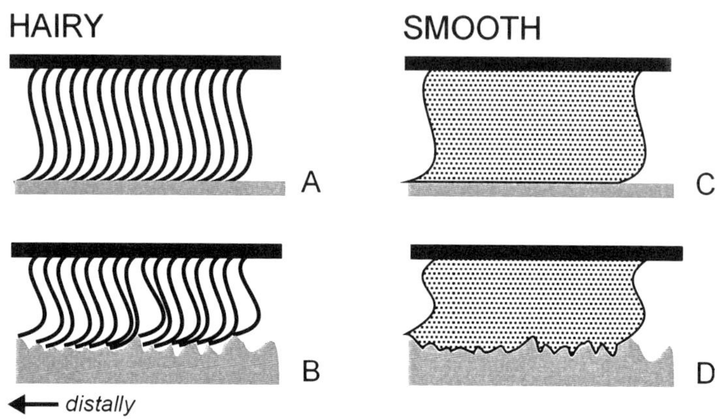

Fig. 9.1. Diagram of the action of «hairy» (A, B) and «smooth» (C, D) pad attachment systems on the smooth (A, C), and structured (B, D) substrata (Beutel and Gorb, 2001). Both systems are able to adapt to the surface profile.

It is remarkable that these highly-specialised structures are not restricted to one particular area of the leg. They may be located on different parts, such as claws, derivatives of the pretarsus, tarsal apex, tarsomeres, or tibia. Therefore, it appears obvious that these structures adapted for attachment have evolved several times, independently (Breidbach, 1980; Schliemann, 1983). However, previous authors have discussed this phenomenon considering only members of few hexapod taxa. Only a phylogenetic analysis of the pad characters, processed together with characters of other organ systems, may aid in resolving the question of the pad evolution (Chapter 13).

Literature data on particular hexapod groups, contains information on leg attachment devices. This data is sometimes almost useless, because of confusing terminology. Authors often refer to any attachment structures as pulvilli, or just as attachment pads. Below, a short explanation of tarsal and pretarsal structures is provided to clarify differences in morphology of attachment devices. The terms used here are largely in accordance with the definitions given by (Dashman, 1953).

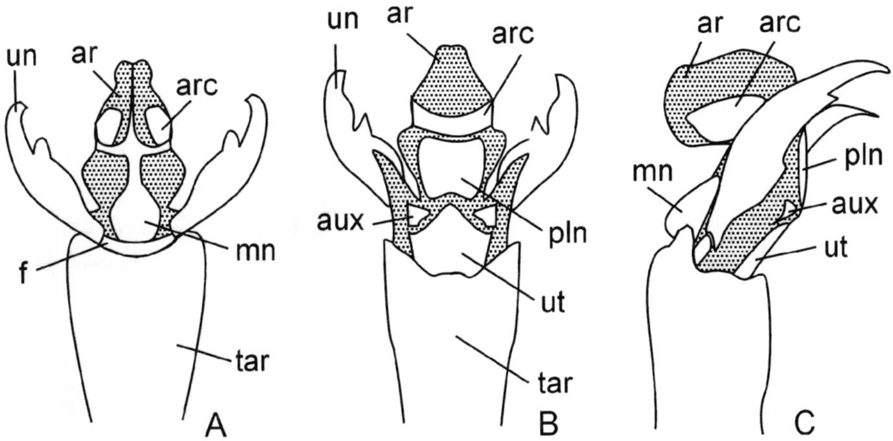

Fig. 9.2. Terminology of arolium structures (*Apis mellifera*). A. Dorsal aspect. B. Ventral aspect. C. Lateral aspect. ar, arolium; arc, arcus; aux, auxillae; f, marginal flange of the terminal tarsomere; mn, manubrium; pln, planta; tar, tarsal segments; un, claw; ut, unguitractor plate.

Tarsus is the distal part of a hexapod or myriapod leg. It consists of 1 to 5 *tarsomeres* or tarsal segments. The proximal tarsomere is often the longest, and is called the *basitarsus*. The most distal tarsomere is often called the terminal tarsomere or *distitarsus*. *Pretarsus* (*posttarsus*) is the terminal part of the hexapod leg, closely associated with the distal end of the terminal tarsomere. It is represented by a complete annulus in entognathous hexapods, and usually consists of two ventral sclerites, the *planta* and *unguitractor* plate, in representatives of other groups. *Planta* is the distal part of an unguitractor plate that is divided into two sclerites (Snodgrass, 1956) (Fig. 9.2). *Unguitractor* (*unguitractor plate*): median basal plate of the ventral surface of the pretarsus. It is usually invaginated into the end of the terminal tarsomere (Fig. 9.2).

Dactylopodite is the simple, claw-like, distal leg-segment existing in most arthropods, including Crustacea. It is controlled by one pair of muscles (*levator* and *depressor*). Dactylopodite-like pretarsi exist in entognathous hexapods, and in larval Lepidoptera and Coleoptera. However, in these taxa it is controlled only by a single depressor-muscle (Snodgrass, 1935). Depressor muscle (adductor, or flexor, of the claw or retractor muscle, according to Dashman, 1953). Its tendon is connected to the median claw or unguitractor.

Claws (*ungues*) are hollow, multicellular, mobile structures, which articulate dorsally at the distal end of the tarsus. The claw-segment is usually reduced to a small median claw, or a lobe-like structure. *Unguifer* is a median process of the dorso-distal part of the terminal tarsomere, to which the ungues are articulated.

Arcus is an elastic, U-shaped band, which embraces the base of the arolium ventrally, with its arms extending distally, in the lateral walls on either side (Fig. 9.2). *Manubrium* (*orbicula* according to Dashman, 1953) is an elongated medial sclerite in the dorso-basal region of the arolium. It is proximally articulated on the end of the tarsus, between the bases of the claws, and it is attached, like a handle, to the base of the arolium by its narrowed distal end (Fig. 9.2).

Arolium is a specialised attachment organ, which originates as the median hollow lobe of the pretarsus. It can be completely membranous or at least partly sclerotized (Figs. 9.2; 9.3 A). *Empodium* is a median process between the pulvilli. It arises from the distal end of the unguitractor plate, is spine-shaped or lobe-like, and is similar in form to the pulvilli (empodial pulvillus) (Fig. 9.3 B, C). The arolium is morphologically and ultrastructurally, completely different from a lobular median empodium. Its surface is smooth and its cuticle consists of rod-like chitin crystallites, oriented perpendicularly, or at some angle to the surface. The median empodium is composed of common layered cuticle, and its surface is covered by acanthae.

Auxiliae or *auxiliar sclerites* (= *basipulvilli* according to Dashman, 1953) are lateral sclerites beneath the bases of the claws (Fig. 9.2). *Pulvilli* are paired attachment-pads, smooth or hairy, originating as lateral membranous lobes, ventral to the claws. They are located on the auxiliae (Fig. 9.3 B, C), which presumably participate in control of pulvilli movements.

Euplantulae are flexible pad-like attachment structures, without hairs, on the ventral side of one or more tarsomeres (Fig. 9.3 H). *Plantar lobes* are specialised, adhesive tarsal thorns (Fig. 9.3 I).

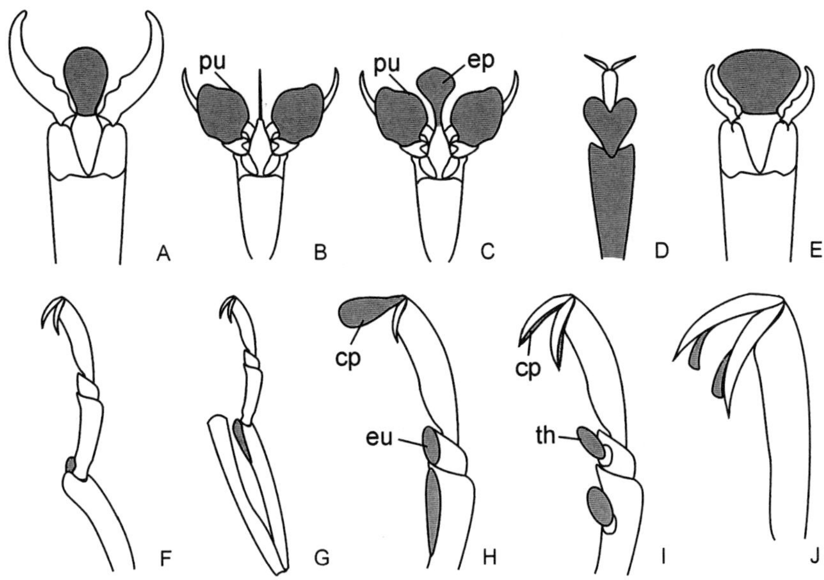

Fig. 9.3. Diversity of the leg attachment devices (grey-colored areas) in hexapods (Beutel and Gorb, 2001). A. Arolium (smooth). B. Pulvilli (smooth or hairy). C. Empodial pulvillus (ep) (hairy). D. Hairy adhesive soles of tarsomeres. E. Eversible pretarsal bladder (smooth). F. Eversible structure between tibia and tarsus (smooth). G. Fossula spongiosa (hairy). H. Euplantulae (eu) and claw pad (cp) (both smooth). I. Tarsal thorns transformed into adhesive structures (th), claw pad (cp) (both smooth). J. Adhesive claw setae.

2. EXTERNAL TARSAL MORPHOLOGY

Protura. The tarsus is undivided, as in all other entognathous hexapods. A very short, ring-shaped pretarsus is present. It bears a single claw, an unmodified empodial seta, and a ventral empodial appendage. The claw is usually knife-shaped. Mid and hind-claws have a navicular shape in protentomids and acerentomids, and are equipped with lateral teeth, which are connected by a membrane in members of the latter group (Janitschek, 1970). Specialised adhesive structures such as arolia, pulvilli, or dense fields of hairs are absent, as in all other apterygote hexapods. However, empodial structures may be potential preadaptations for formation of pulvilli.

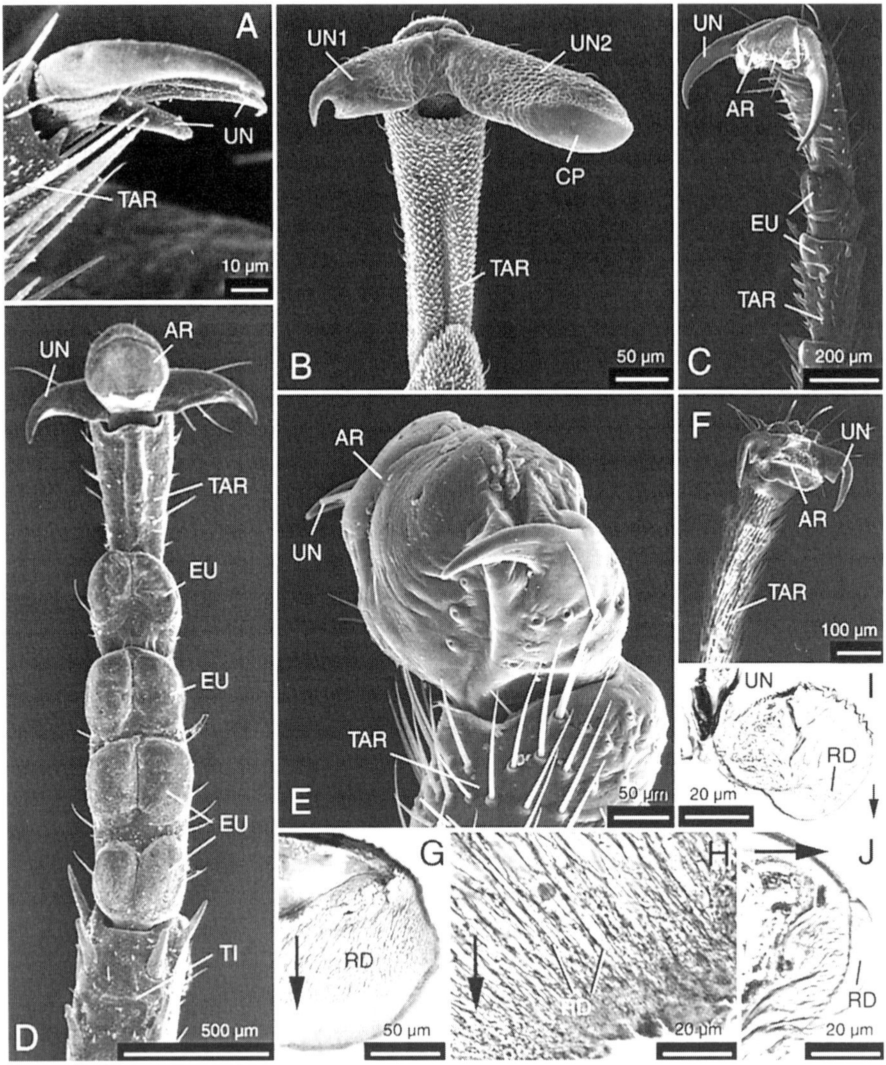

Fig. 9.4. Insect pretarsi. *Lepisma sachharina* (Zygentoma) (Beutel and Gorb, 2001). B. *Paraleptophlebia* sp. (Ephemeroptera). C. *Blatella germanica* (Blattodea). D. *Omocestus viridulus* (Caelifera). E. *Cercopis vulnerata* (Auchenorrhyncha). F. *Chrysoperla carnea* (Planipennia). G, H. *Schistocerca gregaria* (Caelifera), arolium. I. *Brachyptera risi* (Plecoptera), arolium. J. *Cercopis vulnerata* (Auchenorrhyncha), arolium. A-F. SEM. G-J. Sagittal sections; arrows indicate ventral directions. AR, arolium; CP, claw pad; EU, euplantulae; RD, filaments; TAR, tarsal segments; TI, tibia; UN, claw, UN1, hook-like claw; UN2, pad-bearing claw.

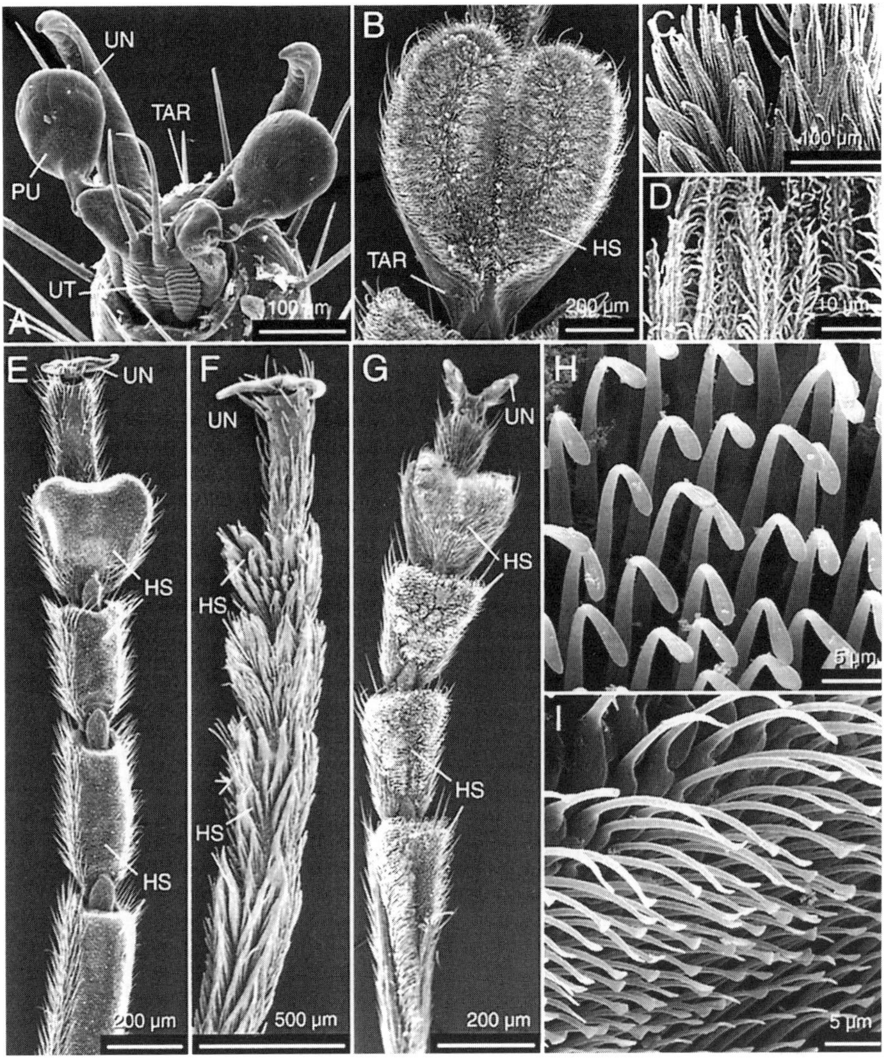

Fig. 9.5. Insect tarsi and pretarsi (Beutel and Gorb, 2001). A. *Coreus marginatus* (Heteroptera). B. *Cantharis fusca* (Coleoptera, Polyphaga), hindleg, 1st and 2nd tarsomeres. C, D, F. *Priacma serrata* (Coleoptera, Archostemata). E, I. *Sialis lutaria* (Megaloptera). G. *Rhagonycha fulva* (Coleoptera, Polyphaga). H. *Cantharis fusca* (Coleoptera, Polyphaga), adhesive setae. HS, hairy soles; PU, pulvilli; TAR, tarsal segments; UN, claw.

Collembola. The pretarsus is a very short and separate leg-section. On the ventral side, opposite the single claw, a spur-like empodial appendage and an empodial tubercle are present, and in some springtails, an accessory empodial appendage. The claw is triangular in cross section with a median edge – sometimes toothed – and two lateral edges. The lateral edges are equipped with one lateral tooth in most species. Sminthurids (e.g. *Sphyrotheca* sp.) are characterised by a membranous claw sheath (tunica) (Schaller, 1970).

Diplura. (Grassé, 1949; Condé and Pagés, 1991). The presence of an empodial median claw and of a pair of larger, lateral claws, is probably a groundplan character of the taxon. This condition is found in *Anajapyx* and *Lepidocampa* species. The median claw is lost in all other diplurans.

Archaeognatha. The tarsus has three or, rarely, two segments (Watson and Smith, 1991). A well-defined, segment-like pretarsus is not developed in archaeognathans or any of the following groups. The reduced pretarsus bears two simple claws. A median claw, or empodium, is not developed.

Zygentoma. The tarsus has three segments in most members of *Zygentoma* (Smith and Watson, 1991), but has five segments in *Tricholepidion gertschi* (Wygodzinsky, 1961). The reduced pretarsus bears one median empodial claw and a pair of lateral claws (Fig. 9.4 A).

Ephemeroptera. The tarsus usually has four or five segments (Peters and Campbell, 1991). The transformation of one of 2 claws into a thickened attachment organ (claw pad), is a characteristic feature found only in ephemeropterans (Figs. 9.3 H; 9.4 B). Other adhesive structures are absent.

Odonata. The tarsus has three segments and the distal segment is longer than the others. The paired claws are usually armed with a hook (Watson and O'Farell, 1991). Specific adhesive structures are absent from dragonflies and damselflies.

Plecoptera. The two proximal tarsomeres of the tri-segmented tarsus bear euplantulae in the larger species (Zwick, 1980). An arolium is present between the strong, paired claws (Fig. 9.3 A). The arolium cuticle bears chitinous threads oriented at an angle to the surface (Fig. 9.4 I).

Dermaptera. The second tarsomere of the tri-segmented tarsus is distinctly shorter than the others in representatives of Dermaptera, excluding *Hemimerus* (Günther and Herter, 1974) (Figs. 9.3 D; 9.7 D, E). The two proximal tarsomeres usually bear a dense brush of hairs on the ventral side. An arolium is found in hemimerines (Günther and Herter, 1974).

Blattodea. Tarsi have five segments. An arolium is either present or absent. Tarsomeres 1-4 are usually equipped with euplantulae (Figs. 9.3 A, H; 9.4 C). Reductions of adhesive structures occur often in cavernicolous species. The arolium cuticle bears chitinous threads oriented at some angle to the surface (Roth and Willis, 1952).

Isoptera. Tarsi have five tarsomeres in *Mastotermes* species, but four tarsomeres in other termites (Watson and Gay, 1991). An arolium is present in alate adults. Euplantulae are not developed.

Mantodea. Tarsi have five segments in almost all species (Balderson, 1991). Euplantulae are developed on the four proximal tarsomeres. An arolium is missing.

Embioptera. Tarsomeres 1 and 2 of the tri-segmented middle and hind leg tarsi, are densely covered with hair. The proximal tarsomere of the proleg is strongly distended due to the presence of silk glands (Ross, 1991).

Ensifera. Tarsi have three or four segments. Proximal tarsomeres are equipped with euplantulae (Figs. 9.3 A, H). An arolium is lacking.

Caelifera. Tarsi have one, two or three segments, with euplantulae on the proximal tarsomeres. The cuticle of the arolium has a characteristic inner structure of chitinous threads, oriented at some angle to the surface (Fig. 9.4 G, H).

Phasmatodea. Tarsomeres 1-4 of the five segments of the tarsus are equipped with euplantulae. An arolium is present.

Grylloblattodea (= Notoptera). The five segments of the tarsi are equipped with euplantulae. An arolium is lacking (Rentz, 1991).

Zoraptera. Tarsi have two segments with an intensely shortened basitarsus. A strong, curved bristle is present between the paired claws. Specific adhesive structures are absent.

Thysanoptera. The presence of an eversible, balloon-shaped, pretarsal adhesive structure at the distal end of a tarsus having one or two segments, is unique to thysanopterans («Blasenfüße») (Fig. 9.3 E). Homology of this structure with the arolium is likely, as was previously demonstrated (Heming, 1970; 1972; 1973).

Psocoptera. The tarsus has two or three segments (Weidner, 1972). Euplantulae and arolium are lacking. A seta, which is slightly extended distally, is present on the toothed or simple claws (Fig. 9.3 J).

Phthiraptera. Tarsi have one or two segments. A claw pad, dissimilar from that of ephemeropterans, is present in some representatives of Phthiraptera (Amblycera, Ischnocera part.) (Fig. 9.3 I). Euplantulae which allow attachment to smooth surfaces, are also present in some mallophagan species.

Sternorrhyncha. The tarsus has two segments in most representatives of Sternorrhyncha, but only one segment in coccoids (Strümpel, 1983). A cushion-like, simple, or bilobed arolium is present in many psylloids. Some representatives of Aphididae possess pulvilli. Other aphid species examined are characterized by an eversible structure, resembling euplantula between tibia and tarsus (Lees and Hardie, 1988) (Fig. 9.3 F).

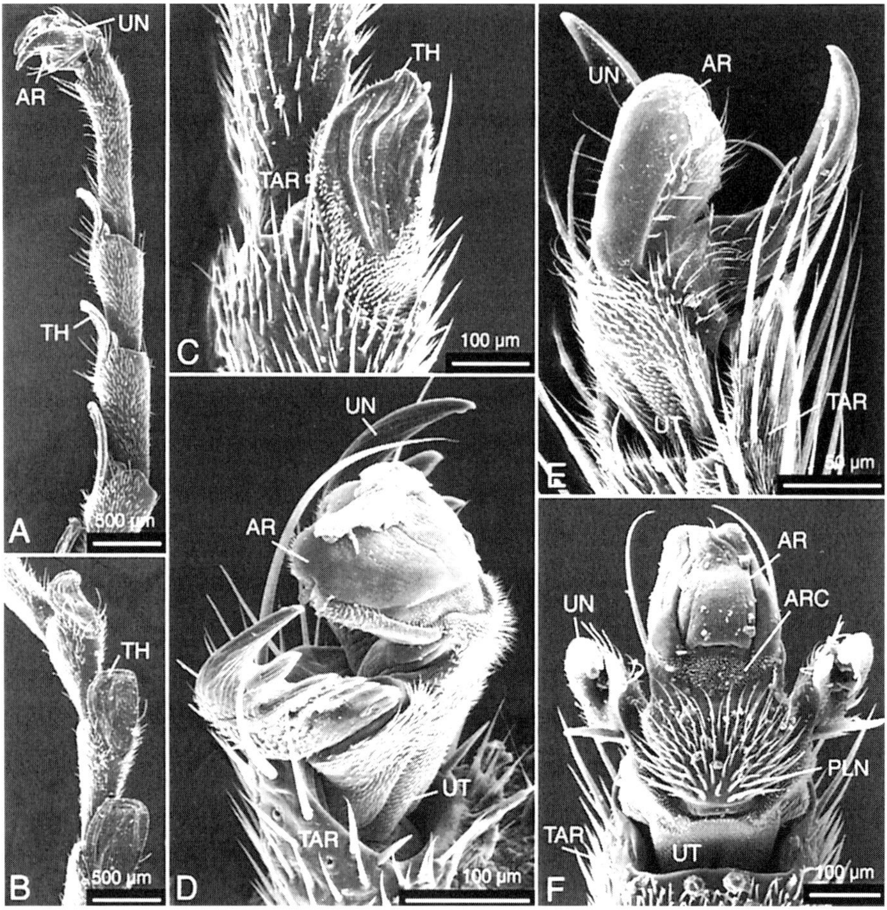

Fig. 9.6. Insect tarsi and pretarsi (Beutel and Gorb, 2001). A, B. *Urocerus gigas* (Hymenoptera), hindleg, tarsus, lateral aspect. C. *Abia seriacea* (Hymenoptera), hindleg, ventro-lateral aspect of the single attachment thorn. D. *Panorpa communis* (Mecoptera). E. *Tipula* sp. (Diptera). F. *Apis mellifera* (Hymenoptera). AR, arolium; ARC, arculus; PLN, planta; TAR, tarsal segments; TH, specialised tarsal adhesive thorns; UN, claw; UT, unguitractor plate.

Auchenorrhyncha. Tarsi have three segments. An arolium has evolved in cicadids and cercopids (Fig. 9.4 E). The arolium cuticle has a characteristic inner structure of chitinous threads, oriented at some angle to the surface (Fig. 9.4 J).

Heteroptera. The tarsus has a maximum of three segments. An arolium is present in slaterellines (Lygaeidae), an arolium and smooth euplantulae in mirines (Miridae), smooth pulvilli are present in members of Coreidae, Pentatomidae, and Pyrrhocoridae (Figs. 9.3 B; 9.5 A), and smooth pulvilli and a fossula spongiosa in reduviids (Carver et al., 1991) (Fig. 9.3 G).

Neuroptera (=Planipennia). The tarsus of the middle and hindlegs almost always has five segments, as in members of other holometabolous groups with the exceptions noted. Modification in number of tarsomeres is only known from the specialised, raptorial forelegs of some mantispids (New, 1991). No specific adhesive structures are present in members of Sisyridae, Myrmeleontoidea, and Ithonidae. An arolium is developed in representatives of other groups (Fig. 9.4 F).

Megaloptera. Tarsomeres with a dense, ventral brush of hair are present in members of Chauliodidae (Maki, 1936), Corydalidae and Sialidae. The fourth tarsomere is expanded and bilobed in *Sialis* species (Theischinger, 1991) (Figs. 9.3 D; 9.5 E, I).

Raphidioptera. The ventral side of tarsomeres 1-3 is densely covered with hairs. The fourth tarsomere is very short and largely concealed between the two lobes of the heart-shaped, third tarsomere (Aspöck and Aspöck, 1971).

Coleoptera. Tarsi have five segments in most groups of Coleoptera. However, reduction in number of tarsomeres occurs in several lineages (e.g. Chrysomeloidea). The ventral side of the proximal tarsomeres is covered with a very dense layer of hairs in males, and in both sexes of many coleopteran species (Fig. 9.3 D; 9.5 B, G, H). This is considered to be a groundplan character of the order. Tarsal hairs are transformed into larger or smaller suckers in representatives of some aquatic Adephaga, such as Gyrininae and Dytiscinae. In *Priacma serrata* (Archostemata), the hairs are branched in a feather-like pattern (Fig. 9.5 C, D, F). Arolium, pulvilli, and euplantulae are absent from adults and most larvae of Coleoptera. An arolium-like structure is present in larvae of Chrysomelidae (Gannon et al., 1994).

Strepsiptera. The full number of five tarsomeres, and double claws, are present in males of Mengenillidae, Mengeidae and Triozocerinae. Tarsomeres are densely covered with hairs (Kinzelbach, 1971). Broadened proximal tarsomeres, and the loss of claws and the distal tarsomere is characteristic in males of Stylopiformia.

Hymenoptera. Tarsi having four or three segments are found only in a few Hymenoptera (e.g. Chalcidoidea part., Platygasteridae, Trichogrammatidae: (Naumann, 1991)). An arolium is generally present (Figs. 9.2; 9.3 A; 9.6 F) and additional, specialised, tarsal adhesive thorns in the non-apocritan groups («Symphyta») (Figs. 9.3 I; 9.6 A-C).

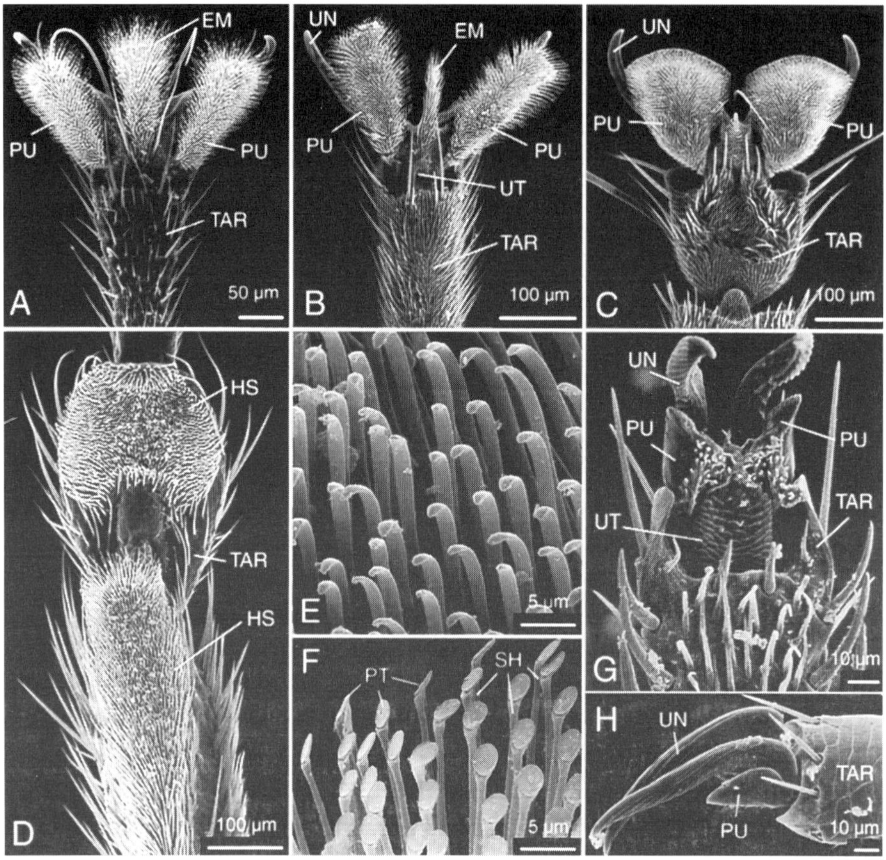

Fig. 9.7. Insect tarsi and pretarsi (Beutel and Gorb, 2001). A, *Bibio nigriventris* (Diptera). B. *Rhamphomya nigripes* (Diptera). C, F. *Episyrphus balteatus* (Diptera). D, E. *Forficula auricularia* (Dermaptera). G, H. *Ceratophyllus* sp. (Siphonaptera), pretarsus, ventral (G) and lateral (H) aspects. EM, empodium; HS, hairy soles; PT, hair plate; PU, pulvilli; SH, hair shaft; TAR, tarsal segments; UN, claw; UT, unguitractor plate.

Trichoptera. Adults of Trichoptera possess either pulvilli or a small arolium.

Lepidoptera. An arolium is usually present in Lepidoptera. Hairy pulvilli are described for *Epiphyas* (Tortricidae) (Nielsen and Common, 1991) and *Micropterix*. This condition is considered to be groundplan feature of the order.

Mecoptera. An arolium is absent in boreids and notiothaumids (Mickoleit, 1971), but present in other mecopterans (Fig. 9.6 D). Other specific, adhesive structures are lacking. Paired claws are present in members of all groups with the exception of Bittacidae, which posses single-clawed raptorial tarsi.

Diptera. Less than five tarsomeres are found in a few representatives of Diptera. An arolium is present in tipulids (Fig. 9.6 E) (Rees and Ferris, 1939; Hennig, 1973) and hairy pulvilli in adults of other groups (Figs. 9.3 B, C; 9.7 A-C, E). A lobe like, pulvilliform empodium is present as an additional adhesive structure in tabanoids (Brachycera) and Nematocera excl. Tipulidae (Fig. 9.3 C; 9.7 A) (Hennig, 1973).

Siphonaptera. Smooth pulvilli are present (Fig. 9.7 G, H). Other attachment structures are lacking.

3. SURFACE CHARACTERISTICS

Hairy systems always contain cuticle protuberances on their surfaces. Protuberances on the hairy pads of Coleoptera, Dermaptera, and Diptera belong to different types. Representatives of the first two lineages have socketed setae on their pads. Setae range in length from a few micrometers to several millimeters. Presumably, in these pads, the sensory cells degenerate in the trichoid complex during development. Dipteran protuberances are acanthae: single sclerotized protuberances originating from a single cell (Richards and Richards, 1969). Acanthae range in length from a few micrometers to, at most, 0.5 mm. The key morphological characteristic of acanthae is their lack of a socket and a sensory cell (Richards, 1965). Both types of structures can be covered with additional, minute outgrowths referred to as microtrichia. Fimbriate setae were found in the beetle *Priacma serrata*, a representative of the basal suborder Archostemata (Beutel and Haas, 2000) (Fig. 9.5 C, D, F). The surface of smooth systems appears smooth under a light microscope. However, it may consist of a pattern of hexagonal structures (area 14.7 μm^2; sd=1.96; n=22) as in representatives of Orthoptera (Fig. 9.8 A). The hymenopteran (Fig. 9.8 E) and mecopteran (Fig. 9.8 D) arolium may be patterned in lines running perpendicularly to the longitudinal axis of the pretarsus. A similar pattern exists on the surface of attachment thorns in some species of non-apocritan Hymenoptera (Fig. 9.6 C). The surface of a tipulid arolium exhibits a complex pattern of microfolds (Fig. 9.8 F).

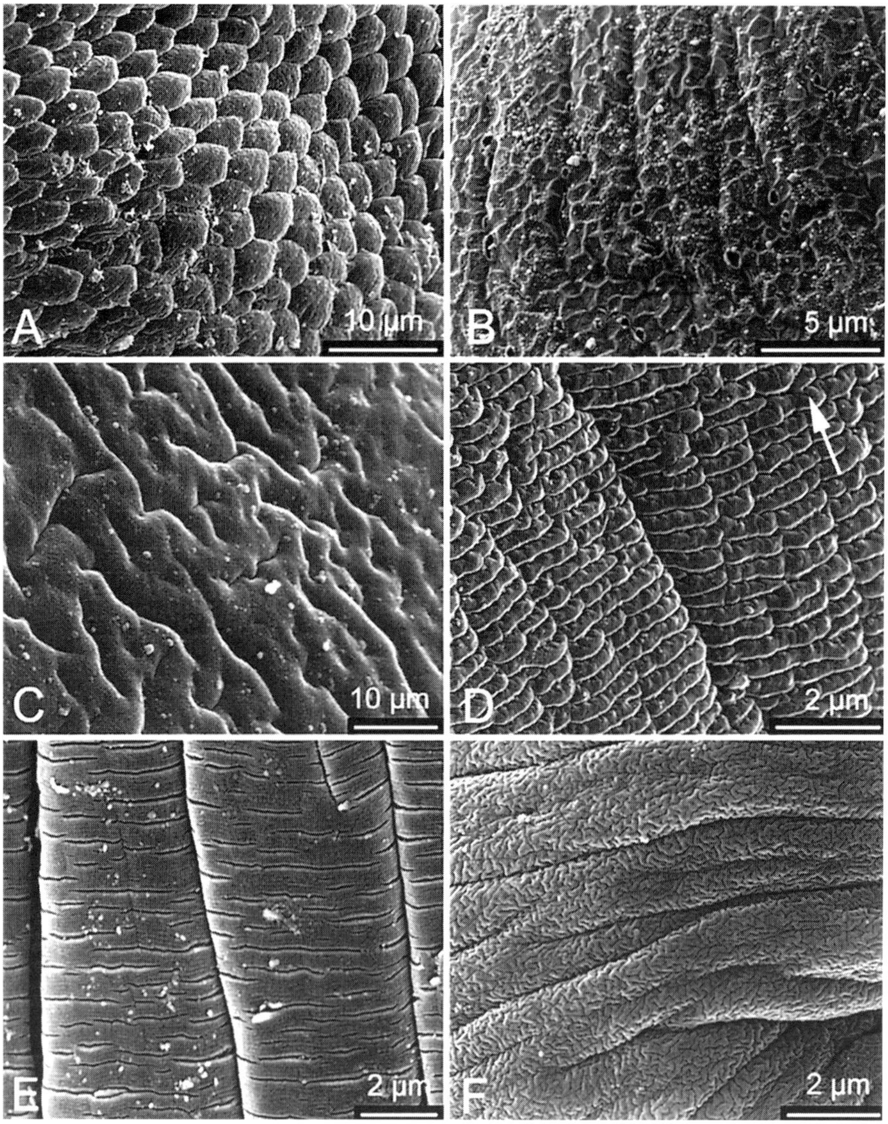

Fig. 9.8. Surface details of smooth systems of hexapod tarsi (Beutel and Gorb, 2001). A. *Tettigonia viridissima* (Ensifera), euplantulae. B. *Apis mellifera* (Hymenoptera), arolium. C. *Urocerus gigas* (Hymenoptera), attachment thorn. D. *Panorpa communis* (Mecoptera), arolium. E. *Vespa crabro* (Hymenoptera), arolium. F. *Tipula* sp. (Diptera), arolium.

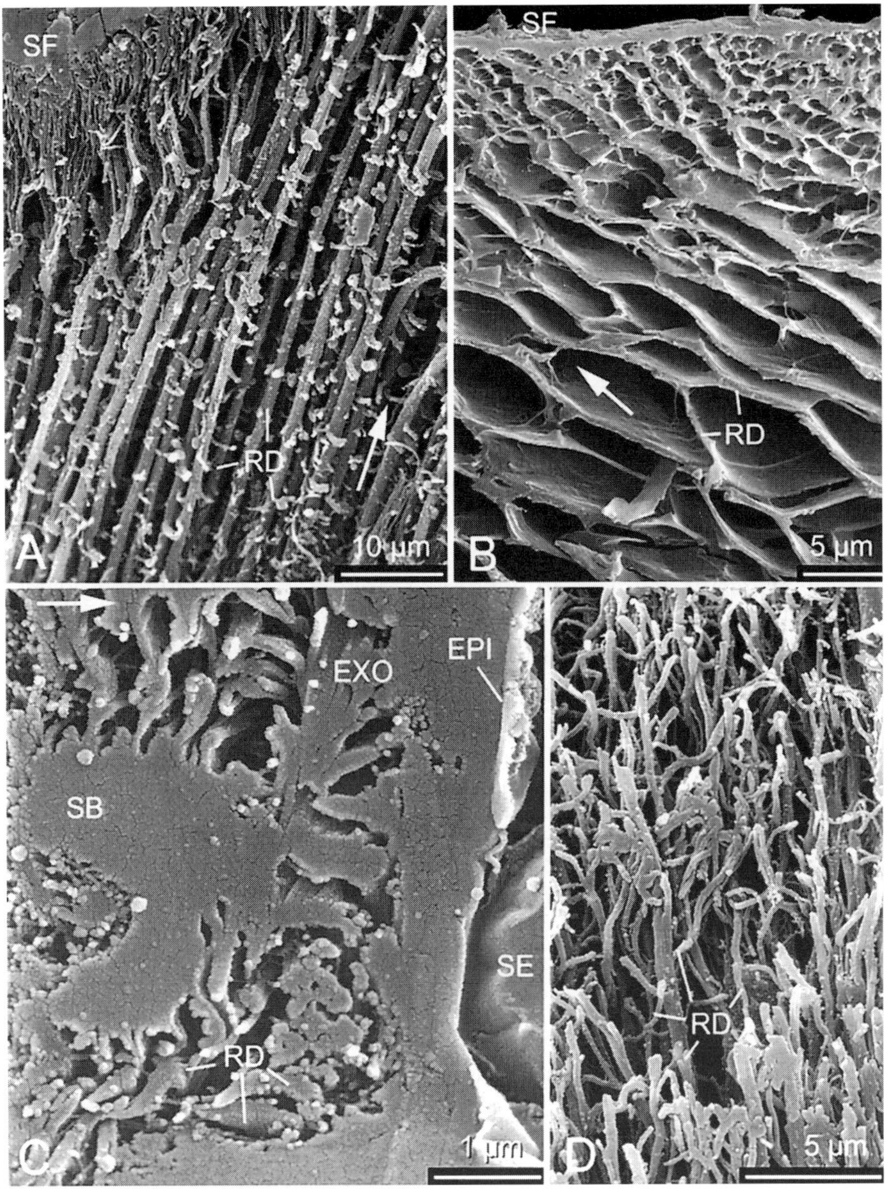

Fig. 9.9. Inner architecture of the cuticle of smooth systems of hexapod tarsi (Beutel and Gorb, 2001). A. *Tettigonia viridissima* (Orthoptera), euplantula. B. *Cercopis vulnerata* (Auchenorrhyncha), arolium. C. *Rhagonycha fulva*, ventral part of the tarsus (Coleoptera, Polyphaga). D. *Apis mellifera* (Hymenoptera), arolium. EPI, epicuticle; EXO, exocuticle; basis of the seta; RD, chitinous threads; SE, seta; SF, surface.

Microfolds found on the surface of attachment thorns in some species of non-apocritan Hymenoptera are much larger (Fig. 9.8 C). In some specimens, residual secretory substances are present on the arolium surface (Fig. 9.8 B). Footprints on glass surfaces were previously reported, for both hairy and smooth systems, as consisting of round droplets (hairy systems: Coleoptera: (Ishii, 1987); Diptera: (West, 1862); smooth systems: Blattodea: (Roth and Willis, 1952); Orthoptera: (Jiao et al., 2000); Sternorrhynchi: (Lees and Hardie, 1988)).

4. ULTRASTRUCTURAL ARCHITECTURE OF PAD MATERIAL

Structural features of fly adhesive-hairs have been previously reported on the syrphid *Episyrphus balteatus* (Gorb, 1998d): the acanthae are hollow inside, and some of them contain pores under the end plate (Fig. 9.7 F). This pore, presumably, delivers an adhesive secretion directly onto the contact area (Chapter 10). Additional pore canals at the base of the shaft may transport secretions to the surface (Bauchhenss, 1979a). The membranous cuticle of hairy pads is made of fibrous composite material with loosely distributed fibres. In coleopterans, the hair bases are embedded in this material (Fig. 9.9 C), which provides flexibility to the hair and helps it to adapt to a variety of surface profiles.

Ultrastructural studies show the cuticle of smooth pads to consist of a natural friction-active material with a specific inner structure. In orthopterans, tiny threads of 0.08 μm in diameter are located just under the epicuticle of euplantulae (Fig. 9.9 A). These filaments are branches of thicker threads, of 1.12 μm diameter, located more deeply in the cuticle (Kendall, 1970; Henning, 1974). It has been shown that these threads can change their shape under loads (Gorb and Scherge, 2000). The arolium of bees contains thinner and longer threads (Fig. 9.9 D), which may provide higher flexibility of the material. Internal architecture of the arolium of cicadas resembles foam, consisting of fluid-filled cells surrounded by flexible cuticle (Fig. 9.9 B). The diameter of the cells increases in deeper layers of the material.

5. HAIRY SYSTEMS

Pulvilli are soft lobes responsible for attachment to smooth surfaces (Walker et al., 1985). They are covered by tenent setae, sometimes termed tenent hairs, which serve to increase the actual area of attachment to the surface. Proximal and distal tenent setae have a different ultrastructure (Gorb, 1998d). Both the shape and size of pulvilli correlate with the weight and biology of the insect (Röder, 1984a; b).

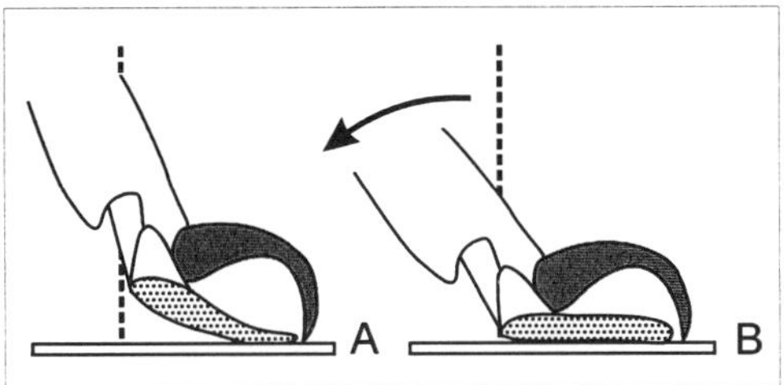

Fig. 9.10. Mechanism of attachment of the fly foot to a surface (Niederegger and Gorb, 2001). Both pulvilli are placed fully onto the surface. Grey area, claw; dotted area, pulvillus.

Tenent setae are relatively soft structures. In *Calliphora* flies, their tips are usually compressed, widened, and bent at an angle of 60° to the hair shaft (Bauchhenss and Renner, 1977). When walking on smooth surfaces, these hairs in flies and beetles produce secretion, which is essential for attachment (Ishii, 1987) (Chapter 10). Some authors have hypothesised that different forces may contribute to the resulting attachment force: capillary adhesion and intermolecular van der Waals forces. The action of intermolecular forces is possible only at very close contact between surfaces. The forces increase, when the contacting surfaces slide against each other. This may explain, why flies sitting on a smooth undersurface always move their legs in a lateral-medial direction (Wigglesworth, 1987). During these movements, pulvilli slide over the surface obtaining optimal contact.

An important role of intermolecular interaction has been shown in experiments on the adherence of beetles (Stork, 1980b), and beetle setae, (Stork, 1983b) on a glass surface. The presence of claws, decrease of air pressure, decrease of relative humidity, or electrostatic forces do not influence beetle attachment on the smooth substrata. In the beetle *Chrysolina polita* (Chrysomelidae), the resulting attachment force directly depends on the number of single hairs contacting the surface. The number of hairs correlates to the body weight. Average attachment force in this beetle species is about 50 times stronger than its weight. In males, it is well correlated with the body weight. For females, such a dependence has not been so clear, because of the different number of eggs that each female had (Stork, 1980b).

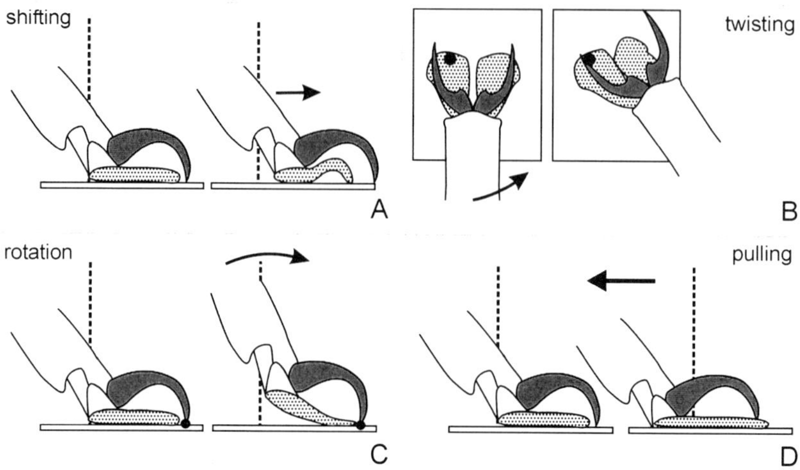

Fig. 9.11. The pretarsus allows the fly to detach from the surface with the use of various mechanisms: (A) shifting, (B) twisting, (C) rotation, (D) pulling (Niederegger and Gorb, 2001). This suggests that the type of detachment movement depends on the position of the foot, and on the walking direction.

The structure of attachment organs and the occurrence of adhesive secretion in walking, has been previously studied in flies (Hasenfuss, 1977a; Bauchhenss, 1979a; Walker et al., 1985; Gorb, 1998d), but little is known about the mechanical aspects of the setae or of the pretarsus kinematics. The following subsections clarify some questions about motion, setae deformability, and gait patterns employed in attachment and detachment of the fly foot.

5.1 Fly-walking: a compromise between attachment and motion?

5.1.1 Attachment/detachment performance

High-speed recordings have revealed that, in attachment, the foot is placed on the surface and the pulvilli are put down flat. Through this movement the claws become spread apart (Fig. 9.10). Four different movements occur during detachment (Fig. 9.11). In *shifting*, the foot is moved forward. This compresses the pulvilli, which in turn leads to a smaller contact area and thus to a reduction of attachment forces. In *twisting*, the leg is turned up to 90°, and pulled backwards. This bends the tenent setae and eventually detaches them from the surface. In *rotation*, the foot is lifted from back to front, which causes the claws to be passively pressed to the surface. With this force, the adhesion is interrupted by peeling. In *pulling*, the foot is simply pulled backward until the pulvilli detach from the surface.

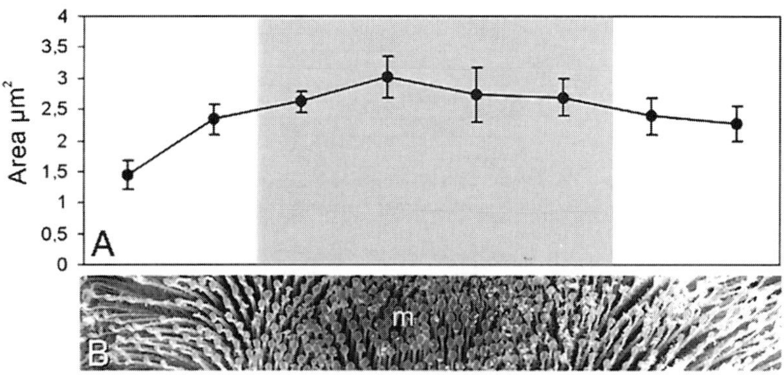

Fig. 9.12. The area of the setae tips in the fly *Calliphora vicina* (Niederegger and Gorb, 2001). Setae are deformed when in contact with the surface. This could lead to an increased contact area and thus to a stronger adhesion of the pulvilli. A. Tip areas of the setae; the grey area indicates the region of the setae that is in contact with the substrate. B. The part of the pulvillus where the contact areas were measured. m, middle of the pulvillus; error bars, standard deviations.

5.1.2 Setae deformation

It was previously suggested that setae are composed of flexible cuticle, and are able to replicate the surface profile (Bauchhenss, 1979a; Stork, 1983b). The results of freezing-substitution experiments show that the area of the setae tips becomes larger when the pulvillus is in contact with the surface. This deformation is best seen in the middle of the pulvillus whereas on the side parts, the setae are often not in contact (Fig. 9.12). The whole pulvillus becomes compressed and elongated while contact is made.

5.1.3 Gait pattern

When a fly walks on a surface, three opposite legs are moved whereas the other three legs remain in a stance phase (Fig. 9.13 A). It is commonly known that flies are able to walk head-down on a horizontal surface. Does the gait pattern of the fly change in this position?

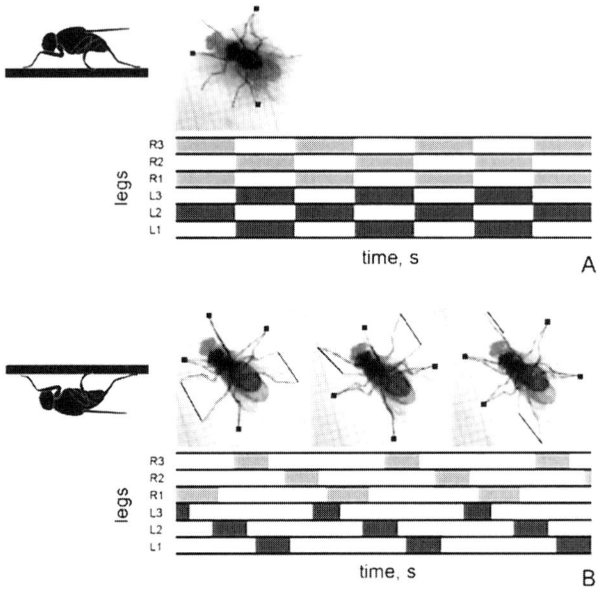

Fig. 9.13. Gait pattern in the walking fly, *Musca domestica*. Two pictures were superimposed to show two succeeding steps. Black squares indicate the feet in the stance phase. The advantage of the specific gait pattern the fly uses when walking under the surface seems to be the additional contacting point, which may result in a larger supporting area. A. Fly walking on a surface. B. Fly walking under a surface; the lines point out the way the free legs moved, during one step.

High-speed camera recordings on walking *Musca domestica* show that the gait pattern of flies, walking on a vertical surface, is not different from the pattern on a horizontal surface. However, the fly walking under a surface moved only two legs, while four legs remained in the stance phase (Fig. 9.13 B). This result led us to presume that the gait pattern may contribute to the fly's ability to walk under a surface, in providing an optimal relationship between the body weight and the supporting contact area.

5.2 Scale effects on the attachment pads and friction force in syrphid flies

Attachment forces mediated by friction, or adhesion, are usually proportional to the area of real contact between two surfaces (Persson, 1998). Because of the flexible material of setae, setose or a hairy mechanism can maximise the possible real contact area with the substratum, regardless of its microsculpture.

Data on the setose pad system of the adult reduviid bug, *Rhodnius prolixus*, led previous authors to suggest that mechanical seizure between adhesive setae and substratum irregularities, is responsible for attachment to the substratum (Gillett and Wigglesworth, 1932). Most authors agree that at least two factors related to pad material can contribute to the attachment force: (1) material flexibility and (2) presence of an epidermal secretion in the contact area.

Although the morphology and ultrastructure of the setose attachment devices have been described in numerous studies, various aspects of the functioning of these systems still remain unclear. Among lizards, which are also able to walk on a smooth surface with the use of «setose pads», the pad area has been shown to be the primary, causal factor influencing clinging ability in geckos, skinks and anoles (Irschick et al., 1996). However, despite the close correlation between pad area and attachment ability, pad area depends less on body mass than attachment ability. When the effect of body size is removed, about 50% of the variation in clinging ability remains unexplained, which suggests that the microsculptural and ultrastructural differences may affect clinging ability. Geckos, skinks and anoles differ in the structure of the setae covering the attachment pads (Ruibal and Ernst, 1965; Ernst and Ruibal, 1967). However, the effect of ultrastructural properties has not been tested.

In the beetle *Chrysolina polita*, attachment force increases with the total number of adhesive setae (Stork, 1980b). The number of adhesive setae can contribute to the attachment force by an increase in the number of single contact points, and/or by an increase in an overall contact area with the substratum. To test the role of constructional and dimensional factors in attachment, a wider number of species has to be tested. Such parameters as pad area, seta density, the area of a single seta tip, and body mass have to be individually measured, to test their effect on the resulting attachment force.

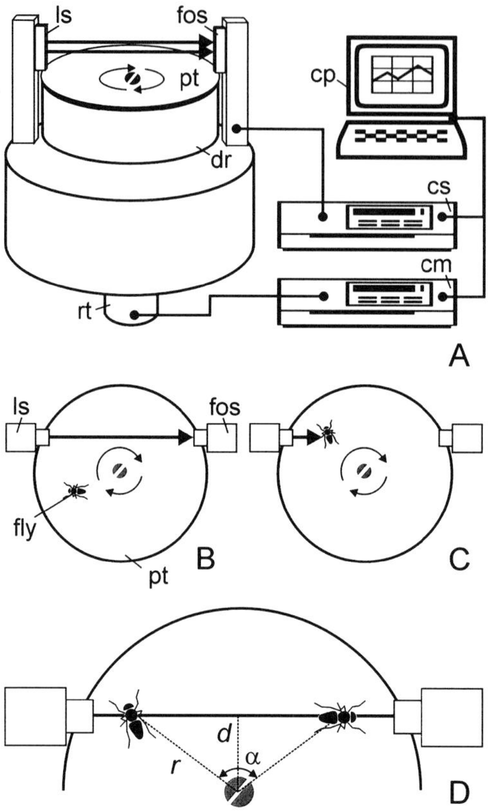

Fig. 9.14. Centrifugal device for measuring friction force. A. Principal diagram of the centrifuge. The metal drum, covered by a PVC disc, is driven by the computer-controlled motor. The fibre-optical sensor is adjusted just above the disc. The sensor signal is monitored by a computer. B-D. Principle of the insect position monitoring. The sensor is shifted to one side of the disc. The fly, rotating to the right, passes the laser beam twice per rotation. This will interrupt the sensor signal twice. Knowing the speed of the motor and the time between two signal interruptions, the fly position (radius, r) on the disc can be calculated (D). cp, computer; cs, sensor control electronics; cm, motor control electronics; dr, metal drum; fly, fly; fos, fibre-optical sensor; ls, light source; pt, PVC plate; rt, rotor of the motor.

This subsection presents a case study on attachment abilities of six species of syrphid flies (*Platycheirus angustatus*; *Sphaerophoria scripta*; *Episyrphus balteatus*; *Eristalis tenax*; *Myathropa florea*; *Volucella pellucens*), which were selected according to their body mass and pad dimensions. The variety of attachment pad design in a number of taxa of the

Syrphidae has been previously described (Röder, 1984a). Most morphological data and measurements given below were collected individually. This allowed direct comparison of the effects of structural properties on attachment ability.

5.2.1 Centrifugal force-tester

The main problem in measuring attachment forces of living insects is the difficulty in the designing experiment, or labour-consuming data processing of the video-recordings. Centrifugal methods of attachment measurement, with an incorporated laser beam system and fibre-optical sensor, can be used to obtain a reasonable amount of comparative data (Fig. 9.14) (Gorb et al., 2001). This method may be applied to test individual attachment performance of a variety of insect species, on a variety of substrata. The main advantage of this method is that almost no prior treatment of the insects is required. This can be very important, especially in the case of small organisms.

The centrifuge method is generally accepted as standard for measurements of friction, and adhesive forces, in a wide range of objects. It was previously used to measure the adhesion strength between starch microspheres and microcrystalline cellulose (Podczeck and Newton, 1995), frictional properties of skin (Highley et al., 1977), strength of barnacle cement (Dougherty, 1990a), and to test insect attachment (Dixon et al., 1990; Brainerd, 1994; Federle et al., 2000). Vertical attachment forces of ants, measured with a centrifugal tester, were significantly higher compared to those measured by a strain gauge force transducer (Federle et al., 2000). This result has been explained by the fact that tethered insects generally continued to move during the strain gauge measurements, and rarely were all six legs simultaneously in contact with the surface. The centrifugal method was applied to compare attachment abilities of syrphid flies.

5.2.2 Body weight, pulvillus area, and frictional force

The body mass of species studied ranged from 4.8 mg to 164.0 mg (Fig. 9.15 A). Only representatives of two species (*Eristalis tenax* and *Myathropa florea*) were similar in their body mass. Other species differed significantly from one another. The area of a single pulvillus does not correspond directly to the body mass, so that, for example, the heaviest species (*Volucella pellucens*) has significantly smaller pulvilli than *Eristalis tenax* and *Myathropa florea* (Figs. 9.15 B; 9.16) (Gorb et al., 2001).

The acceleration at which an insect loses contact with the surface is just termed «acceleration» below. This value was higher in lighter animals, such as *Platycheirus angustatus* (Figs. 9.15 C, 9.17 A). Friction force, which is the resistance force of the insect mass to the sum of centrifugal and tangential forces, was higher in heavier insects, such as *Eristalis tenax*, *Myathropa florea*, and *Volucella pellucens* (Figs. 9.15 D, 9.17 B). This relationship is reflected in the acceleration value. Although lighter species generated lower friction forces, the relation of the friction force to the body weight was higher in smaller species. In other words lighter species demonstrated relatively higher attachment ability.

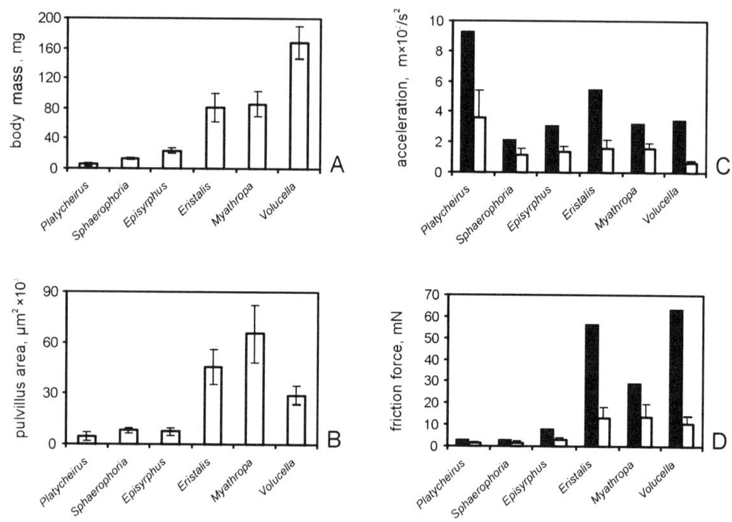

Fig. 9.15. Body mass (A), pulvillus area (B), maximum acceleration (C), and friction force (D) of the six species of syrphid flies studied. Black bars, maximum values; white bars, means; error bars, standard deviations.

The measured horizontal force corresponds to the vertical force in natural situations, when an insect walks on vertical substrata. This situation is completely different from the circumstances of an insect walking under a horizontal surface, due to the set of forces acting on the insect. In such a situation, the insect's weight acts in the direction perpendicular to the surface. In this case, adhesion contributes mainly to insect attachment. In the centrifugal experiment, the insect resisted the force that tried to shift it in a direction parallel to the surface. The insect is able to resist these external forces due to the friction between its attachment pads and the substratum.

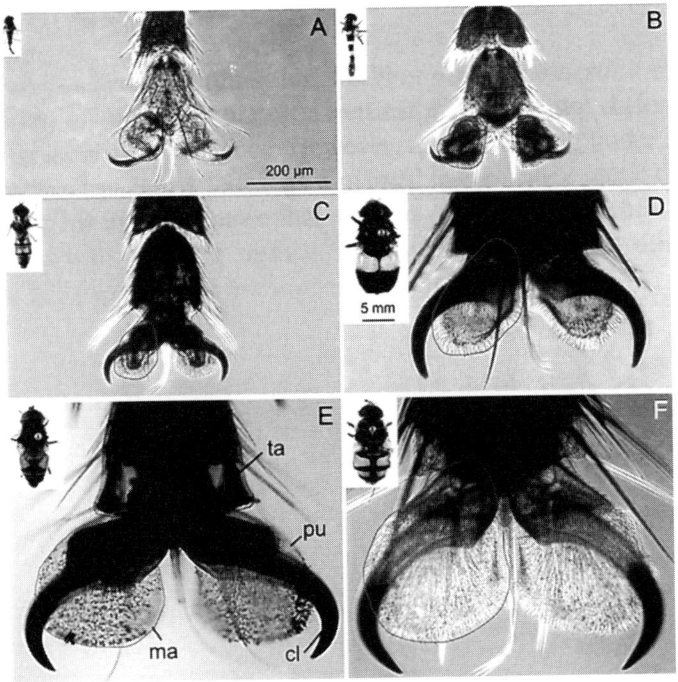

Fig. 9.16. Whole-mounts of pretarsi of syrphid flies. Lines indicate measured areas (ma). Insets show the species used. cl, claw; pu, pulvillus; ta, terminal tarsomere. Scale bar 200 µm is the same for all micrographs, whereas scale bar 5 mm is the same for all fly insets. A. *P. angustatus*. B. *S. scripta*. C. *E. balteatus*. D. *V. pellucens*. E. *E. tenax*. F. *M. florea*.

The acceleration, averaged for all trials with one species was 2-6 times lower than its maximal values. A similar relationship was also obtained for friction force. The relative difference between the maximum measured friction force and the averaged friction force, was smaller in heavier insects. The area of the single pulvillus increases with increased body mass (Figs. 9.16, 9.17 C). Since the pulvillus area is larger and the friction force is higher in heavier animals, it seems that the heavier animals can generate higher friction, due to the larger contacting area with the substratum. Indeed, the measured acceleration decreased, and friction force increased with increased area of attachment organs (Fig. 9.17 D).

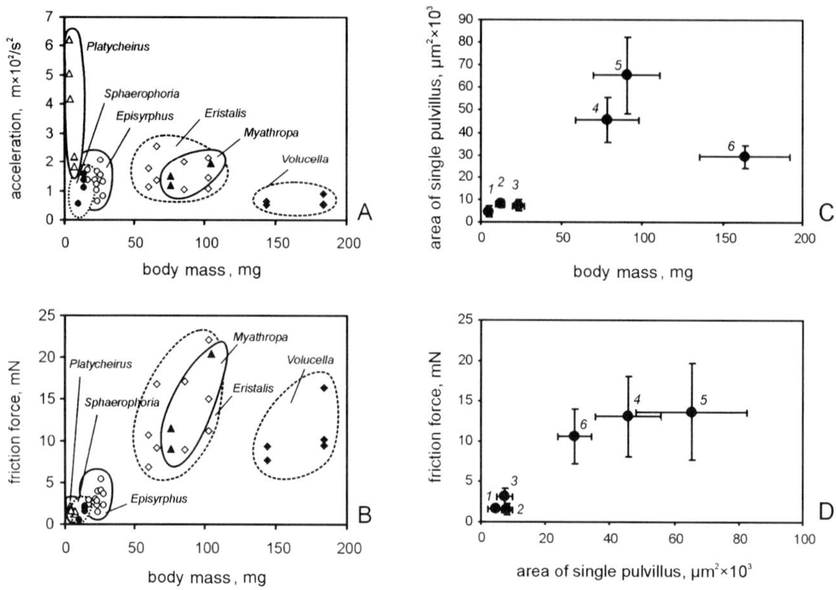

Fig. 9.17. Maximum acceleration (A) and friction force (B) versus body mass. Dependence of the area of single pulvillus on the body mass (C) and friction on the area of single pulvillus (D). 1, *P. angustatus*; 2, *S. scripta*; 3, *E. balteatus*; 4, *E. tenax*; 5, *M. florea*; 6, *V. pellucens*.

5.2.3 Surface characteristics of pulvilli in syrphid flies

Knowing that friction force is dependent on the area of real contact between two surfaces (Chapter 1), it was suggested that surface characteristics of the attachment devices may contribute differently to friction in the species studied. Even qualitative data, obtained from scanning electron microscopy reveals differences in the seta size and density (Fig. 9.18). Two parameters of the pulvillus may be of importance: (1) seta density and (2) area of the seta tip. The setal tip area increased somewhat, with increased body weight (Fig. 9.19 B), whereas seta density slightly decreased with increased body weight (Fig. 9.19 A). The scale effect on the surface characteristics is especially well expressed, when both measured parameters are plotted against each other (Fig. 9.19 C). With an increasing seta tip size, the seta density decreased.

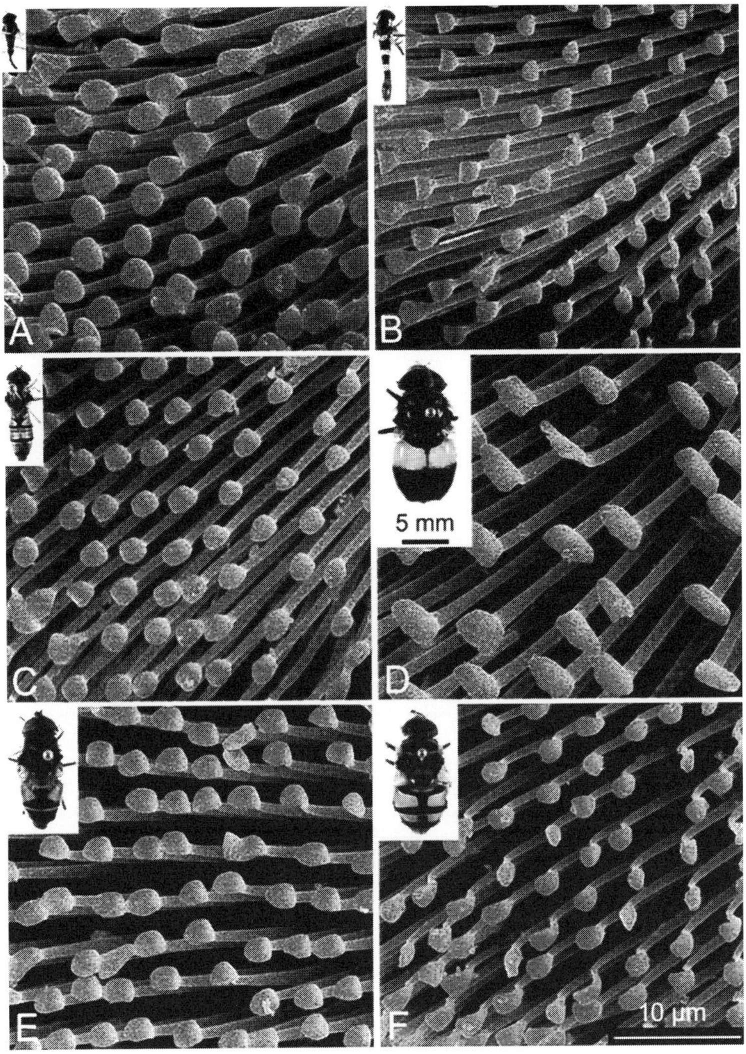

Fig. 9.18. Tenent setae of the species studied. Such images were used to quantify seta density and area of the seta end plate. Scale bar 10 µm is the same for all micrographs, whereas scale bar 5 mm is the same for all fly insets. A. *P. angustatus*. B. *S. scripta*. C. *E. balteatus*. D. *V. pellucens*. E. *E. tenax*. F. *M. florea*.

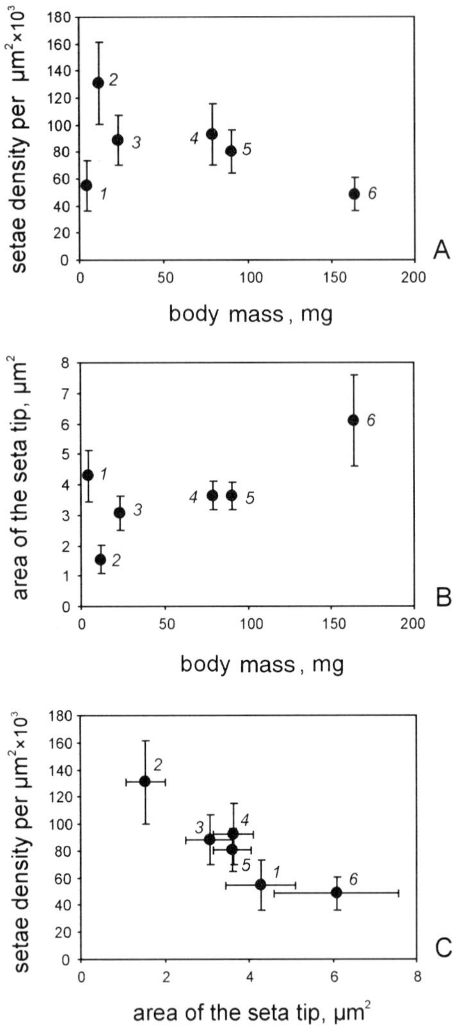

Fig. 9.19. Surface characteristics of the fly pulvilli. A. Dependence of the seta density on the body mass. B. Dependence of the seta tip area, on the body mass. C. Dependence of the seta density, on the area of the seta tip. 1, *P. angustatus*; 2, *S. scripta*; 3, *E. balteatus*; 4. *E. tenax*; 5, *M. florea*; 6, *V. pellucens*.

5.2.4 Lateral tenacity

Tenacity is the adhesion related to the area of apparent contact, between two surfaces. In the case of the fly pulvilli, since friction force is mediated by adhesion, a lateral tenacity of pulvilli was calculated, which represents friction force related to the area of apparent contact. This value ranged from 0.015 to 0.035 $\mu N/\mu m^2$ in species studied, and was not dependent on the body weight (Fig. 9.20 A) or on the pulvillus area (Fig. 9.20 B).

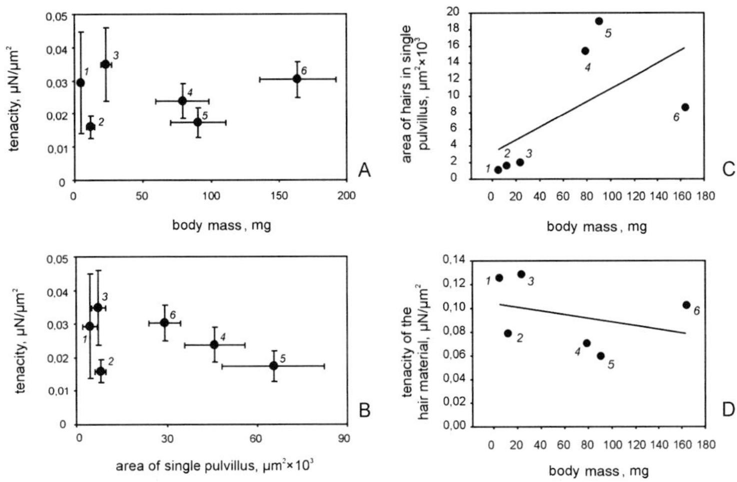

Fig. 9.20. Lateral tenacity of the fly pulvilli and setae material. Dependence of the tenacity of the pulvilli on body mass (A) and on the pulvillus area (B). C. Dependence of the real contact of all setae in a single pulvillus on the body mass. D. Dependence of the tenacity of the material of setae on the body mass. 1, *P. angustatus*; 2, *S. scripta*; 3, *E. balteatus*; 4. *E. tenax*; 5, *M. florea*; 6, *V. pellucens*.

However, the area of pulvillus is not the same as the area of real contact. The area of real contact directly influences the resulting adhesive and friction forces. The area of real contact was evaluated as the aggregate area of all setae, on all pulvilli. It increased with increasing body weight (Fig. 9.20 A). Using this parameter, a lateral tenacity of pulvilli material τ_m was calculated.

Since the evaluated area of real contact was 3-5 times smaller than the area of apparent contact (pulvilli area), the lateral tenacity of material was higher than the previously calculated value of lateral tenacity of the pulvilli.

τ_m ranged from 0.06 to 0.13 $\mu N/\mu m^2$ in the species studied, and was somewhat lower in the heavier species (Fig. 9.20 C, D).

This data shows that the frictional properties of the material of the seta tips is not dependent on the dimensions of the fly species. In other words, friction forces generated by the surface unit of seta tips are similar in the species studied. The same result is obtained for the pad material, as a whole. This means that adhesive properties of secretion, and mechanical properties of the material of seta tips, are more or less constant, and that differences in friction force are mainly related to the real contact area, which can be generated by the pad. This parameter is the sum of broadened seta tips of the pad, contacting with the surface. Real contact area depends on the overall pad area, seta density, and the area of a single seta tip. Although the parameters vary among animals with different dimensions, there is usually a compensation: a smaller area of seta tips is compensated by a higher seta density in the smaller species.

5.2.5 Adhesion-mediated friction

Such factors as the proximity of two surfaces, fluid thickness between surfaces, surface energies, and fluid viscosity, mainly contribute to adhesion. The adhesive setae in syrphid flies *Episyphus balteatus,* are hollow inside and some of them contain pores under the end plate (Chapter 10). The pore, presumably allows passage of the adhesive secretion directly to the contact area. Also, pore canals, located at the base of the shaft, reported from other flies, can contribute to secretion transport to the surface (Bauchhenss, 1979a). Attachment under a horizontal surface is mediated by adhesive forces, which were previously measured on only two species of insects possessing setose attachment pads: the calliphorid fly *Calliphora vomitoria* (2.4 mN) (Walker et al., 1985) and the coccinellid beetle *Epilachna vigintioctomaculata* (2.9 mN) (Ishii, 1987).

The fact that the pad material is designed to provide continuous deliverance of secretion to the contact area, and that adhesive forces are involved in holding an insect under a surface, suggests that adhesive forces may contribute to the friction force (Rabinowicz, 1995) when the insect walks on a vertical surface. Such adhesion-mediated friction can be relatively high, so that surfaces with fluid between them demonstrate a friction coefficient higher than 1. In different species of syrphid flies, the friction coefficient ranged from 7 to 35, which supports the idea of adhesion-mediated friction. Friction between the attachment system of the fly *C. vomitoria,* and a smooth surface, is 8 times higher than adhesion (Walker et al., 1985). If we assume that this relationship is similar among various fly species, the expected adhesive properties of the pad material can be calculated for the species studied.

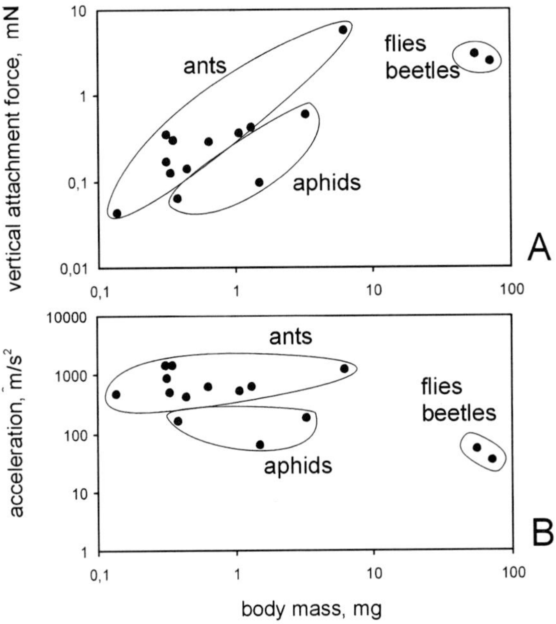

Fig. 9.21. Scale effects on the attachment generated by insect pads. A. Vertical attachment force (adhesive component). B. Maximum acceleration. According to data of different authors (Walker et al., 1985; Ishii, 1987; Lees and Hardie, 1988; Dixon et al., 1990; Federle et al., 2000).

In the case of friction, mechanical properties of material may additionally contribute to the overall attachment force. Material of the fly pulvilli is soft; the membranous cuticle of setose pads is designed as a fibrous composite material, in which the fibers are sparsely distributed. In Coleoptera, seta bases are embedded in this material, which provide high mobility of setae, and their adaptability, on a variety of surface profiles (Fig. 9.9 C). The setae and seta ends are also composed of extremely flexible material. Among setose attachment systems, the flexible nature of seta material has been shown only for beetles, using Mallory's stain (Stork, 1983b), which stains tanned and not-tanned cuticle in different colours (tanned – rot-orange, non tanned – blue).

5.2.6 Scale effects on the friction force

Although heavier species demonstrated higher friction force, the relationship of mass-to-friction was significantly higher in the smallest species (*P. angustatus*), and significantly lower in the largest species (*V. pellucens*). The four other species had a similar mass-to-friction relationship, which was intermediate between the two species named above. Similar relationships between body mass and vertical attachment force (adhesive component of attachment) may be obtained for a broader number of insect species, of both types of pad design (Fig 9.21).

Dependence of friction force on the pulvillus area does not increase linearly; it is logarithmic. A two-fold increase in pulvillus area results only in a 1.5 fold increase in friction force (from 10 to 15 mN). It seems that a further increase of pulvillus area does not improve attachment ability. Possibly, there are some design constraints on a further increase of pad area.

6. SMOOTH SYSTEMS

Smooth systems are composed of cuticle of unusual design (Fig. 9.9). The key property of smooth attachment devices is deformability and the softness of the pad material. Visco-elastic properties were previously suggested (Brainerd, 1994) and were recently, experimentally shown by our studies (Gorb et al., 2000; Jiao et al., 2000). In the first two subsections, a short overview of adhesion, friction and material properties of the pad material is given.

6.1 Friction forces of orthopteran attachment pads

Euplantulae of the great green bush-cricket *Tettigonia viridissima* (Orthoptera, Tettigoniidae) (Fig. 9.22 A), belong to the smooth type of attachment pads (Slifer, 1950). In order to gain detailed tribological information, the friction force of the attachment pads was measured under different loads in living and dead insects, using a micro-tester (Fig. 9.23 A) (for details see Scherge and Gorb, (2001), Scherge and Gorb (2000a)). Profile changes of the surface, and the orientation of cuticle microfibrils, were visualised by means of scanning electron microscopy, followed by freezing-substitution experiments.

In two series of experiments, friction was measured on the euplantulae of the preterminal tarsomere: (1) at different normal forces ranging from 50 μN to 650 μN, at a constant frequency of 0.5 Hz; (2) at different frequencies ranging from 0.05 Hz to 2 Hz, and at constant normal force of 87 μN.

Friction of the pad surface was obtained by oscillating the sample over a distance of 10 μm along the x-axis (distal-proximal), in both directions.

6.1.1 Characteristics of the pad surface and pad material

Our experiments and ultrastructural studies show that the cuticle of orthopteran pads are natural friction-active materials, with a particular inner and outer structure. Each tarsus of *T. viridissima* contains 3-4 euplantulae (Fig. 9.22 A, B). The preterminal tarsomere has the largest euplantula (1.7-2.5 mm wide). Under the light microscope, its surface appears smooth. In reality, it is composed of hexagonal structures (Fig. 9.22 C) (area ca 15 μm^2) with underlying, tiny rods of 0.08 μm in diameter. These rods are branches of thicker rods of 1.12 μm diameter, located deeper in the cuticle. In sections and fractures, the uppermost layer resembles a thin film with a thickness of 180 nm. Non-fibrous structure corresponds to the epicuticle, the outermost layer of insect integument.

6.1.2 Frictional properties of the pad

By oscillating the sample over a distance of 10 μm along the x-axis (distal-proximal), the frictional properties of the pad surface were measured under low normal forces. The dependence of the friction force on load is given in Fig. 9.23 C. The friction behaviour of the pad changes with the velocity. Minimum friction force occurred at 0.5 Hz (10 μm/s) and is higher at slower, and faster, velocities (Fig. 9.23 D). The experiments revealed that the static friction during proximal movement was larger and more stable, compared with distal movement (Fig. 9.23 B). During distal movement, friction slowly increases. This effect is reflected in the rising part of the curve. The anisotropy slightly increases with increasing load.

The friction force remains low, as long as a critical load of about 600 μN is not exceeded. Under loads over 600 μN, pad and substrate remained in contact. Insect footprints are mainly composed of long-chained hydrocarbons with hydrophobic properties (Kosaki and Yamaoka, 1996). This suggests that an attractive force between a water covered surface, such as a silicon surface, and hydro-carbons can be induced by dispersion forces. Dispersion forces represent the most effective type of van der Waals interaction, arising from quantum mechanical effects (Israelachvili, 1992). Hydro-carbon liquids are held together by dispersion forces only, and it is assumed that the force is strong enough to attract the water covered silicon sample.

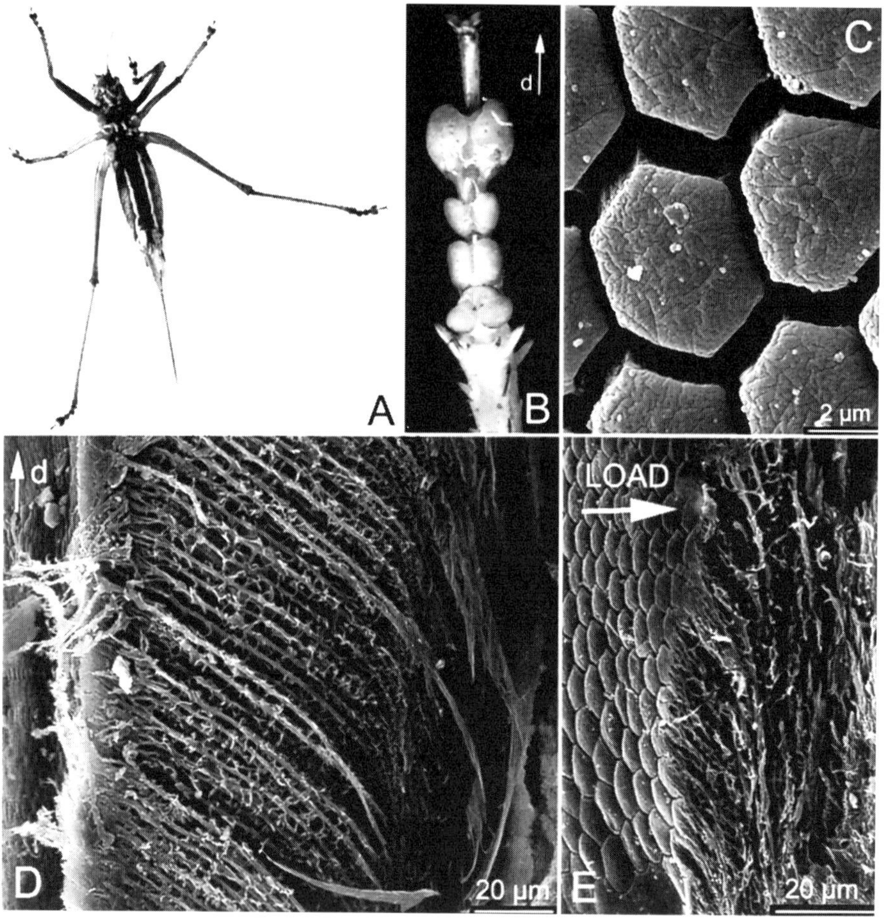

Fig. 9.22. Euplantulae of *Tettigonia viridissima* (Orthoptera, Tettigoniidae) (Gorb and Scherge, 2000). A. Animal on a vertical glass surface. B. Tarsus of the third leg with four euplantulae. C. SEM image of the pad surface. D-E. Shock-frozen pad cuticle after substitution and fracture. D. Free pad that has not been in contact with the substratum. Rods are sloped in a distal direction at an angle of about 45°. E. Pad that has been pressed against the substratum. Rods are sloped in a distal direction at an angle of about 5°. d, distal direction.

The ultrastructural study shows that the inner and outer architecture of pads provide stability and, simultaneously, extreme flexibility. This allows the pad material to adapt to different substrate roughness, which is unpredictable for mobile insects. Through particular orientation of stiff components in the composite material, the material is optimised for maximum friction in one direction. Another interesting feature of the system is that the differently sized rods seem to be adapted to different scales of roughness (micro and meso scale roughness). Shock-freezing experiments revealed that, under loads, rods tend to bend, adapting to the surface profile (Fig. 9.22 D, E).

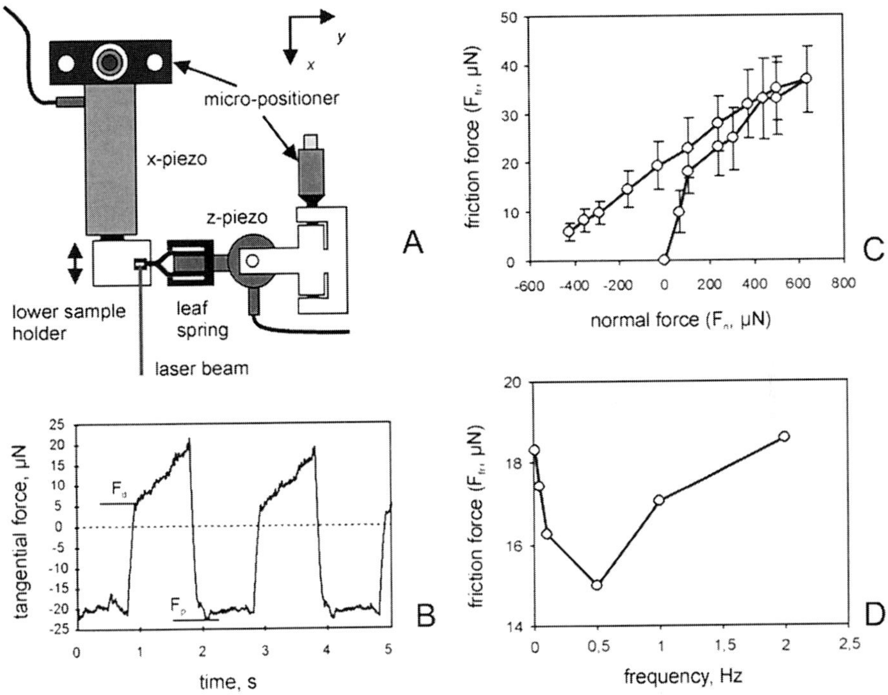

Fig. 9.23. Friction measurements (Gorb and Scherge, 2000). A. Micro-tester set up. The oscillatory motion is provided by an x-piezo. The pad is attached to the x-piezo. A silicon plate, attached to a glass spring, served as an upper sample. A laser beam reflected by a mirror, attached to the spring, was used to detect deflection of the spring. In the z-direction, a z-piezo is attached to adjust the normal force. B. Friction behaviour of the pad in different directions. F_d, friction force to the distal direction of the pad; F_p, friction force to the proximal direction of the pad. C. Friction force versus normal force at a frequency of 0.5 Hz. The lower part of the curve was obtained with an increasing normal force, the upper part was recorded with a decreasing normal force, after full contact between the pad and silicon surface was reached. D. Friction force versus frequency at a normal force of 87 μN.

6.2 Adhesive properties of attachment pads

The force tester used (Fig. 9.24 A) included three main parts, a platform, a glass spring, and a fibre optical sensor. The lower sample was mounted on the platform, the upper sample was attached to the glass spring, and the interacting force was detected by the optical sensor. To measure adhesion between the two samples, the lower sample was approached to the upper sample, then retracted once they remained in contact (remaining time), under a certain load force (normal or applied force).

A typical force-versus-time curve is shown in Fig. 9.24 B. Three types of experiments were carried out. (1) Adhesive properties of the pad secretion. The adhesion of secretion deposited on a silicon chip (the lower sample), was measured by a sapphire ball of 1.47 mm diameter (the upper sample). (2) Adhesion of the pad of a leg of living insects. One tarsus of a leg was firmly fixed on the platform, a silicon chip was attached to the spring as the upper sample. (3) Adhesion measured on the pad of a freshly severed tarsus.

For the first two measurements, applied force in the range of 0.1 to 2.5 mN was used, considering 1 g as the average weight of *T. viridissima*. The indentation on the pad (Fig. 9.24 C) was obtained by comparison of the spring deflections caused by the pad, and a hard lower sample (for example the sapphire ball). Treating the pad like part of a sphere (Fig. 9.24 A), the indentation can be modelled by the Hertz theory (Fig. 9.24 C).

Depending on the location selected, the adhesive force could monotonically increase or decrease, or remain constant with increasing applied force, and in a few cases was oscillating randomly. The adhesive force depended strongly on the location on the silicon surface.

Adhesive force, measured on the fresh pad, increased at smaller applied forces, and remained constant when the applied force exceeded a certain value. In the case shown in Fig. 9.24 D the saturated adhesive force of 1.1 mN was reached at an applied force of 0.8 mN. In different individuals studied, the saturated adhesive force ranged from 0.7 to 1.2 mN at applied forces of 0.7-1.5 mN.

To study the influence of the change of pad material on pad adhesion, measurements on a freshly severed leg were carried out. The freshly cut sample had the same adhesive force as a living pad. However, the adhesive force decreased to 0 after 70-75 min of the experiment. This effect can be explained by results of another experiment, showing that the pad cuticle has higher evaporation abilities than a normal leg cuticle. About 3 h after the start of the experiment, the pad lost about 50% of its initial weight due to water evaporation.

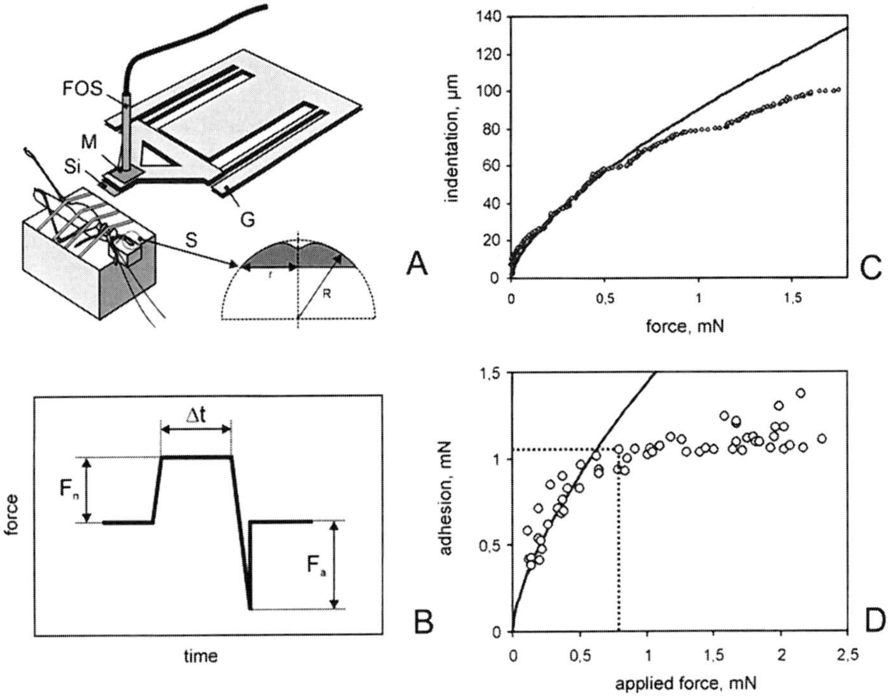

Fig. 9.24. Adhesion measurements A. Micro-force tester. The lower sample (S), the pad of a living insect, is fixed to the platform, and the upper sample (Si), a silicon chip, is attached to the spring. Driven by a motor, the platform moves the lower sample to make contact with, and retract from the upper sample. The deflection of the glass spring (G) is monitored by the fibre-optical sensor (FOS) through the mirror (M). Inset shows a cross section of the pad outlined by a semi-circle. R, radius of pad curvature; r, pad width. B. Diagram of a typical curve of the interacting force recorded versus time. F_n, applied force, F_a, adhesive force, Δt, remaining time. C. The pad indentation versus applied force (dotted line). The solid line is the indentation data fitted, according to the Hertz theory. D. Dependence of adhesion on applied force. At smaller applied forces the adhesive force increased with increasing applied force, and was saturated at an applied force over 0.8 mN (dotted line). The saturated adhesive force was about 1.1 mN. Based on the assumption that the adhesive force is proportional to the contact area, caused by applied force, the data was fitted in accordance to the Hertz theory.

The adhesive force of secretion on the silicon chip, detected by the sapphire ball, showed location dependence. It was due to the unequal distribution of the secretion, and the small contact area between the hard silicon surface and the hard ball. In the case of the whole pad, where the contact area is larger due to the pad deformation, location effect would presumably no longer play a role.

The saturated adhesive force of 1.1 mN was reached at an applied force of 0.8 mN, and the corresponding indentation was found to be about 75 μm (Fig. 9.24 C). The indentation of 75 μm indicates that the pad reached maximum contact with the silicon surface (thickness of pad ca. 100 μm). The contact area did not increase with greater applied force after the maximum contact. This suggests that the adhesive force F_a can be expressed as

$$F_a = K_a A,$$

where K_a is a constant, A is the contact area at separation, which is caused by the applied force. At applied force < 0.65 mN the deformation of the pad could be well described by the Hertz theory

$$\delta = \frac{1}{K^{2/3} R^{1/3}} F_n^{2/3}$$

where δ is an indentation; F_n, applied force; K, effective elastic modulus of the pad; R, radius of pad curvature (Fig. 3), the adhesive force in Fig. 4 agrees with

$$F_a = K_a A = \pi \left(\frac{R}{K} \right)^{2/3} K_a F_n^{2/3}$$

where ($a^2 = R\delta$; a, contact radius).

It is concluded that the attachment ability of the *T. viridissima* pads to the smooth surface, strongly depends on the contact caused by the mechanical deformation of the pad. This seems to be a general principle involved in the function of smooth attachment pads.

6.3 How does the bee release its arolium?

The most complex question is concerning attachment-detachment of the pad system. Both actions take a very short time, and structures involved are too small to study these mechanisms without special equipment. The mechanisms of such an action vary in different insect lineages, and most tarsal mechanical systems are not fully understood. Bees and other hymenopterans have unusually complex and specialised structures of the pretarsus. The pretarsus ends with a soft attachment pad (arolium). The mechanism of the extension, and the flattening of the arolium on the smooth surface, have been previously demonstrated in a dead bee. The spreading movement of the pad results automatically from the pull of the claw-flexor-muscle, and the downward pressure of the tarsus on the base of the pretarsus, when claws fail to grasp the support (Snodgrass, 1956). Here, some data is provided to clarify the mechanism of arolium detachment.

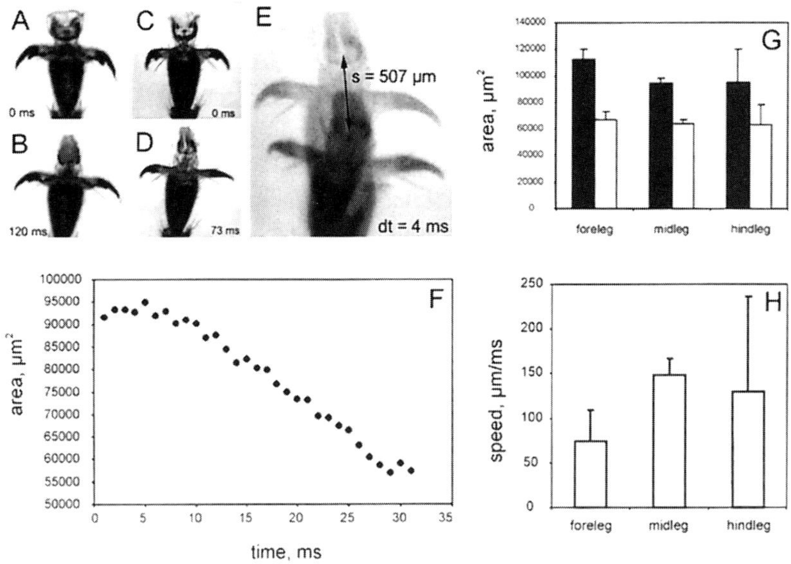

Fig. 9.25. Detachment dynamics of the arolium of *Apis mellifera* (Baur and Gorb, 2001). A-D, F, G. Changes in the area of the spread arolium under a glass surface. A, B. Initial full-spread and final positions of the arolium just before contact loss, ventral view. C, D. Initial full-spread and final positions of the arolium just before contact loss, dorsal view. F. Changes in the area of the spread arolium during leg detachment. G. Initial (black columns) and final (white columns) positions of the arolium just before contact loss. E, H. Speed-measurements. E. An example of two overlaid frames (just before and when contact is lost) used in speed measurements. H. Detachment speed of fore, mid, and hindleg. A-E. Single frames of high-speed videorecordings, 1000 fps. F-H. Data from high-speed videorecordings.

The bee-pretarsus consists of membraneous regions, of which the arolium is the most striking (Figs. 9.2, 9.6 F). The ungues (claws) are articulated with a small plate (flange) at the end of the tarsus. The U-shaped arcus surrounds the arolium. The aroliar manubrium is articulated between the marginal tarsus-flange and the arolium. The planta and auxilary sclerites are located on the ventral side of the pretarsus. The retractable unguitractor is joined with the tendon of the flexor muscle. The folding and flattening behaviour of the arolium is in part a result of the specific material structure of the soft distal part (Fig. 9.9 D).

Action of the arolium during leg detachment was recorded using a high-speed videocamera on bees, walking on a glass plate under a dissecting microscope. It was revealed that the bee releases its arolium by a change of the arolium contact area, with the undersurface. During this action, the soft part of the arolium is driven by the arculus, lateral parts of which are pressed together. This action usually takes about 30 - 40 ms, and results in a change of the contact area to 20% of the fully-spread condition. After the contact area is reduced, the leg jumps from contact with a speed which ranges from 0,024 m/s to 0,24 m/s. Presumably, the folding movement of the pad results from an upward pull of the tarsus on the base of the pretarsus, and/or from a decrease of internal pressure within the arolium.

7. ATTACHMENT PADS IN OTHER ARTHROPODS

Among terrestrial arthropods, setae with the attachment function used for locomotion on the smooth substrata, are known from the scopula in different species of spiders (Hill, 1978; Müller, 1981; Miller et al., 1988; Rambla, 1990; Perez-Miles, 1994). Scopula setae are very complex structures due to specialised joints on the base of each seta, and microtrichia coverage (Fig. 9.26). Setae of the scopula are responsible for maximum contact during locomotion on a smooth substrata.

Since no specialised glands were found at the bases of the scopula setae, the spiders' attachment is presumably achieved by forces of capillary adhesion, or in other words, by use of an extremely thin water-film on the substratum (Homann, 1957; Roscoe and Walker, 1991). This hypothesis is supported by experiments on Teflon-covered substrates, where the water film is absent. Spiders fall off of such substrata.

In large spider species, the ventral surface of the tarsus and metatarsus are additionally covered with setae of a structure similar to that located on the scopula. Therefore, they have the best attachment abilities. As has been previously reported, large spiders, such as *Cupiennius* (Ctenidae), could hold about ten times their own body weight, when sitting on a vertical glass plate (Foelix, 1982). Scopula is absent in the web spiders.

The two alternative designs of leg attachment systems, used in locomotion, seem to appear convergently in Chelicerata. Setose pads have been found in Aranei, whereas smooth pads have been reported from Solifugae and Pseudoscorpiones (Beier et al., 1941).

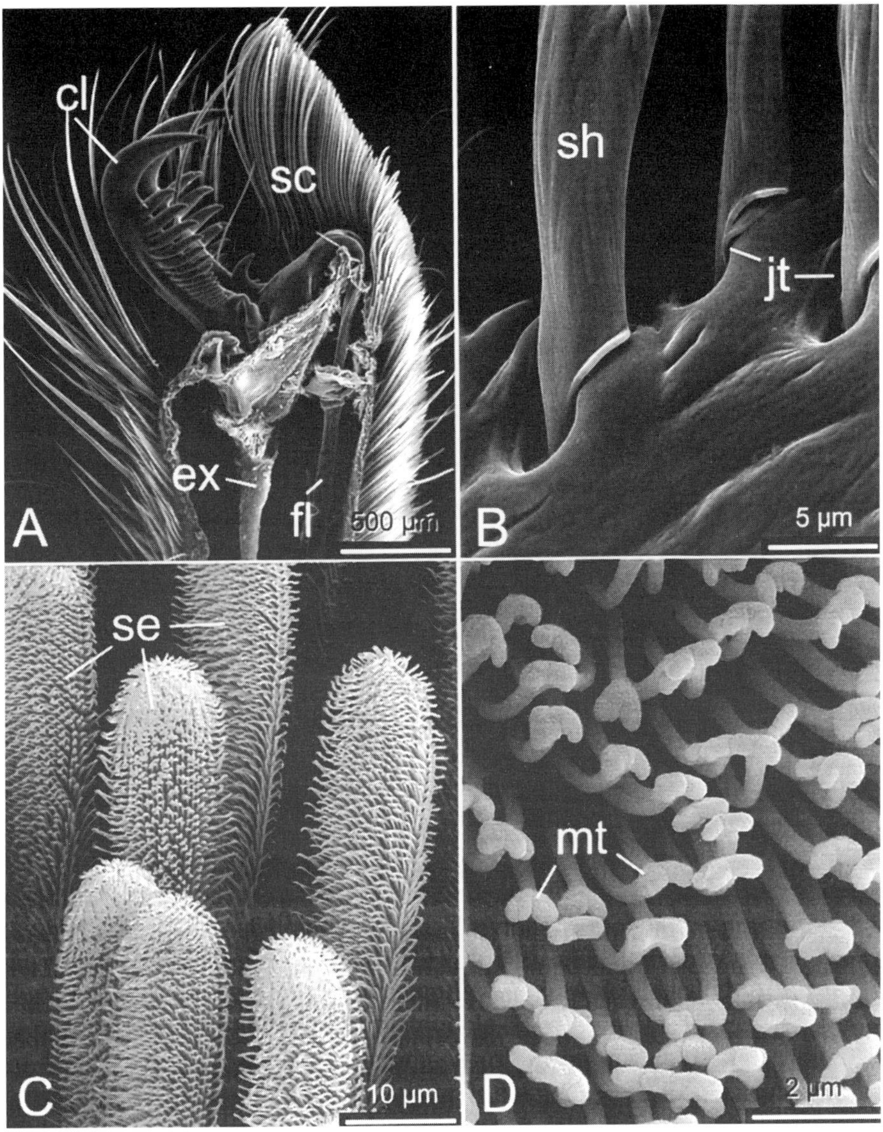

Fig. 9.26. Scopula in the wandering spider, *Cupiennius salei* (Ctenidae). A. Lateral aspect, the segment wall is removed to expose tendons. B. Bases of scopula setae. C. Tips of scopula setae. D. Microtrichia of the scopula setae. cl, claws; ex, tendon of the claw-extensor-muscle; fl, tendon of the claw-flexor-muscle; jt, joints of the setae; mt, microtrichia; sc, scopula; se, setae; sh, hair shaft.

Chapter 10

Secretions in frictional systems

Cuticular fixation systems may be designed according to several fundamental principles: hooks, lock or snap, clamp, spacer, sucker, expansion anchor, friction, and adhesive secretions (Chapter 4). A real system is usually based on two or more of these principles. The odonate head-fixation system, wing-locking devices in Coleoptera and Hymenoptera, and insect tarsal pads are often based on friction-active surfaces supplemented by epidermal secretions.

Light-microscopical studies have shown the presence of vesicles in epidermal cells underlying the microtrichia-covered areas in the attachment systems, based on two complementary surfaces. Later investigations have revealed numerous pore canals penetrating the cuticle of these devices, and their microtrichia. Secretory substances have also been revealed in attachment systems adapted for a variety of surfaces. It is well known that insects, walking on smooth surfaces, leave microscopically small fluid prints.

Material preventing separation of two surfaces may be defined as an adhesive. The question arises, why are adhesives used for attaching materials to surfaces? There are a variety of natural attachment devices based on mechanical principles (Nachtigall, 1974). There are at least three reasons for the use of adhesives: (1) they join dissimilar materials; (2) they improve stress distribution in the joint; and (3) they increase design flexibility (Waite, 1983). These reasons are relevant to the evolution of natural attachment systems, and man-made joining materials.

What are the origin and nature of epidermal secretions existing in frictional systems? What is contained in epidermal vesicles and transported to the contact area? What are the chemical composition and properties of these secretions? All these questions will be discussed below, clarifying the functioning of some attachment devices.

1. SYSTEMS OF TWO COMPLEMENTARY SURFACES

1.1 Secretory epidermis and nature of secretions

In semi-thin sections stained with toluidine blue, the cuticle of most attachment devices is viewed as consisting of two layers: an external one (exocuticle, deeply blue stained) and an internal one (endocuticle). In frozen sections stained with Sudan Red, endocuticle is reddish. Epidermal cells of the postcervical sclerite of the dragonfly arrester system (Chapter 6) and those underlying TDM, TAL, and TPL fields of the beetle wing-locking devices (Chapter 8), contain dense cytoplasm with rough endoplasmatic reticulum, many mitochondria, and large electron-lucent vesicles of diverse shapes. The vesicles, which are located in the part of the cell facing the inner layers of the endocuticle, form large junctions resembling a layer filled with secretion. These junctions are connected to numerous pore canals of endocuticle. This area is presumably associated with the release of the vesicle content into the pore canals.

There are several sources of evidence of the lipid nature of the vesicle fluid. In TEM preparations, the vesicles are electron lucent (Fig. 10.1 A, B). In the semi-thin sections treated with Maxwell solution, the vesicles appear as cavities in SEM (Fig. 10.1 D). In the semi-thin sections stained with toluidine blue, the vesicles are viewed as transparent structures embedded in deeply-stained cytoplasm. In the histochemical assays, Sudan Black stains vesicles black, Sudan Red stains them orange.

The comparisons of the cuticle surface dehydrated, and critical point dried, with the same surface after air drying, often revealed the presence of amorphous coverage of the microtrichia in air-dried specimens. A comparison of surfaces of the shock-frozen microtrichia of the odonate arrester, which were dehydrated in ethanol and acetone, with those air-dried, shows that the surface of dehydrated microtrichia is smooth and clean, whereas non-dehydrated microtrichia is covered with non-volatile bloom. The bloom contained little flakes, 0.1-0.4 μm long and 0.05-0.50 μm wide. The density of these flakes is about 5-15 in μm^2. From these observations, it can be concluded that the material stored in vesicles and liberated on the external surface of the microtrichia, is presumably a non-volatile lipid. Similar results have been obtained in experiments with microtrichia fields of the wing-lock in Coleoptera (Fig. 10.2).

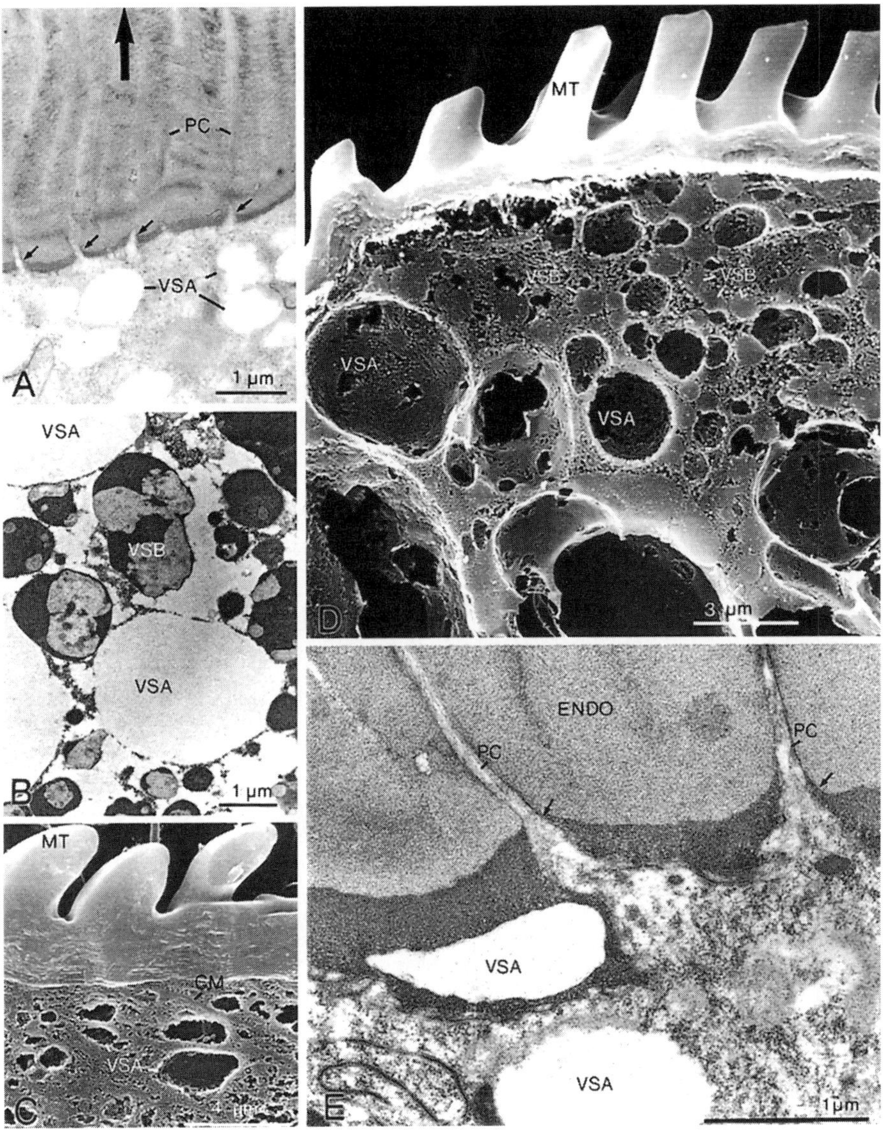

Fig. 10.1. Elytra-to-body locking devices in Coleoptera. TEM pictures (A, B, E) and SEM pictures of the semi-thin sections (C, D) of the area underlying the cuticle of the TAL field, in beetle species *Tenebrio molitor* (A-C, E) and *Tribolium castaneum* (D). CM, cell membranes; ENDO, endocuticle; MT, microtrichium; PC, pore canals; VSA, electron-lucent vesicles; VSB, electron-dense vesicles; large arrow indicates preferred directions of the canals; small arrows indicate cell connections to the canals.

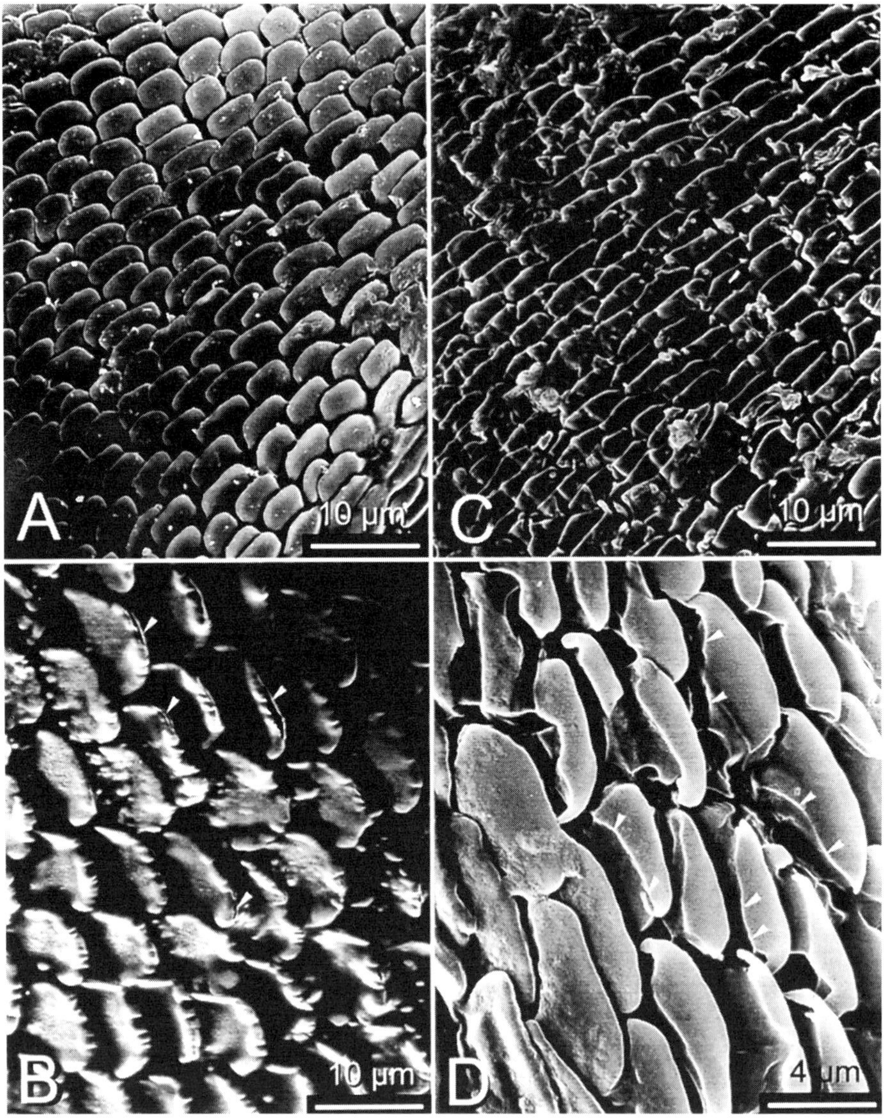

Fig. 10.2. SEM pictures of the dorso-medial field of the elytra-to-body locking devices in the tenebrionid beetle *Tribolium castaneum* (Gorb, 1999b). A. Specimen fixed in glutaraldehyde, dehydrated in ethanol and acetone. B. Air-dried specimen. C, D. Dental wax casts of the surface of a living insect. White arrowheads, secretory substances.

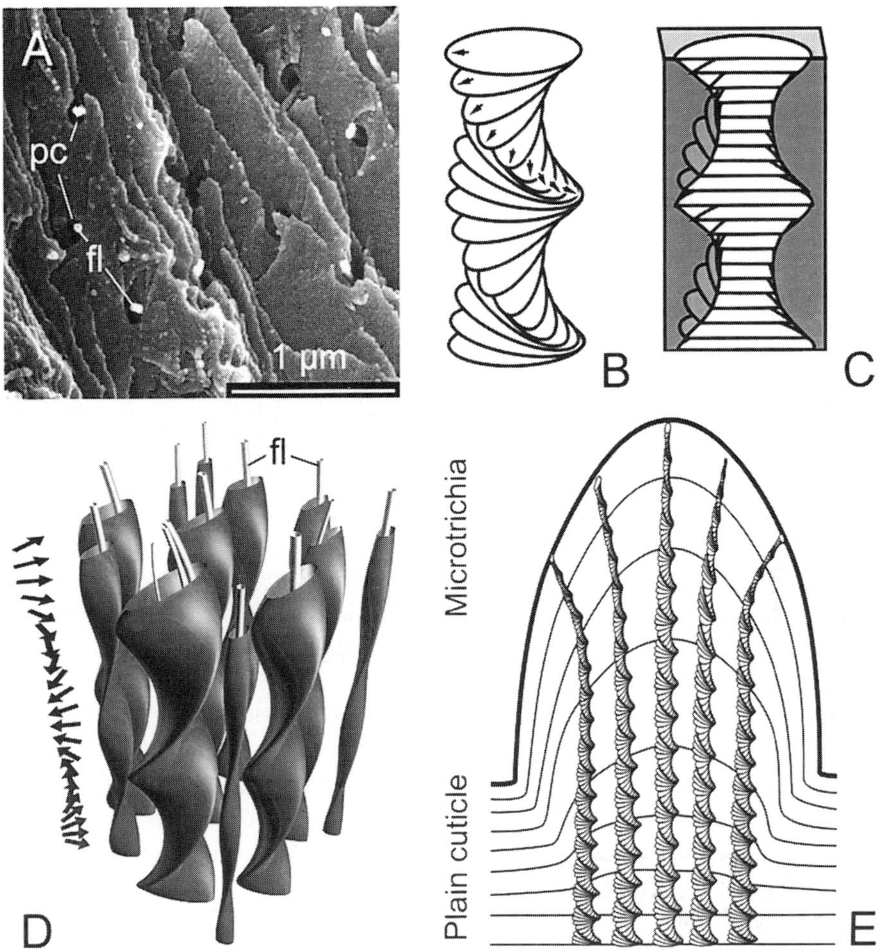

Fig. 10.3. Pore canals (Gorb, 1997b; 1998b). A. Oblique fracture of the cuticle in the head-arrester system of the damselfly *Lestes barbarus*. B-E. Models of the design of pore canals in the flat cuticle (B-D), and within the microtrichia cuticle (E). B. Three-dimensional ribbon-like organisation of the canal as it appears in the serial cross sections. C. Longitudinal section of model shown in B. D. Computer-based model of pore canals with filaments inside. E. Directionality of canals in the underlying cuticle and within the microtrichium. Deviations of the canal depend on the orientation of the cuticular layers. Arrows in B and D indicate preferred orientation of microfibrils within the layer.

1.2 Cuticle transporting systems

1.2.1 Pore and terminal canals

The content of vesicles is transported through the cuticle of microtrichia by the pore canal system (Chapter 2). Large surface pores have been found only in cockroach pulvilli (Arnold, 1974), in the microtrichia of the mesothoracic part of the wing-lock in water bugs from families Nepidae, Naucoridae, and Nepidae (Chapter 8), and in the coxal locking mechanism in cicada (Chapter 7). Such large pores were not found on the microtrichia surface of the majority of other frictional systems. A pore canal is a tube of an elliptical cross section, which is twisted regularly along its length so as to form a twisted ribbon (Neville et al., 1969; Neville and Luke, 1969b; Barth, 1970; Compère and Goffinet, 1987a; b; Halcrow and Powell, 1992; Cribb and Jones, 1995). The shape of the pore is helicoidal and presumably results from the forces associated with crystallisation of the forming cuticle (Fig. 10.3 B, D). The density of canals in the damselfly endocuticle is about 6-7 μm^{-2}. In exo and endocuticular layers, pore canals are usually oval, 0.3-0.4 μm at the larger diameter and 0.1-0.3 μm at the smaller one. In the endocuticle, the area of a pore canal in a cross section varies from 0.01 to 0.15 μm^2.

Pore canals contain up to five canal filaments with a diameter (probably together with residual secretions) of about 0.08-0.10 μm (Fig. 10.3 A, D). Usually, a canal contains one filament, but canals located in the deeper layers of the endocuticle often have from two to five single filaments. The filaments were observed both in the intact cuticle and in the cuticle enzymatically treated with chitinase. In the cuticle treated with NaOH, the canal filaments were absent. Thus, it can be hypothesised that the canal filament presumably has protein-like components, but does not contain chitin. The extremely small size of the canals, together with its twisted-ribbon shape and the presence of filaments, make up a kind of fluid-transporting system that must have particular fluid-mechanic properties.

In microtrichia, two types of canals mainly exist. Canals of the first type are larger, 0.1-0.2 μm in diameter. They are electron lucent, located in the endocuticle and in the exocuticle as well, and usually contain canal filaments (Fig. 10.3 A). In some cases, canals can bifurcate, especially in deep layers of the endocuticle. Some canals of this type reach the surface of the cuticle (Fig. 10.3 E). In the external layers of the endocuticle and in the exocuticle, the canals tend to branch into smaller electron-denser canals of the second type, called terminal canals that are 0.01-0.10 μm in diameter. These canals form a dense network within each microtrichium. The density of these canals within external layers of exocuticle is about 10-30 μm^{-2}.

Terminal canals are quite different from pore canals. They do not have filaments inside and do not penetrate the thin epicuticular layer. It seems that secretion reaches the epicuticle through terminal canals and goes through the epicuticle without any canal structures. The diameter of such a termination is approximately 0.01 μm. It can be suggested that they may transport substances not only physically, due to capillary forces, but also chemically, due to active enzymatic transport. Soft cuticle of the damselfly postcervical sclerite, which is densely penetrated by pore and terminal canals, resembles a sponge-like tissue that can easily release its contents under pressure (Gorb, 1997b).

1.2.2 Orientation

Early ultrastructural studies of the pore canal system in representatives of Arthropoda, based on polarised microscope observations and on TEM images following ultra-thin sectioning (Neville, 1967; 1970; Neville et al., 1969), suggested a twisted-ribbon model; this has also been supported by later studies (Neville, 1975; Compère and Goffinet, 1987a). Pore canal orientation has been used as an indirect method for studies of the macromolecular architecture of the cuticle (Neville and Luke, 1969b). Canal shape is oval in cross section, and the orientation of the larger radius of the oval is strongly dependent on the preferred orientation of cuticle microfibrils within the lamella. This is the reason for a characteristic zig-zag formation of pore canals in their longitudinal sections (Fig. 10.3 C). The shape of the pore canal is defined by the macromolecular organisation of the chitin-protein microfibrils. The long axes of the canals are orientated parallel to the axis of the preferred orientation of the cuticular microfibrils. The preferred direction of the fibres is rotated in each successive layer, which is about 10-30 nm thick, at an angle varying about 5-10°.

Generally, pore canals are directed perpendicularly to the cuticular lamellae. Vertical pore canals pass from the epidermal cells through all cuticular layers to the surface of the cuticle. The change in direction of the pore canals within the microtrichia depends on the orientation of the cuticle lamellae. In the plain cuticle, layers show a parallel orientation. When the cuticular lamella protrudes to form a microtrichium, the canals remain orientated perpendicularly to each of the lamellae, and are consequently curved within the microtrichium. Numerous canals pass through a microtrichium and most of them end at its side wall (Fig. 10.3 E).

1.2.3 Functions of canals

The transporting function of the cuticle is well known. Cuticular waxes covering an insect body are deposited on the cuticle surface using the pore canal system (Shvanvich, 1949; Locke, 1960a; Gorb, 1995b). Therefore, pore canals are usually interpreted as a transport system providing deposition of new epidermal materials into/onto the cuticle and, in addition, provide reabsorbtion paths for the transference of organic products (Neville and Luke, 1969a; Compère and Goffinet, 1987b). There are also other functions of pore canals. It has been suggested that pore canals constitute structural devices offering resistance to cuticle splitting (Mutvei, 1974). Apparently, canals can prevent fracture propagation under mechanical loads, and such design property in tanned cuticle may provide an increase in stability in the joint regions. Also, the cuticle of apodemes usually contains large pore canals (Gorb, 1996a). In insect chemo and hygroreceptors, pore canals provide stimulus conduction to the sensory cell (Altner and Prillinger, 1980; Cribb and Jones, 1995).

Pore canals, described in the cuticle of the odonate head-arrester and beetle locking device, provide transportation of lipid-containing secretions to the surface of the cuticle. The elastic cuticle of the postcervical sclerite, which is densely penetrated by pore canals, resembles a sponge-like tissue that can easily free its contents under pressure. It is suggested here that mechanical contact of the sclerites with the microtrichia fields of the head, results in the liberation of epidermal secretions into the contact area. The high capillarity of the surface covered with microtrichia, can provide quick spreading of the secretions within the region of contact.

1.3 Functions of secretion

Attachment devices, which are designed as a pair of co-opted surfaces, usually consist of one solid and one flexible part. Such a difference in material properties results in an increase of frictional forces in the contact area (Bowden and Tabor, 1986). When corresponding fields are pressed together, deformations of a flexible part increase the real contact area between corresponding structures, and consequently increase frictional forces. Additionally, structures may interlock mechanically.

Microtrichia deformations result in the liberation of epidermal secretions, which may be surface-active and serve to increase adhesion at the contact area. There is an alternative hypothesis presuming that epidermal secretions may aid in the lubrication of contacting microtrichia. This may help an interlocked system to be easily released in only one particular direction. The exact function of the secretions remains unknown.

2. SECRETIONS IN ATTACHMENT PADS

Insect adhesive pads are designed as flexible areas of cuticle (Chapter 9) supplemented with cuticular secretions. Secretion has been demonstrated in both alternative designs of the pads; hairy and smooth. Pad fluids have been previously reported from hairy adhesive pads of reduviid bugs (Edwards and Tarkanian, 1970), flies (Bauchhenss and Renner, 1977; Bauchhenss, 1979a; Walker et al., 1985), coccinellid beetles (Ishii, 1987), and from the smooth pads of cockroaches (Roth and Willis, 1952), aphids (Lees and Hardie, 1988; Dixon et al., 1990) and bugs (Hasenfuss, 1977a; 1978; Ghasi-Bayat and Hasenfuss, 1980a).

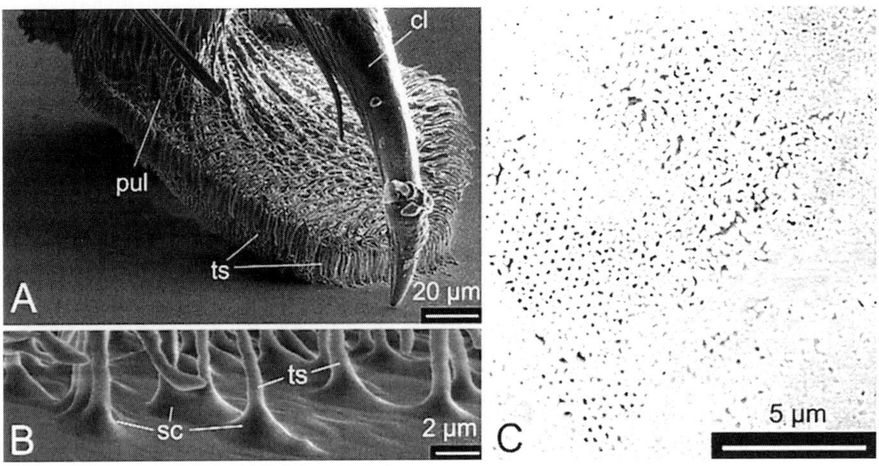

Fig. 10.4. Secretion in the fly attachment system. A. Pulvillus of the syrphid fly *Eristalis pertinax,* in contact with the glass surface. B. Tenent setae tips surrounded by secretion. C. Footprints on a glass surface. Note the regular pattern of droplets. cl, claw; pul, pulvillus; sc, secretion; ts, tenent setae.

In the adhesive pads of insect legs, epidermal cells with secretory activity usually contain at least two types of vesicles: electron-lucent and electron-dense (Bauchhenss and Renner, 1977; Bauchhenss, 1979a; Lees and Hardie, 1988). A possible explanation for this may be that adhesive pads are adapted to provide attachment to a variety of substrates and, therefore, need to liberate a mixture of different adhesive components. However, this hypothesis needs additional studies on the chemical composition of secretion and its interaction with various substrata.

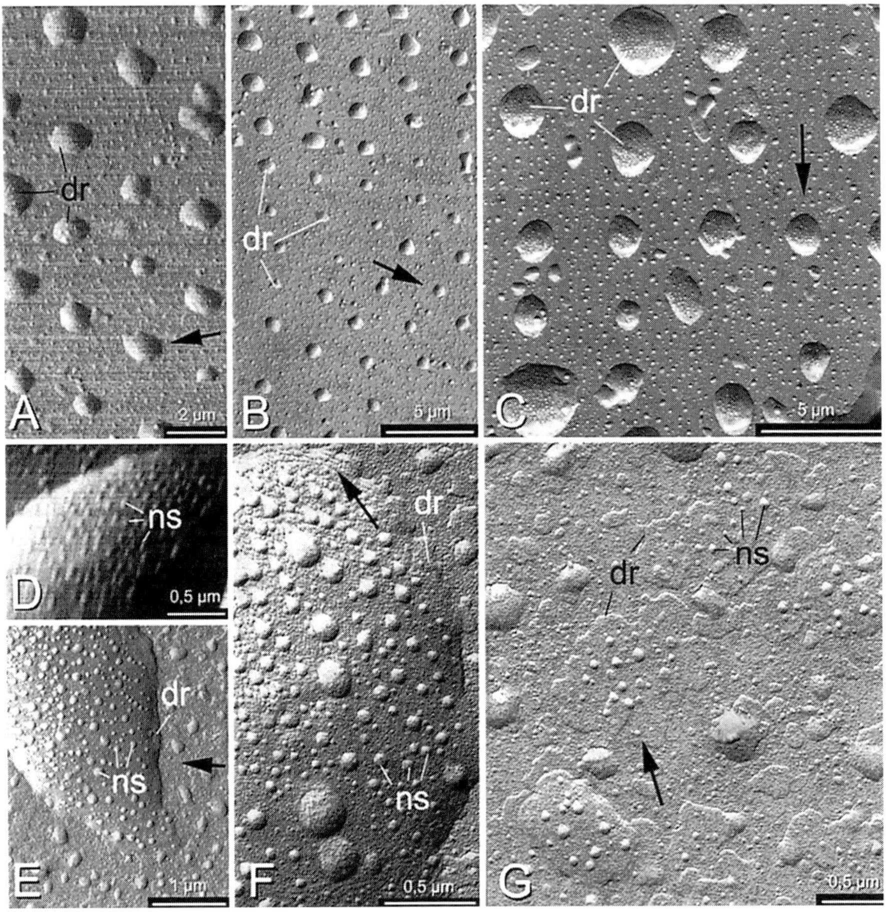

Fig. 10.5. Carbon-platinum replicas of the frozen footprints of the calliphorid fly *Calliphora vicina,* in SEM (A, D) and TEM (B-C, E-G). A-F. Fresh prints. G. Dried-out prints. dr, droplets; ns, nanostructures. Black arrows indicate direction of coating. Preparation of Y. Stierhof (MPI of Biology, Tübingen).

Footprints can be easily observed with the light microscope, especially under phase contrast (Fig. 10.4 C). In hairy systems, they usually have a regular pattern corresponding to the pattern of tenent setae. In the smooth systems, prints have a randomly distributed pattern (Jiao et al., 2000). At a higher magnification, prints contain an additional pattern of nanostructures, 30-100 nm in diameter, on their surface. These structures remain on the surface even when the content of the main droplet partly evaporates, after several hours (Fig. 10.5). This observation supports the hypothesis about multi-phasic composition of the secretion, and that at least one part of the secretion is non-volatile.

2.1 Composition of the secretion

The pad secretion of diverse hexapods contains a non-volatile, lipid-like substance that can be observed in footprints, under-water (Fig. 10.9 A, B), or in footprints stained with Sudan Black. It has been shown with the use of thin-layer chromatography that in lady bird beetles (Coccinellidae) the chloroform-soluble part of the pad secretion consists mainly of hydro-carbons, fatty acids, and alcohols (Ishii, 1987). Gas chromatography has revealed that pad adhesive secretions consist of hydro-carbons and true waxes (Kosaki and Yamaoka, 1996), which correspond well to the composition of the cuticle coverage.

Some observations were made on tarsi of the beetle *Hemisphaerota cyanea* (Chrysomelidae, Cassidinae), submerged in water. It was noted that the contact surface of the tarsi was water-repellent. Although in places where water did seep into the spaces between clustered bristles, the clusters were seen to be moistened by a fluid that did not mix with water (Eisner and Aneshansley, 2000). Similar observations were made on the smooth pads of the grasshopper *Tettigonia viridissima* (Jiao et al., 2000). Chemical extracts of *H. cyanea* tarsi, or of glass surfaces to which they had clung, yielded mixtures of saturated and unsaturated linear hydro-carbons of C_{20} to C_{28} chain length, with (Z)-9-pentacosene as the principal component. These results led previous authors to presume that the fluid is an oily substance.

Measurements made with the use of the Atomic Force Microscope (see subsection 2.3 of this Chapter) demonstrated a low contact angle of the droplets on hydrophilic substrata: height-to-diameter ratio of the droplets was about 1/50 (Fig. 10.6 A). After several hours, droplets evaporated by up to 15% (Fig. 10.6 B). The residues remain on the surface for a long time. These results indicate the presence of water or other solvents in the secretion. However, identity of the water-soluble fraction of the fluid, remains unknown.

2.2 Design of the fly's tenent setae

The design of adhesive setae is adapted for the release of adhesive substances close to the area of contact. Here it is shown that tenent setae located on the distal part of the pulvillus are responsible for secretion release very close to the contact area, between the seta tip and the substratum. Setae located on the basal part of the pulvillus do not have such a mechanism. The two types of setal design may be an adaptation for attachment to a variety of substrata.

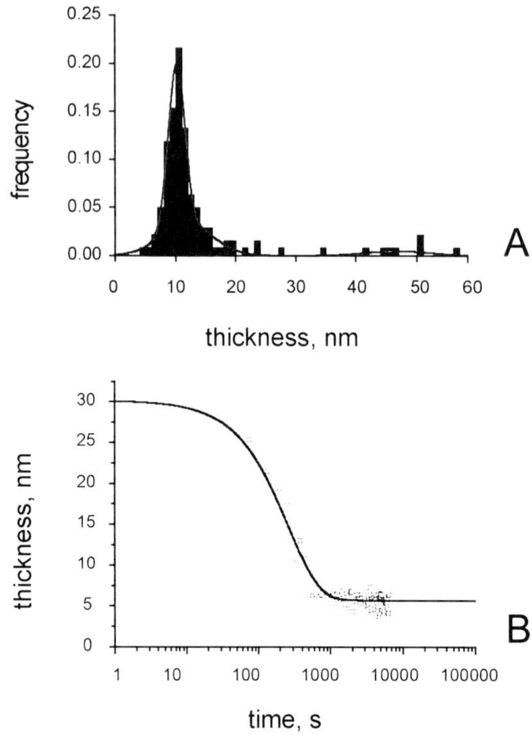

Fig. 10.6. Droplet thickness of the secretion in the fly *Calliphora vicina* (AFM-measurements). A. Frequency histogram of the droplet thickness. B. Time dependence of the droplet thickness. Line indicates exponential decay.

In *Calliphora* flies, the non-volatile lipid secretion is produced by large cells, located at the base of each pulvillus, and stored within a «spongy» layer of cuticle (Bauchhenss, 1979a). A well-developed system of pore canals has also been described at the base of the tenent setae, located on the basal part of the pulvillus. It has been hypothesised that the canals are responsible for the secretion release onto the surface of the seta bases. The loss of secretion could be minimised by its re-absorption back into the cuticle when the fly detaches, due to the high capillarity of the system. The corrugations of the dorsal surface of the adhesive pads of flies and bugs, have been previously interpreted as fluid-transporting structures (Hasenfuss,

1977a; 1978; Ghasi-Bayat, 1979; Ghasi-Bayat and Hasenfuss, 1980b). However, this hypothesis was not experimentally supported.

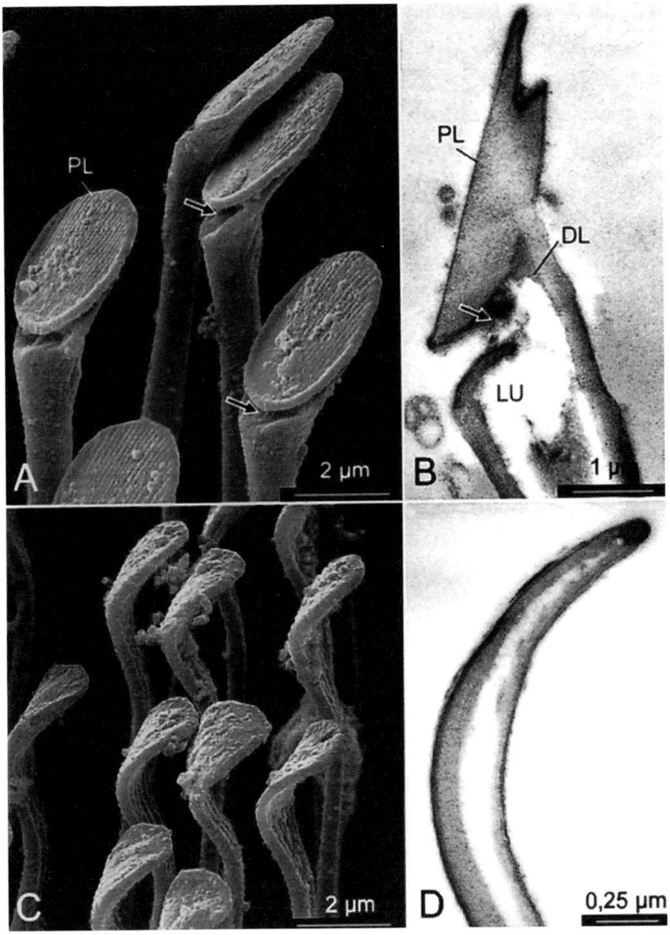

Fig. 10.7. TEM and SEM micrographs of the two types of tenent setae in the syrphid fly *Episyrphus balteatus* (Gorb, 1998d). Setae of type 2 (A, B) are located distally in the pulvillus, whereas setae of type 1 (C, D) are located at the base of the pulvillus. DL, dense layer; LU, lumen; PL, end plate.

In spite of these findings, some investigators have indicated that the secretion is mainly located at the spatulate ends of the tenent setae (Walker et al., 1985), which means that it may be released in the area located close to the distal tip of such setae. These authors have doubted the pore canal hypothesis of secretion liberation, but did not propose any alternative ideas about secretion release.

There are two types of tenent setae existing on the ventral side of each pulvillus (Fig. 10.7). Setae of the first type exist on the basal part of the pulvillus. They are spoon-like, with an intensely compressed tip region. Setae of the second type are located on the distal part of the pulvillus. They are characterised by the ellipsoid plate located at the distal tip of each seta. The ventral and the dorsal surfaces have small grooves running parallel to the longitudinal axis of the hair shaft. The type 2 seta has been termed tenent hairs, or adhesive setae, in the literature (Walker et al., 1985).

Both types of setae are tube-like. In longitudinal and cross-sections the central parts of the setae are electron-lucent, as viewed in TEM (Fig. 10.7 B, D). These electron-lucent areas suggest lipid-containing substances, which were dissolved during the electron-microscopic preparation, due to the use of ethanol and propylene oxide as solvents. No distal openings are found in the distal part of type 1 setae. However, for type 2 setae there is a transition between the central lumen of the shaft and the outside (Fig. 10.7 B). This area is located in the posterio-distal part of the seta, between the seta thickening and the distal plate, where large canals (0.1-0.4 μm wide) occur.

The ventral grooves of a seta tip are often covered with amorphous substances, as observed in air-dried preparations of pulvilli, attached to a cover-slip and released after drying procedure. In such preparations, the distal parts of intact tenent setae were often covered with secretion, close to the region of canals. In preparations, washed in a variety of organic solvents, the distal seta surface was almost clean.

SEM-study of preparations of pulvilli attached to cover-slips shows that lipid substances are not spread over the whole contact area between pulvillus and substratum, but as discrete droplets just under the distal end-plate of the type 2 setae (Fig. 10.4). Moreover, the surfaces of a seta base remained clear after attachment. Setae were very seldom glued to each other, and space between them was usually not covered by the secretion. Such observations allowed the conclusion that the adhesive, lipid-containing substances are targeted directly to the contact area between the distal plate of a single tenent seta and the substratum. However, the fact that tenent setae are adapted to release secretions distally does not exclude the possibility that the pore canal system is involved in secretion release.

The position of the setae on the substratum is always the same when a fly holds onto an inclined surface. It seems that the end of the tenent seta, together with the seta orientation, is adapted to release secretion when a pulling force is directed along the substratum surface to the proximal part of the pulvillus (Fig. 10.8). When force is directed in the opposite direction, both types of setae can detach from the substratum, and openings of tenent setae can probably be closed by such an action.

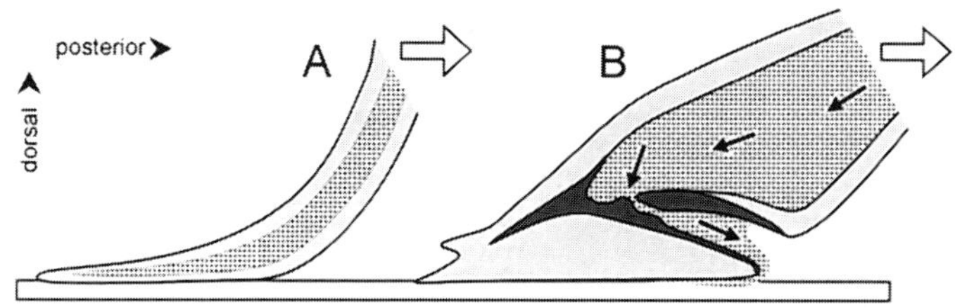

Fig. 10.8. Diagram of position of type 1 (A) and type 2 (B) setae on the substratum. Dotted area indicates lipid-containing secretion. Black arrows indicate the route of secretion release. White arrow indicates direction of pulling force.

An interesting point is the origin of the distal openings. Two hypotheses have been previously suggested (Gorb, 1998d). The first is that the distal opening is a sort of pore canal penetrating the cuticle of the tip of each seta. In this case, the space within the tenent seta is part of the inside of the body of the fly (canal hypothesis). Evidence for this is that the lumen is directly connected with inner layers of cuticle of the pulvillus, in the deepest portion of the seta (personal observation). An alternative hypothesis is that the distal opening is an invagination. In this case, the lumen within the tenent seta is «external space» (invagination hypothesis). Possible evidence for this is the presence of an electron-dense layer covering the walls of the lumen. This layer may be interpreted as epicuticle. Each of these hypotheses needs additional evidence.

In the smooth system of the leg pads of aphids, the cavity of the adhesive organ is connected with the external surface through fine epicuticular filaments. These are presumably sites of elaboration of the final secretory product, and are presumably places for the extrusion of the adhesive secretion (Lees and Hardie, 1988).

2.3 Properties of the secretion

There are several theories explaining how insects hold onto a smooth surface. Walker, with co-workers (1985) summarised them as follows: (1) the purely mechanical hold of setae, (2) setae acting as suckers, (3) electrostatic forces between setae and substratum (Maderson, 1964), (4) seizure theory (Gillett and Wigglesworth, 1932). Data on the hairy pad system of the adult reduviid bug *Rhodnius prolixus,* let previous authors suggest that mechanical seizure between adhesive setae and substrate irregularities is responsible for attachment to the substratum (Gillett and Wigglesworth, 1932). In the hairy system of the chrysomelid beetle *Chrysolina polita*, two main forces contribute to overall attachment: (1) molecular adhesion between setae and substratum, and (2) surface tension in a thin layer of epidermal secretion.

Molecular adhesion requires very close approximation between contacting surfaces. Surface tension requires the presence of secretion in the contact area. Indeed, pad secretion is an essential factor for attachment to smooth substrata. For example, the attachment performance of the reduviid bug *Rhodnius prolixus,* was reduced when pads were treated with lipid solvents (Edwards and Tarkanian, 1970). The aphid *Aphis fabae,* looses its ability to attach to smooth surfaces after walking on silica gel for a certain period of time (Dixon et al., 1990).

This fluid adheres well to both hydrophilic and hydrophobic surfaces, such as wax (Roth and Willis, 1952) and silanised glass (Dixon et al., 1990). It has been suggested that the fluid contains a kind of surfactant, similar to Triton X-100, which would make adhesion less sensitive to properties of the substrate (Dixon et al., 1990).

Measurements carried out on the pad secretion of the grasshopper *Tettigonia viridissima,* with the use of a sapphire ball revealed that adhesive forces varied from 0.1 to 0.6 mN (Fig. 10.9). Depending on the location selected, the adhesive force could monotonically increase or decrease, or remain constant with increasing applied force ranging from 0.1 to 2.5 mN (Fig. 10.9 C, D). In this case, the increase or decrease in adhesive force typically ranged between 0.1-0.2 mN. In a few cases, the adhesive force oscillated randomly. The adhesive force strongly depended on the location on the silicon surface (Fig. 10.9 E, F). Most likely the secretion is highly viscous, and the measured adhesive force is mainly a result of the viscosity of the secretion. It is well-known that the pulling force needed to separate the two surfaces decreases, with increasing thickness of the fluid film between two solids (Bowden and Tabor, 1986). The indirect evidence of viscous force is, that in some cases, stronger adhesive force was measured on single droplets compressed by the sapphire ball. However, in the case of the pad, secretion droplets would presumably spread well on the pad surface even under low forces, due to the soft material of the pad (Jiao et al., 2000).

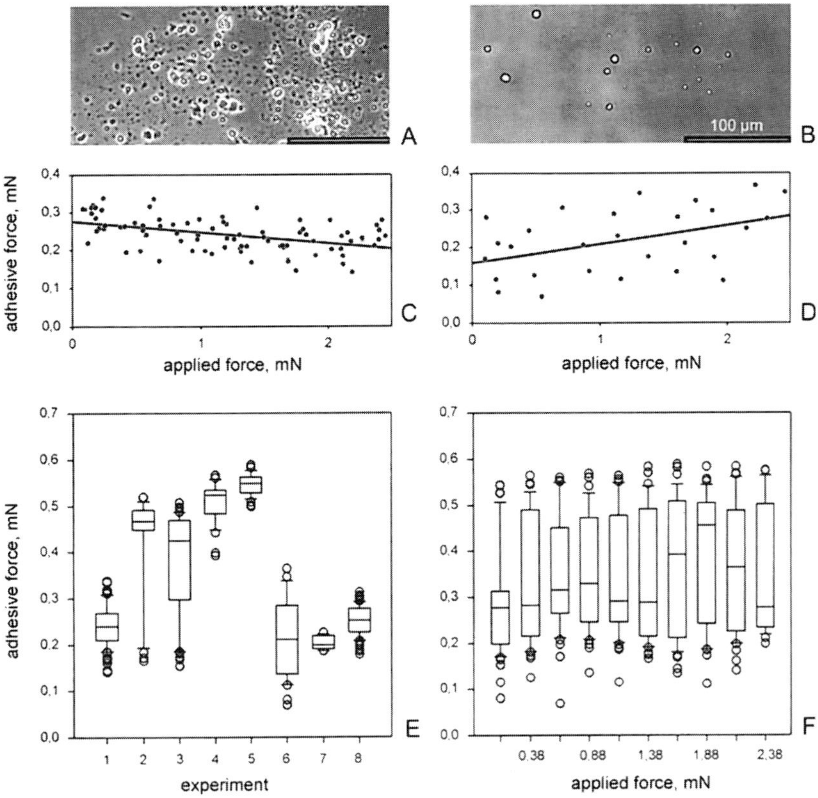

Fig. 10.9. Footprint secretions of the grasshopper *Tettigonia viridissima* (Jiao et al., 2000). A. The unequal distribution of the secretion on a glass surface (cover slip). B. Droplets occurring in footprint preparations, mounted in distilled water. C-F. Adhesive properties of the secretion. Due to the unequal distribution of secretion on the surface, the adhesive force decreased (C), or increased (D) with increasing applied force. E. Average adhesion at eight locations. F. Adhesive force data obtained at eight locations pooled together and plotted versus the applied force.

For measurements of small forces in the nano and subnanonewton range, an Atomic Force Microscope (AFM) can be used. This method is non-destructive and has the advantage of a high three-dimensional resolution (Fig. 10.10). It was employed to measure forces directly inside the droplet area (Wallentin et al., 1999). Secretion droplets were located by directing the probe at different spots on the samples and checking the shape of the acquired force-distance curves (Fig. 10.11). Force-distance curves were recorded at a zero lateral scan size in a vertical scan range of 1 µm, with a constant maximum deflection of 50 nm (corresponding to a maximum load of about 35 nN). To obtain information about the viscous properties of the fly secretion, the retracting speed of the cantilever was varied during the measurements. The vertical cantilever velocity ranged between 20 nm/s and 20 μm/s.

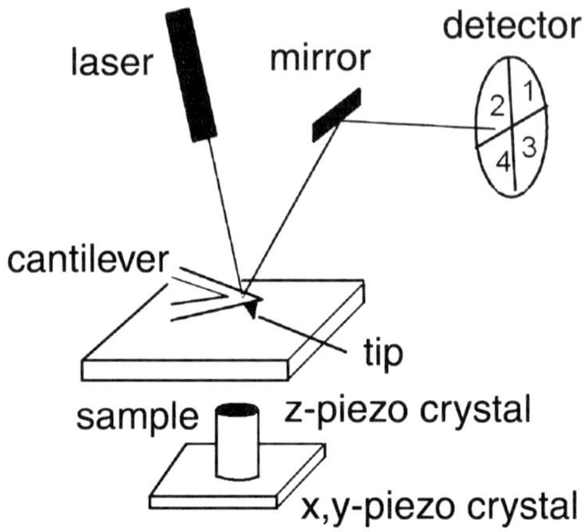

Fig. 10.10. Schematic set-up of an Atomic Force Microscope (AFM). The bending of the cantilever (length about 100 μm) caused by distance dependent forces, is monitored by a laser deflection system with a split photodiode. The sample can be moved by piezo crystals with 0.1 nm resolution. For surface imaging (contact mode), the sample is scanned in an xy-direction and the cantilever deflection is compensated by a feedback movement of the sample in a z-direction, which is recorded as topography. For force-distance measurements, the xy-position is held constant while the sample is scanned in the z-direction, and the resulting deflection of the cantilever is recorded.

The secretion droplet prevents a direct jump from *F* to *H'* (Fig. 10.11 B). In contrast to the cycle on the condensed water layer, the retracting curve in the secretion additionally contains the line *F* to *G*. In this region, the tip stays inside the liquid, but the contact to the surface is broken. Because of high viscosity of the liquid, the capillary meniscus can be pulled out until it finally breaks (jump *G* to *H*). This is not possible with the low viscosity of pure water.

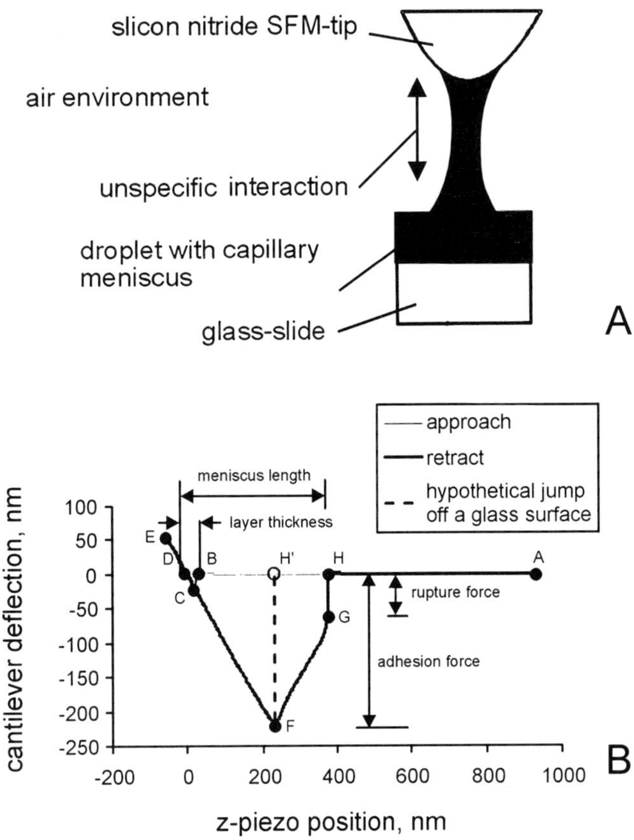

Fig. 10.11. A. Diagram of the viscous, liquid bridge holding the cantilever tip inside the fluid. B. Typical force-distance cycle in a fly's secretion. Such parameters, as layer thickness (*B-D*) on the approaching part of the cycle, meniscus length (*D-H*), adhesion force (*F-H*), and rupture force (*G-H*) on the retracting part, were extracted. Note that the tip can be separated from the sample without a sharp drop-off (line *F-G*).

A characteristic curve shape (Fig. 10.11 B) was found throughout the complete set of experiments. It was used to consider if the force-curve was actually taken inside the secretion droplet. The adhesion force corresponds to the total interaction between the tip and the surface in the fluid, while the rupture force combines interactions between the tip and surface, or the internal interaction in the fluid. These interactions depend on the relationship between viscosity and fluid-surface tension. Probably, the interaction between the fluid molecules themselves mainly contributes to the interaction. As it can be confirmed by theoretical calculations, the liquid breaks inside the meniscus, and not at the contact area between fluid and surface.

Meniscus lengths, rupture forces and adhesion forces correlate linearly with the fluid thickness, which indicates the capillary nature of the interaction and confirms the hypothetical interaction. There is direct velocity dependence of adhesion forces and the meniscus length (Fig. 10.12). Higher adhesion and rupture forces and an increased meniscus length at higher retracting speeds, might be due to the secretion viscosity.

The shape of the force-distance curves, and the correlation of mean forces and meniscus lengths with the layer thickness, points at a capillary attraction mechanism between tip and fly secretion. The velocity-dependent forces and meniscus lengths indicate the high viscosity of the fluid (Wallentin et al., 1999).

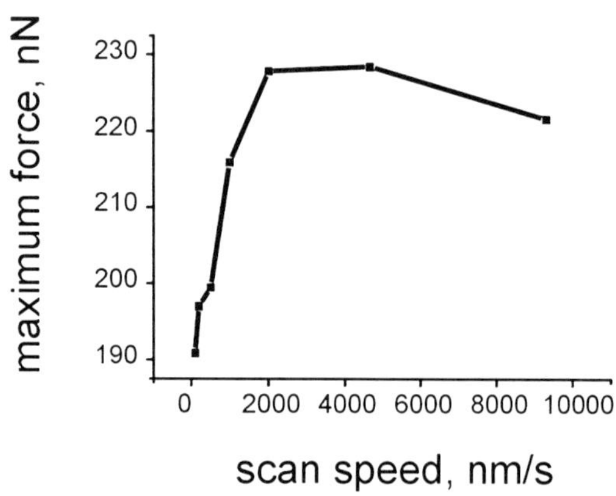

Fig. 10.12. AFM-measurements carried out on single droplets of the footprints in the fly *Calliphora vicina*. The diagram shows velocity dependence of the pulling force acting on the cantilever tip, contacting with the droplet.

However, quantitative evaluation of the rheological parameters of the secretion seems to be impossible, because of the unknown geometry of droplets, exact shape of liquid bridges, and contact areas between liquid and the two hard surfaces. The forces measured on single fly droplets with the use of an AFM cantilever, are comparable with interactions between the single seta and the substratum, due to similar size of the cantilever tip to the fly's seta.

Chapter 11

Protuberances and their fields

Origin and shape of cuticle protuberances involved in the design of attachment devices varies among different insect groups (Chapter 3). For example, leg attachment systems of the hairy type have appeared several times in insect evolution (Chapter 13), and therefore, they may consist of setae with sockets (Coleoptera, Dermaptera) or of acanthae without sockets (Megaloptera, Diptera). Microtrichia shape also varies enormously among frictional systems (Fig. 11.1). However, in systems composed of two corresponding surfaces, one could expect some constraints in shape and size of co-opted structures. Detailed information about the structure and dimensions of protuberances, together with material properties of the underlying cuticle, is crucial to understanding behaviour of surfaces in contact.

Fig. 11.1. Diversity of the shape of protuberances from different functional systems. Two views of each protuberance are shown.

This chapter combines data from different functional systems in order to understand which characters of the frictional system are of great importance for a particular function. Questions raised in this chapter are not intensively discussed in the literature. How do different parameters, such as length, width, density, and distance between protuberances, correlate within the same field and between corresponding fields? How are protuberances directed in functionally corresponding fields? Are there differences between sexes in the design of microtrichia and their fields?

Cuticle protuberances existing in frictional systems are usually 0.5-30.0 μm long. Knowing that dimensions, surface, volume, and mass, scale differently, it is interesting to compare the size of protuberances and their fields in closely related animal species of different size. For example, the density of hairs in the leg attachment devices sharply increases with increased body weight. Since the area of attachment devices cannot be increased proportionally to the body weight, there is another mechanism to increase attachment ability by increasing the number of single contacting points, and real contact area in larger animals (Fig. 11.2). Such data may also be useful for surface prototyping for industrial applications in systems of a larger scale.

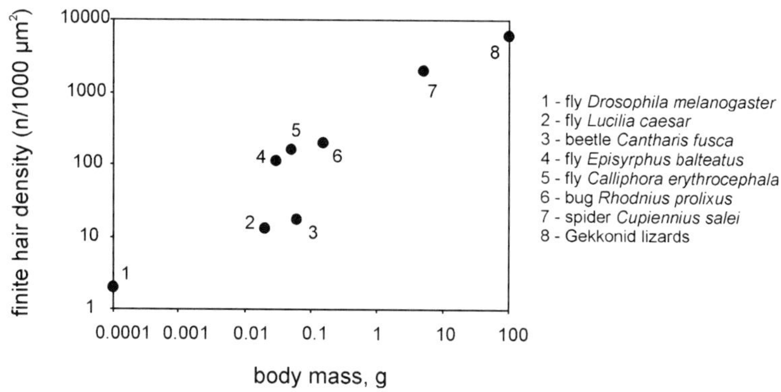

Fig. 11.2. Dependence of the finite hair density of the attachment pads on the body mass, in hairy pad systems of representatives from diverse animal groups.

1. SCALE EFFECTS ON THE MACROSCULPTURE

In insects, the relationship between the head width and the width of its articulation with the neck, usually ranges from 2 to 4. In adult Odonata, this relation varies from 8 in the Petaluridae to 16 in the Hemiphlebiidae. When the head is arrested to the neck on two microtrichia fields (MFH) by a pair of neck sclerites, the head obtains two additional points of contact with the neck area. This arresting system provides greater stability of the head (Chapter 6). The microtrichia fields differ in their size and location on the rear surface of the head in representatives of different families. It is plausible that these parameters may depend on the construction of the head capsule. To test this hypothesis, morphometric characters of the head in 227 representatives of 26 odonate families were studied (Gorb, 1999a) (Fig. 11.3). The analysis shows that the arrester basis correlates well with the head width; the relationship is linear (Fig. 11.4). However, the slopes of these lines evaluated for seven odonate families were different. This can be explained by constructional differences of the head capsule (Steinmann, 1967; Gorb, 1998b).

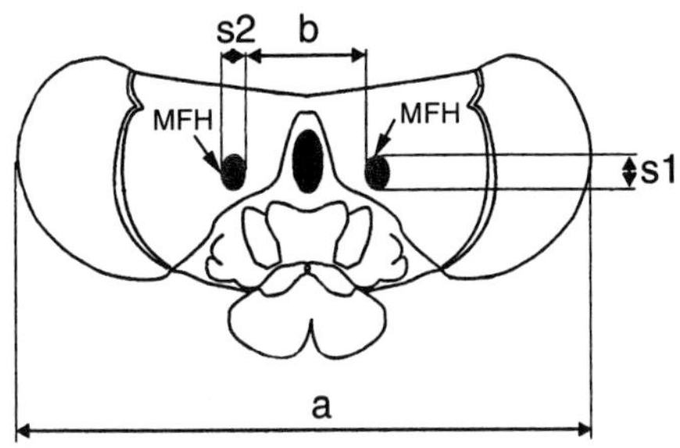

Fig. 11.3. Measurements of some arrester characters: b, arrester base or the distance between microtrichia fields (MFH) on the rear side of the head; s1, height of the field; s2, width of the field.

Surprisingly, the size of microtrichia fields is scarcely dependent on the head width (Fig. 11.4 C). In other words, the arrester base, scales in a different way than field dimensions. Presumably, the leverage of head arresting, and not the area of contacting surfaces, plays a crucial role in the fixation to the neck. An additionally important parameter, which can influence scaling properties of arrester characters, is the shape of the triangle connecting microtrichia fields and the head-neck joint. However, it is very difficult to quantify it experimentally.

Many attachment devices, such as leg attachment pads, also scale with an increase of the animal size. However, due to differences in the scaling of the mass and area, the relationship between attachment ability and body weight decreases very rapidly (see Chapter 9). Since resulting attachment force depends very much on the surface properties, we have to consider scale effects on the microsculpture of attachment devices.

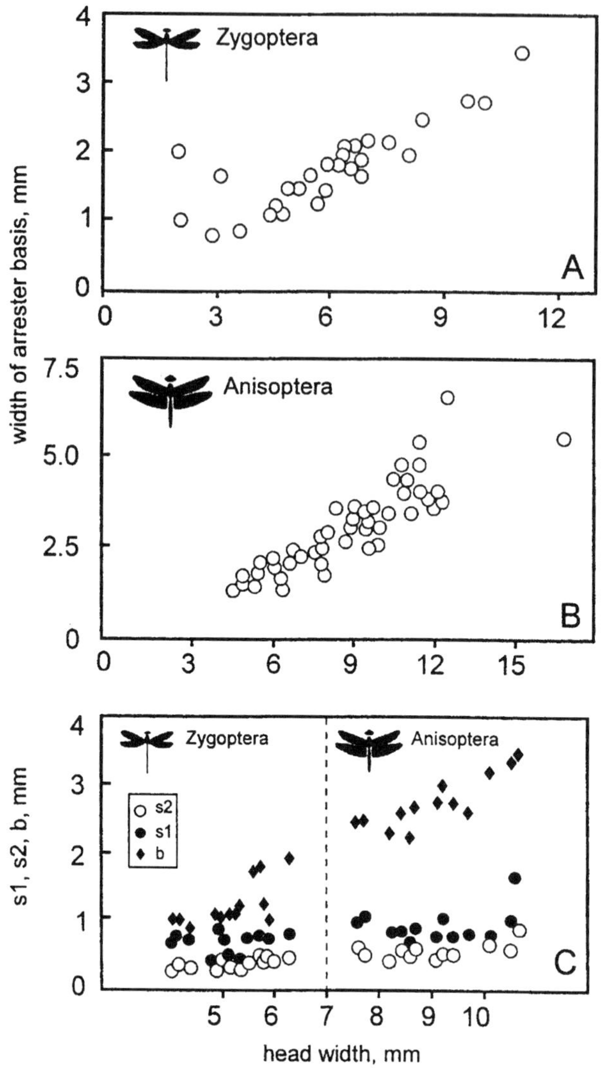

Fig. 11.4. Scaling of macrostructures of the dragonfly arrester (Gorb, 1998b). A-B. Dependence of the width of arrester basis in Zygoptera (A) and Anisoptera (B), on the head width. C. Width of arrester basis, height and width of the microtrichia fields of the head. Graphs are based on data obtained from representatives of seven odonate families (Gomphidae, Aeshnidae, Corduliidae, Libellulidae, Calopterygidae, Lestidae, Coenagrionidae).

2. SCALE EFFECTS ON THE MICROSCULPTURE

In the literature, it has been reported that diverse characters of the surface microsculpture depend on the animal size. Such scale effects occur in Hymenoptera. Setae-like hooks, serving as a sliding coupling between forewing and hindwing, are larger in larger species (Schneider and Schill, 1978).

Scale effects have been quantified in detail, in the wing-locking device of three species of tenebrionid beetles (Chapter 8). The five-fold increase in the body size results in the four-fold increase in length of microtrichia, and in two-fold increase of the width of microtrichia (Gorb, 1998a). However, a five-fold decrease in density of the protuberances occurred. The dependence of microtrichia length and width on the size of an insect is linear, whereas the dependence of the microtrichia density on the insect size is exponential (Fig. 11.5). A calculation of the areas in simple geometrical models shows that, in the case of cone-like structures, an increase in both the length and density, and a decrease in the width is advantageous (Gorb, 1998a).

Armoured membranes of Diptera (Chapter 7) interlock their folds due to surface protuberances. In these systems, the complexity of membrane armature is also dependent on the dimensions of the species (Gorb, 1997a). Relatively small species, such as *Musca domestica* and *Drosophila melanogaster,* have only short, ungrouped microtrichia or micropapillae on membranes of various articulations of the body. Large flies have the highest degree of grouping, the largest microplates, and the longest microtrichia (Fig. 11.6).

This result can be explained in terms of relative degrees of loading. In larger flies, the membrane can experience higher loading during, for example, head or proboscis movements, or even during resting. The head loading is the relationship between head volume (or mass) and the occiput area. These parameters for flies having different head width are following: *D. melanogaster* (0.88 mm), *Lucilia caesar* (3.26 mm), and *Tabanus bovinus* (6.50 mm). Occiput areas, head volumes, and relationships between them are approximately as follows: 0.10 mm^2, 0.17 mm^3, 1.7 (*D. melanogaster*); 0.40 mm^2, 9.00 mm^3, 22.5 (*L. caesar*); 0.60 mm^2, 71.90 mm^3, 119.8 (*T. bovinus*). Thus, for larger species, such as *L. caesar* and *T. bovinus,* the head loading should be 10-60 times higher than for smaller ones. To fixate the perturbed head of *T. bovinus*, the frictional forces generated by the membrane in the neck-head articulation must be significantly higher than that of *D. melanogaster*.

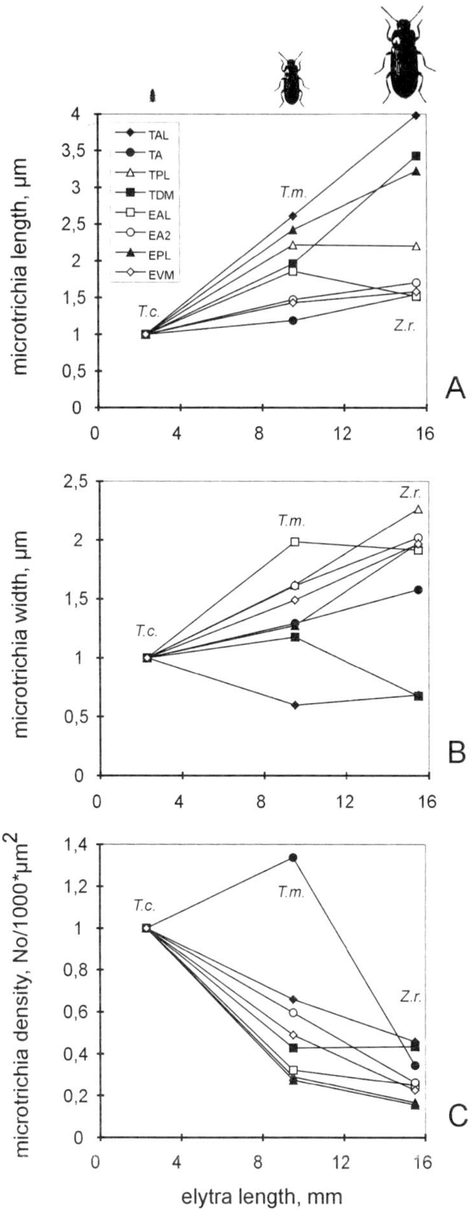

Fig. 11.5. Dependence of the length (A), width (B), and density (C) of microtrichia of eight microtrichia fields on the elytra size in the elytra-locking system of tenebrionid beetles (Gorb, 1998a). All parameters are recalculated relative to those in the smallest species studied. *T.c.*, *Tribolium castaneum*; *T.m.*, *Tenebrio molitor*; *Z.r.*, *Zophobas rugipes*.

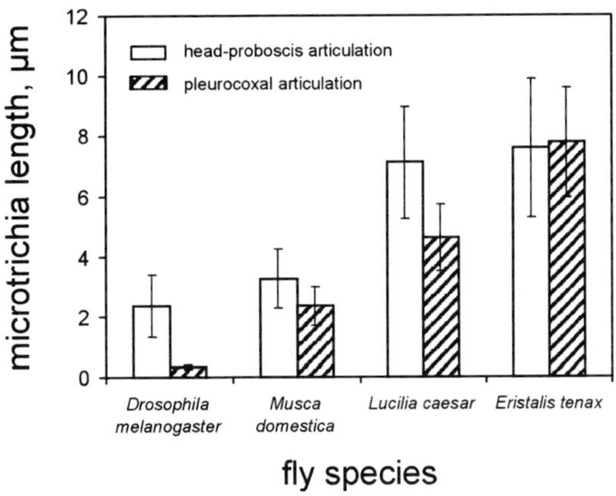

Fig. 11.6. Length of microtrichia in the armoured membranes of flies. Fly species are arranged according to an increase in size. Error bars are standard deviations. Data from Gorb (1997a).

In the dragonfly head arrester, the microtrichia fields on the rear surface of the head and the postcervical sclerites are functionally co-opted structures correlating in the shape, size and density of microtrichia. The absolute area of the microtrichia fields in extremely large pseudostigmatids is larger than in very small representatives of the family Protoneuridae. However, the size of the microtrichia is usually larger in large species, and their density is lower (Fig. 11.7 A, B). Density of microtrichia is usually higher on the postcervical sclerite (Fig. 11.7 D). The difference in the microtrichia density on the corresponding structures is more distinct in the larger species (Fig. 11.7 C).

An increase in size of single microtrichia in larger species may be explained by possible correlation with the cell size. It is more difficult to explain why the difference in the microtrichia density on the corresponding structures increases in larger insects. As was mentioned above, the density increase is advantageous, because of the increase of the real contact area between corresponding structures. Since length to width ratio of elongated protuberances influences the spring constant of protuberances, differences in the microtrichia length to width ratios may be a mechanism by which material of the same basic mechanical properties, may be modified into a softer or stiffer layer of microtrichia coverage.

Thus, in closely related species, microstructures scale with an increase in size and a decrease in their density. If we consider scaling effects among species of different, high-ranking taxa, the size of the microstructures usually decreases and their density increases. The most interesting question, which remains unresolved for the majority of frictional systems, is how do scale effects on the microstructure influence frictional properties of surfaces and resulting attachment abilities? The first attempt to test these relationships was made in an investigation of the attachment system of pulvilli in syrphid flies (Chapter 9).

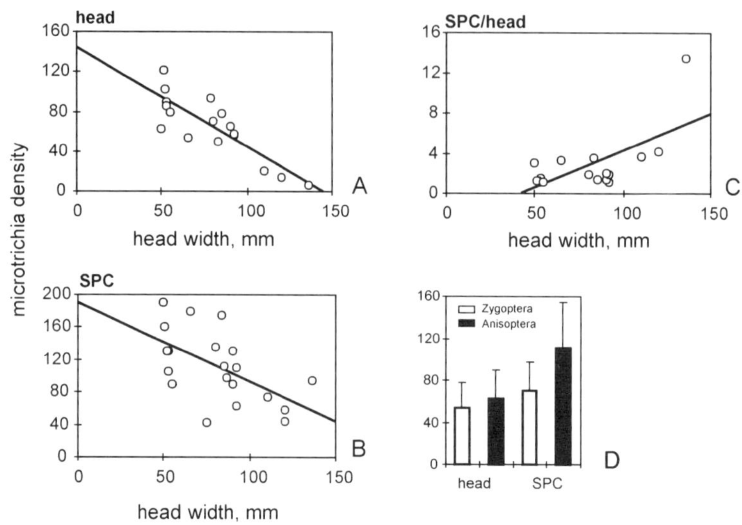

Fig. 11.7. Microtrichia density of corresponding fields in the head arresting system of 16 odonate species. A. Head. B. Postcervical sclerite (SPC). C. Relationship between microtrichia density on the head and postcervical sclerite. D. Means and SD of the data on the microtrichia density pooled together for representatives of Anisoptera and Zygoptera.

3. DIRECTIONALITY OF PROTUBERANCES

In extremely rare cases, cuticle protuberances are perpendicularly oriented to the underlying cuticle. Usually they are sloped at some angle in one particular direction. In some cases, their orientation pattern is more complex. Sloped surface irregularities generate different frictional forces in different directions (Chapter 1), and in the case of natural frictional devices, they may be responsible for preventing sliding in one particular direction. Even more complex is the case of two corresponding fields. Below, we provide a comparison of the microtrichia directionality in the coverage of two corresponding parts of the head-arresting system of dragonflies (Fig. 11.8).

On the rear side of the head, the protuberances are oriented in the dorsal or dorso-lateral direction. If two discrete fields of microtrichia are present (Coenagrionoidea and Anactini), the structures are directed ventrally in the upper field, and dorsally in the lower one (Fig. 11.8 G). On the corresponding surface of the postcervical sclerite, microtrichia are usually oriented in the opposite direction (compare A and B in Figure 11.8). However, the postcervical sclerite, corresponding to the two discrete microtrichia fields on the head, is covered by microtrichia directed ventrally in the dorsal part of the sclerite, and dorsally in its ventral part.

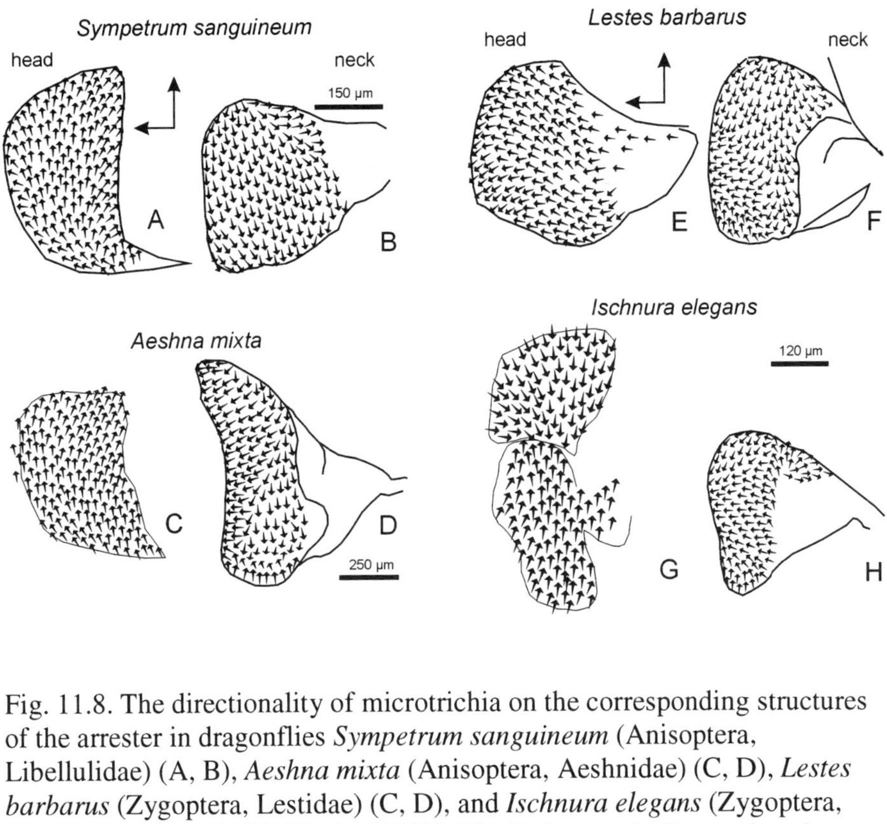

Fig. 11.8. The directionality of microtrichia on the corresponding structures of the arrester in dragonflies *Sympetrum sanguineum* (Anisoptera, Libellulidae) (A, B), *Aeshna mixta* (Anisoptera, Aeshnidae) (C, D), *Lestes barbarus* (Zygoptera, Lestidae) (C, D), and *Ischnura elegans* (Zygoptera, Coenagrionidae) (G, H) (Gorb, 1998b). Vertical arrow indicates dorsal direction. Horizontal arrow indicates lateral direction. Directions were defined from SEM pictures.

In the beetle elytra-locking devices, the microtrichia composing a field are usually oriented in one preferred direction. The map of local directions of microtrichia together with the preferred direction of the field is given in

Figure 11.9. According to the directionality of the microtrichia, the fields can be described using their preferable direction along each of three spatial axes. For example, if the microtrichia field directed posteriorly (X-axis), medially (Y-axis), and dorsally (Z-axis), its index of directionality should be *PMD*. In a field directed anteriorly (X-axis), laterally (Y-axis), and ventrally (Z-axis), its directionality index should be *ALV*.

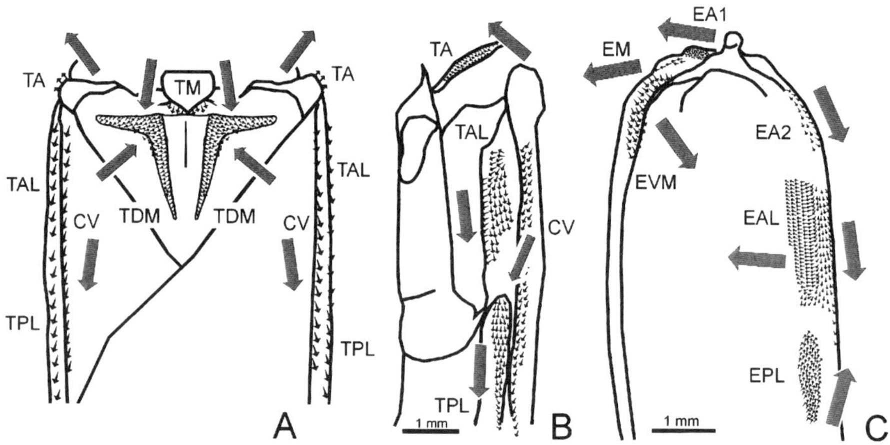

Fig. 11.9. Directionality of microtrichia (MT) in the fields of the elytra-locking devices in the tenebrionid beetle *Tribolium castaneum* (Gorb, 1998a) A. Meso and metahorax, dorsal aspect, elytra removed. B. Meso and metahorax, lateral aspect, elytra removed. C. Ventral aspect of the anterior half of the left elytra. Small arrows indicate local directions of MT; large arrows indicate preferred directions of microtrichia in the field. CV, EA1, EA2, EAL, EM, EPL, EVM, TA, TAL, TDM, TM, TPL, microtrichia fields. For nomenclature see Chapter 8.

Using these rules, the fields can always be divided into two groups, depending on their preferable directionality along one particular spatial axis. The set of fields of the thorax contain two functional groups preventing movements of elytra along the longitudinal axis (X-axis): 1) *the posteriorly-directed group* presumably preventing anterior shift (CV, TAL, TPL, TM; and 2) *the anteriorly-directed group* preventing posterior shift (TDM, TA). Similar groups can be recognised for two other spatial axes. Y-axis: 3) *the laterally-directed group* preventing medial shift (TA, TAL, TPL, TM, CV); 4) *the medially-directed group* preventing lateral shift (TDM). Z-axis: 5) *the dorsally-directed group* preventing ventral shift (TPL, TM, TDM); 6) *the ventrally-directed group* preventing dorsal shift (TA, TAL). On the elytra, there are six similar groups: 1) the EA2, EAL, EM, EVM, CV; 2) the EA1,

EPL; 3) the EA2, EPL, EVM, CV; 4) the EA1, EAL, EM; 5) the EPL; 6) the EA1, EA2, EAL, EM, EVM, CV. However, co-opted fields can be directed in the opposite directions (TDM - EVM, TPL- EPL) as well as in the same direction (TAL - EAL).

The directionality of the microtrichia within one particular field, presumably depends not only on the directionality of microtrichia of the corresponding, but also on the microtrichia arrangement in the neighbouring fields. It seems that microtrichia directionality prevents elytra from shifting along all three spatial axes. There are at least two fields responsible for both directions along one particular spatial axis. Such a construction can only be actively released from the locked state. Presumably, the elytral muscles are also involved in such an action.

4. MECHANICAL PROPERTIES OF MATERIALS IN CORRESPONDING FIELDS

The friction force between two surfaces is mainly dependent on both the microsculpture of each surface, and the force pressing them together. It is well-known about solids that the friction is dependent on the stiffness of corresponding surfaces (Bowden and Tabor, 1986). The pair of surfaces contacting each other with the largest number of contact points, has the largest friction coefficient (Chapter 1). When one of two contacting surfaces is composed of flexible material, the number of contact points between them is larger than in the case, when both materials are stiff.

For biological structures, it seems to be an important rule as well. The general conclusion for frictional systems, based on co-opted microtrichia fields, is that at least one of two surfaces, composing a frictional device, is elastic or flexible. Direct observations of the cuticle of living animals under a dissecting microscope, and the study of the ultrastructure of the cuticle, revealed the differences in material properties of the cuticle between microtrichia fields of the wing-locking mechanism in «Symphyta» (Schrott, 1986), and the dragonfly head-arresting system (Gorb, 1990a). Such a design may also prevent abrasion of microtrichia. Moreover, biological devices are generally designed to be flexible in their attachment-detachment performances, so that one surface does not have to be positioned exactly onto the corresponding surface.

It was also previously hypothesised that the notal and pleural thoracic fields of the beetle elytra-locking devices (TAL, TPL, TDM) consist of elastic cuticle, whereas corresponding elytral fields are much less flexible (EAL, EPL, EVM). The differences in material properties of corresponding microtrichia fields were quantitatively studied on one functional pair of fields (TAL, EAL).

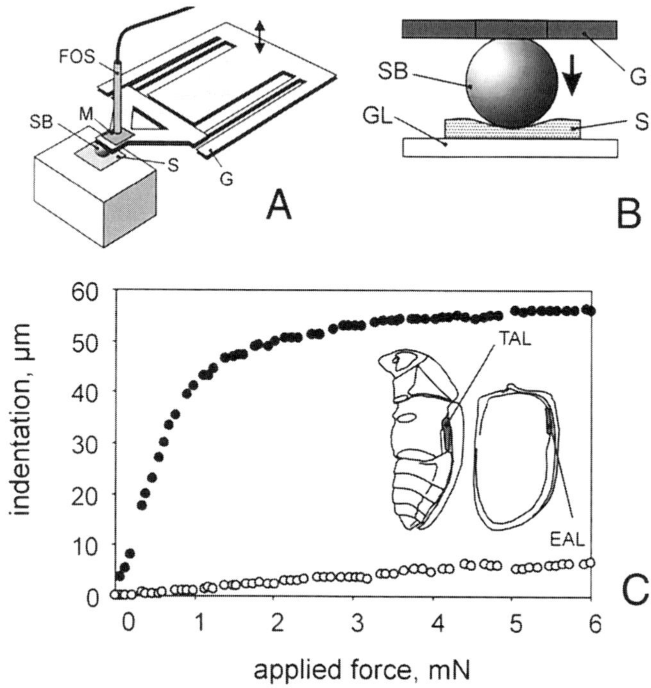

Fig. 11.10. Material properties of corresponding fields in the dung beetle *Geotrupes stercorosus*. A. Diagram of the force tester. B. Indentation process. C. Data on cuticle indentation properties of two corresponding fields of microtrichia of the elytra-locking device (TAL, EAL). FOS, fibre-optical sensor; G, glass spring; GL, glass plate, M, mirror; S, cuticle sample; SB, sapphire ball.

In the experiment, the force tester (see also Chapter 9) was used as an indentor (Fig. 11.10 A). A sapphire ball with 0.5 mm diameter was glued to the glass spring with a known spring constant (230 N/m). Driven by the piezo-drive, the spring was capable of moving up and down. During the downward movement, the material sample was compressed by the sapphire ball (Fig. 11.10 B), and the spring deflection was monitored by a fibre-optical sensor. The difference between signals, obtained with a hard sample (bare glass) and cuticle sample, resulted in an indentation curve (Fig. 11.10 C). Such curves can then be used to calculate the elastic modulus by fitting them, according to the Hertz theory (Hertz, 1881; Israelachvili, 1992).

Results show clearly that material of the elytral field (EAL) is approximately 10 times stiffer than material of the corresponding field on the thorax (TAL).

Such experiments, together with quantitative data on microtrichia shape and size, may help to clarify behaviour of protuberances in contact. In Figure 11.11, taking into account microtrichia shape and differences in mechanical properties of fields, possible behaviours of co-opted pairs of fields are given. The flexible fields are able to adapt to the shape of outgrowths located on a stiffer, corresponding surface.

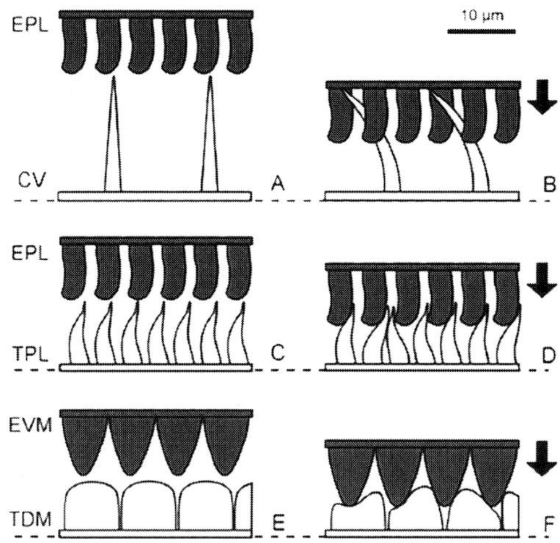

Fig. 11.11. Hypothetical interlocking mechanisms between microtrichia covered surfaces in the elytra-to-body locking mechanism (Gorb, 1998a). A, C, E. Shape of structures when they are released. B, D, F. Contact deformations. Light-grey coloured parts are more elastic and mobile; black parts are stiffer and immobile. The dimensions of microtrichia and their arrangement, correspond to those measured in *Tenebrio molitor*. Arrows indicate direction of loading force. CV, costal vein; EPL, posterio-lateral field of the elytra; EVM, ventro-medial field of the elytra; TDM, dorso-medial field of the thorax; TPL, posterio-lateral field of the thorax.

5. INTERDEPENDENCE OF MICROTRICHIA PARAMETERS

The size and shape of microtrichia within the same field and between fields, cannot vary without limitations. Possible constraints for the geometric properties of protuberances may be the size of underlying epidermal cells, and the geometry of structures of the corresponding field. To provide a quantitative description of the geometry of microtrichia, protuberances of 13 fields of elytra-locking devices, in three species of beetles, were measured from SEM micrographs (Fig. 11.12).

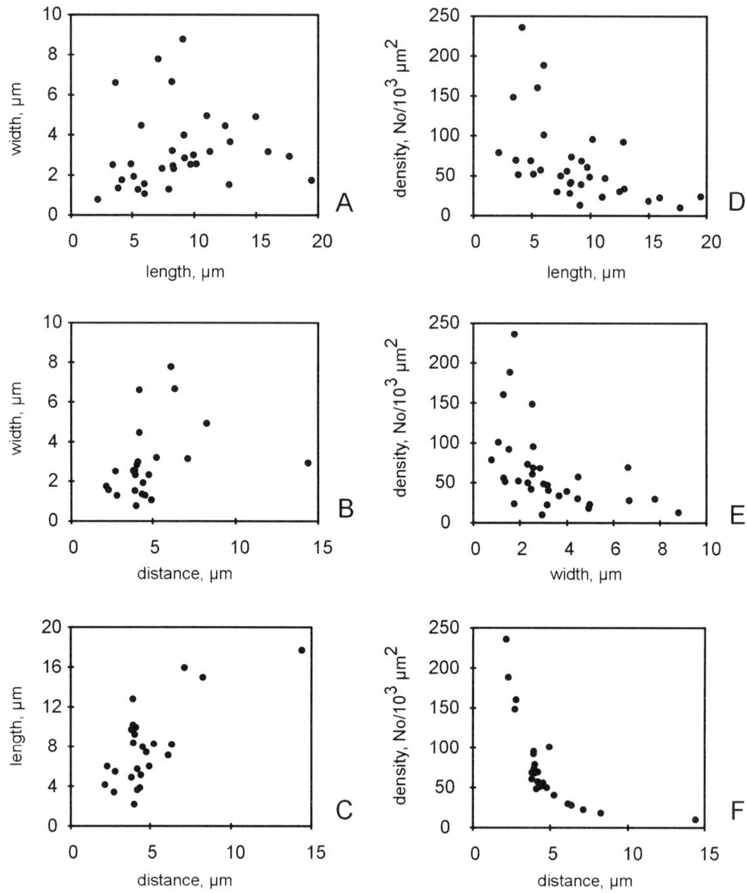

Fig. 11.12. Interdependence of microtrichia parameters measured in three species of tenebrionid beetles (Gorb, 1998a). The data sets used are averages of 13 fields of measured microtrichia. The value of the slope of the linear regression line is given in the upper-right corner of each graph.

The density of surface irregularities is one of the most important parameters for the description of a frictional system. It is correlated with the number of contacting points, which has a direct effect on the mechanical interlocking between surfaces. At the same time, the density of irregularities is correlated to the length, width, and especially the distance between single protuberances, so that in biological frictional systems, the longest and widest protuberances are usually sparsely distributed within the field (Fig. 11.12 D-F). However, there is only a weak correlation between length and width of microtrichia (Fig. 11.12 A). In other words, longer microtrichia are not always wider. Distances between longer and wider microtrichia are greater (Fig. 11.12 B, C). Interplay of these parameters may result in a very different shape of microtrichia and their assemblages.

As mentioned above, co-opted fields were subject to co-evolution in their historic development. Most microtrichia parameters, such as length, width, distance, and density of one field, correlate well with corresponding parameters of the co-opted field. The direct interdependence of the width is revealed for most functional pairs of fields. Length-to-width ratios in elytral fields are usually, remotely correlated with those of thoracic fields: the relatively long microtrichia sometimes correspond to the short and wide microtrichia. The distances between microtrichia on the elytra surface directly depend on those of the thorax. Distance-to-width ratio in one surface slightly increases with an increase in this parameter in the co-opted surface. The density of microtrichia in co-opted fields is usually quite different. On a large scale, however, data shows direct dependence between microtrichia densities in thoracic and elytral fields.

6. SEXUAL DIMORPHISM

Sexual dimorphism in the morphology of attachment devices is usually correlated with the use of these organs, in courtship and/or copulation. The best-known example is a tarsal sucker, present only in male dytiscid beetles (Chapter 4). A similar phenomenon is also known for frictional systems. The area of attachment pads in the males of chrysomelid and cerambycid beetles is usually larger than that of the females. There are also dimorphisms in the structure and density of cuticle protuberances. Males of the colorado potato beetle *Leptinotarsa decemlineata* (Coleoptera, Chrysomelidae), bear specialised setae on their tarsi for long-term fixation on the smooth female elytra (Chapter 5). Such setae are absent in females (Pelletier and Smilowitz, 1987). Many other insects, which do not display obvious sexual dimorphism, may be different in terms of size and design of attachment organs in both sexes.

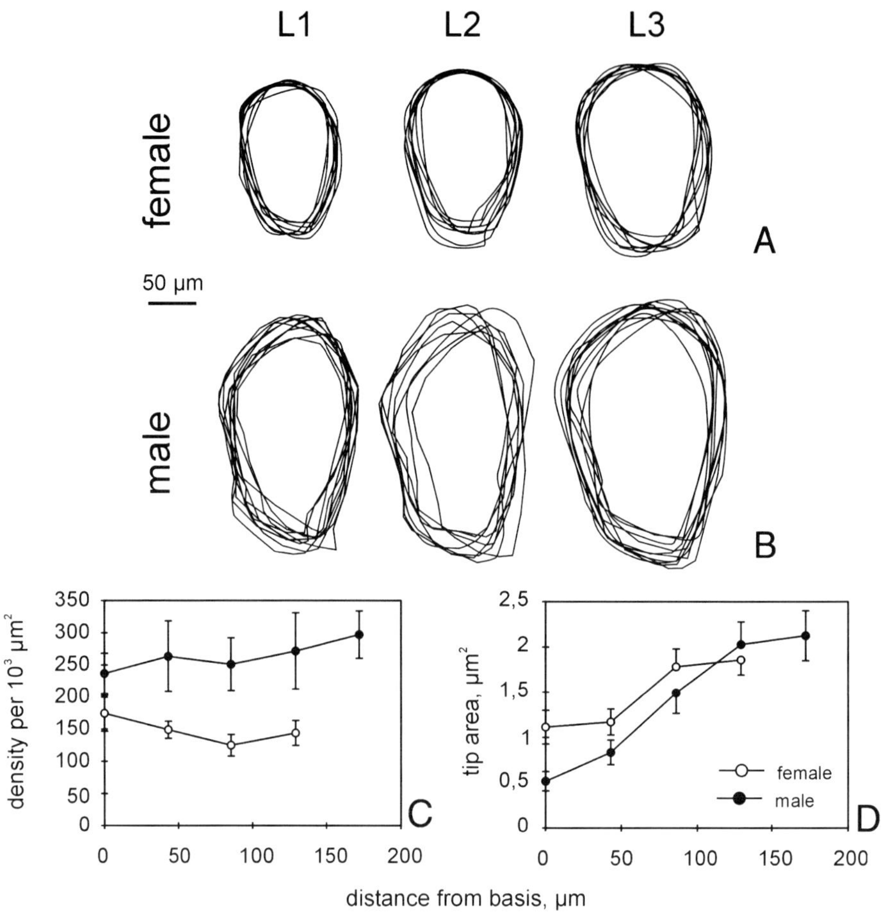

Fig. 11.13. Sexual dimorphism in the macro and microstructures of the fly *Calliphora vicina*. A-B. Shapes of single pulvilli of the fore (L1), mid (L2) and hindlegs (L3), in females (A) and males (B). C. Density of the tenent setae. D. Tip area of the tenent setae.

Males of *Calliphora vicina* flies are significantly smaller than females. However, their attachment pads in all three pairs of legs are significantly larger (Fig. 11.3 A, B). Although the tip areas of single tenent setae, covering pulvilli, do not differ in either sex (Fig. 11.3 D), those of the males have a higher density (Fig. 11.3 C). The presence of sexual dimorphism in fly pads, which were thought to be a purely locomotory adaptation, suggests that these devices may represent multi-functional structures, like the majority of biological systems.

Chapter 12

Sensory equipment

An important constructional property of attachment devices is their sensory equipment, which is responsible for monitoring the contact and probably forces in the contact area. The sensory organs are mainly represented by hair (trichoid) and campaniform sensilla. Hair sensilla detect the presence of the contact by monitoring deflection of the hair shaft, whereas campaniform sensilla are able to measure strains in the cuticle. Through this, they can detect both the contact state and forces acting on the material in the contact area. Usually, sensilla are located on both functionally corresponding surfaces, but occasionally only one part of the system is equipped with sensors. Sensilla of two distinct structures working as frictional devices, are usually convergently projected in the same regions of the central nervous system. Using as an example, the head-arresting system of Odonata (Chapter 6), this chapter shows variability of sensilla distribution, together with some general conclusions about the functions of sensory equipment.

1. CENTRAL PROJECTIONS OF SENSILLA

If an attachment system is responsible for fixating body parts together, then sensilla, monitoring the condition of such a system, are located on two different segments. To co-ordinate information from both sensory fields, sensory projections should converge in the central nervous system. However, there were no specific studies done on the central sensory projections of sensilla, from functionally corresponding parts of an attachment system. One example, which is partly related to this problem, is known in the damselfly ovipositor. Median (upper valvulae) and anterior ovipositor gonapophyses (lower valvulae) are structures responsible for unidirectional egg displacement in the posterior direction, by the use of frictional coverage on their inner surface (Chapter 5).

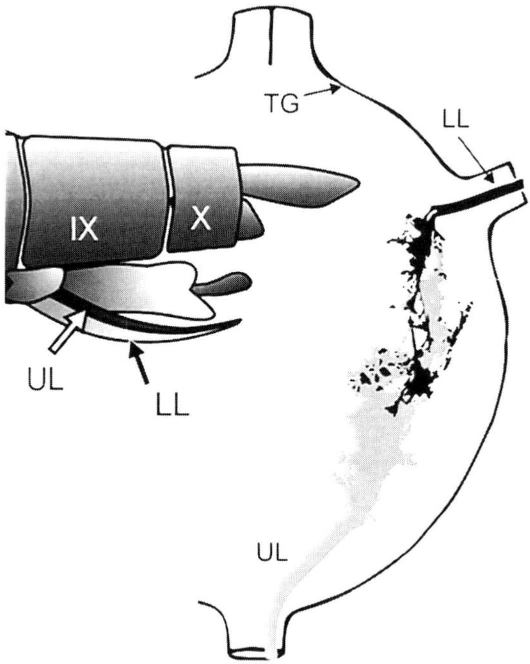

Fig. 12.1. Central projections of sensilla of the lower (LL) and upper (UL) leaves, or valvulae, of the endopytic ovipositor in the damselfly *Sympecma annulata* (Zygoptera, Lestidae) (Gorb, 1994). Inset shows the lateral aspect of the ovipositor. IX-X, 9th and 10th abdominal segments, respectively; TG, terminal ganglion.

These structures belong to different abdominal segments and have rich sensory equipment, which is involved in the oviposition control. Abdominal segments 8-11 and their appendages are innervated by a terminal ganglion (Tillyard, 1917), which is, in fact, a synganglion of the last three abdominal ganglia fused together. Backfilling of afferent fibres using anterograde cobalt fills, revealed that the main fibre bundles of both valvulae end in the lateral part of the ganglion, while entering through different nerves (Fig. 12.1). The convergence of sensory projections in the same CNS region is the morphological basis for sensory integration of co-opted structures, belonging to different abdominal segments (Gorb, 1994).

2. TOPOGRAPHY OF SENSORY ORGANS

Sensilla distribution is very specific in every attachment system. In the beetle elytra-to-body locking devices (Chapter 8), only trichoid sensilla exist. Some are located directly within microtrichia fields. They were found within the TPL field (Fig. 12.1 B) and in the vicinity of the TAL and TPL fields. The lateral sides of meso and metathorax, and the abdomen are covered with hair sensilla, whose location corresponds to the lateral margin of the elytra. In the transitional area between the TA and TAL, there is a compact sensory field consisting of 4-10 sensilla. The posterior part of the meso-thoracic tergite is covered with hair sensilla, which are presumably responsible for the monitoring of contact to the elytra (Gorb, 1998a). Presumably, in this attachment system, sensilla are only responsible for the monitoring of contact between microtrichia fields.

Odonate arrester (Chapter 6) is a more complex system as to its sensory equipment, because it is responsible not only for monitoring contact between frictional surfaces, but it is also involved in the sensory loop controlling flight (Gorb, 1991a; b). Electrophysiological recordings show that the loss of contact between the head and neck parts of the arrester, elicits activity of the wing muscles. In experiments in which the head was pulled away anteriorly with a micromanipulator, and contact between microtrichia fields was lost, the resting dragonfly had initiated flight.

2.1 Odonate arrester sensilla

The reflexes of starting and ending flight are under the control of both campanifom, and hair sensilla, of corresponding structures of the head and neck. The postcervical sclerite is differentiated into a more sclerotized base and a softer distal region, covered by microtrichia. Microtrichia of the sclerite and rear surface of the head are non-innervated cuticular protuberances. However, innervated structures, such as the trichoid (hair) and campaniform sensilla were found. These sensilla belong to two different types of cuticular mechanoreceptors. Trichoid sensilla are innervated hairs (setae) projecting out from the cuticle, whereas campaniform sensilla are round or oval-shaped areas of thin, flexible cuticle surrounded by thick, hard cuticle (Barth, 1986). The sclerite has both trichoid and campaniform sensilla (Fig. 12.3). The microtrichia field on the head is equipped only with trichoid sensilla (Fig. 12.2 A) that may exist as either single mechanoreceptors, or as sensory fields.

Trichoid sensilla. On the heads of representatives of the family Libellulidae, the field of trichoid sensilla may be located laterally or medially to the microtrichia field. Trichoid sensilla were not found within the field of microtrichia. However, the marginally located sensory fields can slightly overlap the microtrichia field by as much as 10-15 μm. The setae from the marginal sensory fields are usually directed towards the microtrichia field. Most representatives of Libellulidae have micropapillae covering the proximal region of the postcervical sclerite. In this area, trichoid sensilla are sparsely distributed. In rare instances, a few sensilla may be located at the margin of the pronotum (*Leucorrhinia pectoralis*), the setae of which are in contact with the postcervical sclerite (TS1, Fig. 12.4). Detailed, comparative data on sensilla distribution in the head-arresting system of representatives of different odonate families is given elsewhere (Gorb, 1991b).

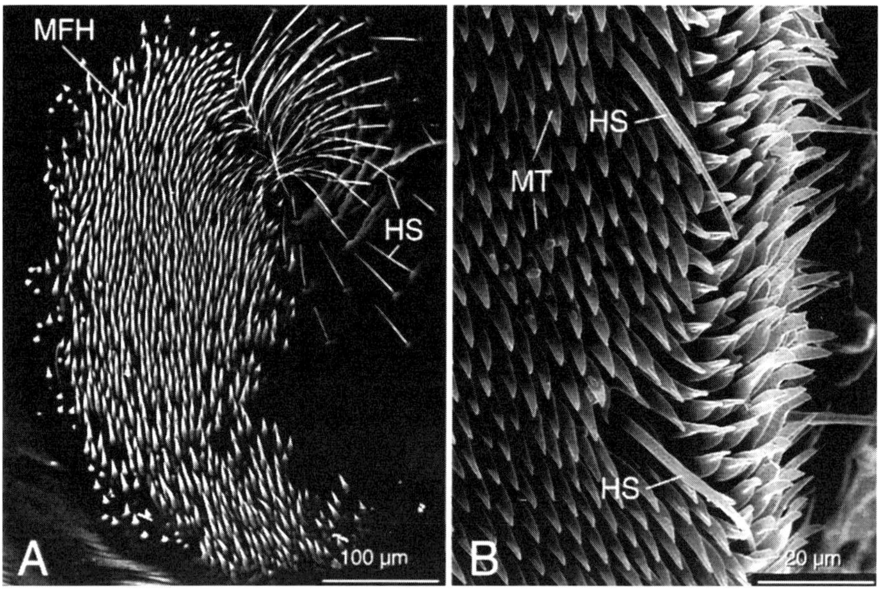

Fig. 12.2. Sensilla in frictional devices. A. Microtrichia field on the head (MFH) in the dragonfly *Idionyx safronata* (Anisoptera, Libellulidae). B. Hair sensilla within the microtrichia field of the elytra-locking device in the beetle *Tenebrio molitor* (Tenebrionidae). HS, hair sensilla; MT, microtrichia.

Campaniform sensilla. Campaniform sensilla are located only in the distal region of the postcervical sclerite (Libellulidae, Aeshnidae, Gomphidae), or in the transitional area between the proximal and distal regions of the postcervical sclerite (Calopterygidae, Coenagrionidae) (Fig. 12.3). It is difficult to localise campaniform sensilla within the microtrichia field using the scanning electron microscope, because they are shadowed by surrounded structures. However, microtrichia located close to the sensillum are usually, slightly sloped away from it.

Usually, campaniform sensilla are, more or less, evenly distributed at an average distance of 15-50 μm from each other. In representatives of Coenagrionidae (*Ischnura*), the cap of the sensillum is round in shape, or slightly lengthened in the dorso-ventral direction (5.0 × 5.5 μm). Its surface is rather smooth. In species of Calopterygidae (*Vestalis spp., Hetaerina spp.*), about seven intensely elongated sensilla are located at the ventral margin of the postcervical sclerite, in the transitional area between its proximal and distal regions. In species of Libellulidae (*Sympetrum decoloratum*), the sensilla are also elongated in the dorso-ventral direction (2.0 × 5.0 μm) and slightly plunged into the cuticle.

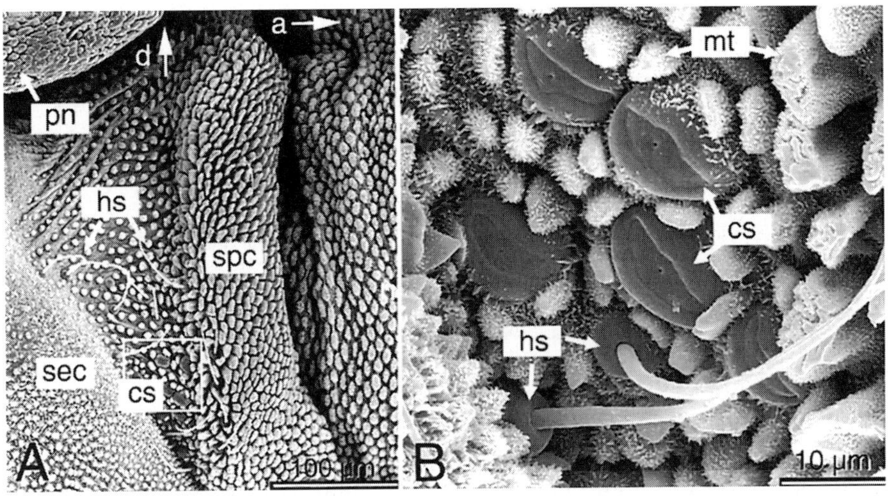

Fig. 12.3. Hair (hs) and campaniform sensilla (cs) from the postero-ventral part of the postcervical sclerite (spc) of the head arrester system in the damselfly *Calopteryx splendens* (Zygoptera, Calopterygidae) (Gorb, 1998b). a, anterior direction; d, dorsal direction; mt, microtrichia; pn, pronotum; sec, eucervical sclerite.

2.2 General rules regarding the distribution of arrester sensilla

In spite of the diversity in sensilla location, there are some general rules on their distribution in the arrester system (Fig. 12.4).

1. Usually, each of the arrester parts are provided with trichoid sensilla. When one part of the arrester (either the head or neck) has scarcely-developed trichoid sensilla or lacks them completely, such sensilla exist in a higher number on the corresponding structure. The presence of the field of trichoid sensilla on one part of the arrester, may be enough to monitor contact between corresponding structures.

2. Trichoid sensilla are located in the most projected areas of the hardened cuticle, or in the regions contacting the most projected areas of the corresponding structure. Such a location provides the highest probability that the sensilla would touch the corresponding structure, when the arrester system is in the fixed condition.

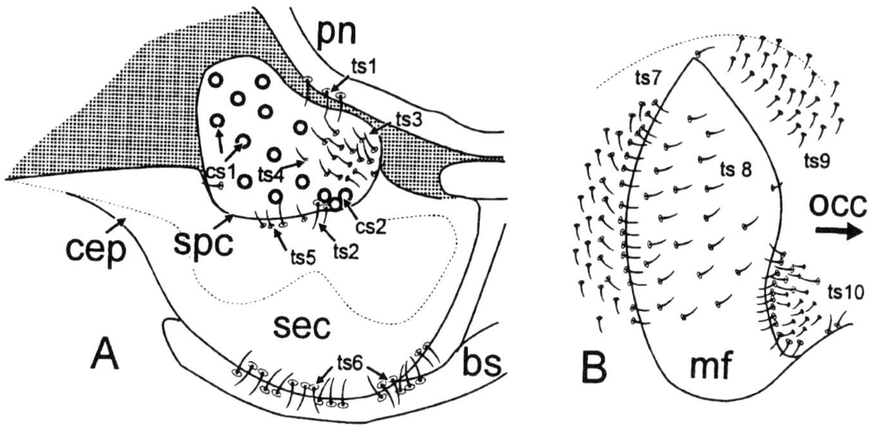

Fig. 12.4. Diagram of the sensory equipment of the head-arresting system in dragonflies (Gorb, 1998b). It summarises data obtained from a variety of representatives of different families of Odonata. A. Lateral aspect of the left side of the neck. B. Posterior aspect of the microtrichia field on the left side of the head. bs, basisternum; cep, cephaliger; cs, campaniform sensilla; mf, microtrichia field; occ, occiput; pn, pronotum; sec, eucervical sclerite; spc, postcervical sclerite; ts, trichoid sensilla.

3. Trichoid sensilla are usually located in the marginal areas of the microtrichia fields of both the head and postcervical sclerite. On the sclerite, setae of sensilla are directed toward the field of microtrichia. Trichoid sensilla, which are located within the microtrichia field, usually have long setae (*Cordulegaster, Epallage*) or they are present in a very limited number (*Epallage*).

4. In representatives of Anisoptera, a total number of trichoid sensilla, and the area covered by them, are larger than those of Zygoptera. This may be explained by the more complex design and larger sizes of arrester parts. The highest complexity of the arrester in representatives of the family Libellulidae, is reflected in the diversity of microtrichia types and in the sensory equipment of the arrester as well. Sensilla were found at almost all locations in which they have been found in other odonate species, with two exceptions, TS1 and TS9 (Fig. 12.4).

5. Fields of trichoid sensilla were also reported in the neck membrane, connecting sclerites (Fig. 12.5). The movements in the head-neck region can be transmitted by the flexible membrane and monitored by the trichoid sensilla of the membrane (Gorb, 2000). Presumably, these sensory fields are also involved in the sensory system, controlling movements of neck sclerites and thus the head-arresting system.

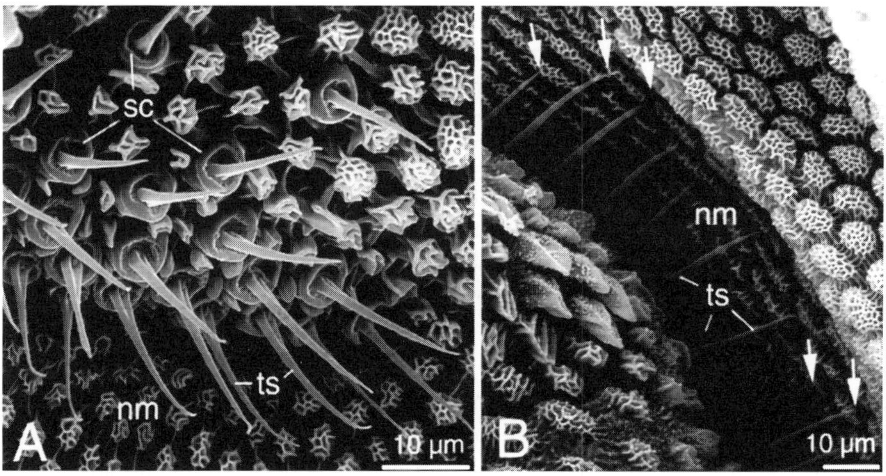

Fig. 12.5. Sensory fields of the odonate neck membrane. A. *Hypolestes clara* (Zygoptera, Lestidae). B. *Ischnura elegans* (Zygoptera, Coenagrionidae). Arrows indicate contact of the hair tips with the folded neck membrane. nm, neck membrane; sc, sockets; ts, trichoid sensilla.

2.3 Possible functions of different sensory fields

Head movements are involved in many behaviours such as food intake, gaze stabilisation, and object tracing. It has been shown by physiological experiments on a variety of insects that the sensory fields of the neck region take part in the control of gravity (Markl, 1962), or in control of the head movement in general (Strausfeld et al., 1987; Liske, 1989; Hengstenberg, 1991). Similar functions of the mechanoreceptors of the head-neck region have also been shown for Odonata (Mittelstaedt, 1950). The rich sensory equipment of the arrester provides not only a morphological basis for monitoring head fixation and release by neck sclerites, but is also involved in some flight reflexes.

The majority of the trichoid sensilla fields are probably involved in the monitoring of contact between corresponding surfaces (TS3, TS4, TS6, TS7, TS8), whereas both the TS5 and TS2 can monitor movements of the postcervical sclerite relative to the eucervical sclerite (Fig. 12.4). The TS1 can monitor movements of the postcervical sclerite relative to the pronotum. Since the postcervical sclerite is connected to the eucervical sclerite, and can be displaced by its movements, its movement can be additionally monitored by sensory fields located on the ventral edge of the eucervical sclerite, and dorsal edge of the basisternum (TS6).

Campaniform sensilla monitor cuticle deformations of the bases of the postcervical sclerite during the contact of its distal part, to the microtrichia fields on the rear side of the head. This may detect the torque, generated by working mouth-parts, or by the male pulling the female's head with anal appendages, as in anisopteran tandem flight.

The question of how the male and female communicate while in tandem is still not resolved (Utzeri, 1989). It is suggested that the arrester campaniform sensilla may play a crucial role in the control of directionality of the female during tandem flight (Beckemeyer, 1997). When postcervical sclerites are in contact with the microtrichia fields of the head, campaniform sensilla may monitor the strains in the elastic cuticle in response to male movements. In flight, the male's turns result in an asymmetry of strains in the cuticle of the right and left postcervical sclerites of the female. This asymmetry may indicate to the linked female, the direction of the male turn. In Sympetrinae (Libellulidae), swinging movements of the male during oviposition may also provide triggering information to the female about when an egg must be released.

Chapter 13

The evolution of frictional systems

The appearance of cuticle was a necessary preadaptation for the further development of surface outgrowths of different size and shape. The most important feature of the evolution of frictional systems is their multiple origin in different insect lineages, and in different functional systems. For example, there is no doubt that attachment systems consisting of a pair of microtrichia-covered surfaces appeared independently in the head-arresting system of Odonata, coxal-locking devices of Neuroptera, and the elytra-locking system of Coleoptera. The possibility of the parallel appearance of similar solutions for the same function, makes it difficult to understand evolution of one particular attachment system. Also, it is often difficult to decide which character-state was the initial one, because some attachment devices have gradually disappeared together with the loss of other functions. For example, in the case of loss of flight capability, and reduction of elytra-interlocking mechanisms.

In this chapter we discuss several examples to give the reader an impression about possible pathways of evolutionary changes of attachment devices. The first two subsections are dealing with the odonate arrester and ovipositor valvulae as model cases of diversification, and a gradual increase of design complexity. These examples show the system's evolution but do not reflect odonate phylogeny. The last subsection presents a study on the evolution of a functional system analysed together with a complex of other characters. Such an approach is based on cladistic method, allowing to recognise parallel evolution of one particular system on the basis of group phylogeny.

Since the odonate arrester is not a highly, structurally-variable system, its character states usually reflect levels of families and subfamilies. Other attachment devices, such as tarsal pads, are very adaptable structures.

However, the adaptability may cause changes in only a few subordinate characteristics, but not influence the principal design of the system. Many aspects of adaptability of attachment devices, especially tarsal pads of the legs, are discussed in Chapter 14.

1. EVOLUTION OF THE ODONATE ARRESTER

Functional morphology of the odonate head-arrester is presented in Chapter 6. The diversity of arrester design is reported below, together with an interpretation on the directionality of the system's evolution. The present chapter also provides information for further phylogenetic reconstructions. Different attempts to clarify interrelationships of odonate subgroups were based on characters of different body parts and organs; characters of wings and genitalia being the most widely used (Tillyard, 1917; 1928; Martynov, 1924; Fraser, 1957; Kiauta, 1968; Pfau, 1971; 1986; 1991; Pritykina, 1980). Nevertheless, the phylogeny of the Odonata is mainly based upon wing-characters or on those of the genitalia. An attempt to compare several head-characters in representatives of a few odonate families has been previously made (Steinmann, 1967).

Some arrester-characters, such as the shape of the postcervical sclerites (SPC) and the morphology of the microtrichia fields on the rear side of the head (MFH), demonstrate a rather low variability at both family and subfamily levels. For the scheme of arrester evolution the following characters have been taken into account: (1) the presence, number and location of the microtrichia fields on the head, (2) the shape of the SPC, its separation from the eucervical sclerite (SEC), and (3) the shape, size, density and location of microtrichia on the SPC and in the microtrichia fields of the head.

1.1 Zygoptera

The head is medio-laterally widened. There are three basic types of MFH design. The coenagrionid-type is characteristic of all Coenagrionoidea, and for some members of families of the Calopterygoidea (Polythoridae, Chlorocyphidae). Lateral to the occiput there are two isolated MFHs, ventral and dorsal, on each side. The lestid-type, consisting of one clearly isolated MFH on each side of the occiput, is characteristic for members of Lestoidea and Hemiphlebioidea, and for some families of the Calopterygoidea (Euphaeidae, Diphlebiidae, Dicteriastidae). The calopterygid-type was found only in Calopterygidae. On the rear surface of the head there is no clearly isolated MFH, the entire surface being covered by micropapillae or short microtrichia. Additionally, calopterygids have small, slightly-separated SPCs. Fixation probably takes place in the region of small depressions, lateral to the occiput. It can be hypothesised that these characters represent the groundplan condition of design of the arrester-system of Zygoptera, and that the calopterygid-type is ancestral for Zygoptera. Comparative data on the arrester design may support the idea that the calopterygid head capsule is quite basic among the Odonata (Steinmann, 1967).

In calopterygids, the rear surface of the head is completely covered by scattered micropapilla. These might serve for short-term head fixation at any point of the rear surface of the head. High-speed movies, taken in the field by G. Rüppell and D. Hilfert (University of Braunschweig, Germany), have clearly demonstrated how *Calopteryx* males crash into each other during flight to remove another male from tandem position with a female. It was previously hypothesised that aggressive behaviour of males may be an additional situation, during which the arrester system is used for the short-term fixation of the attacking male's head to prevent mechanical damage. This explains the design of microtrichia coverage of the head in Calopterygidae rather well. Large areas of micropapilla on the rear surface of the head seem to be better adapted to perform an extremely quick fixation, or release of the head.

Among the Zygoptera, two main directions in the evolution of arrester design can be distinguished from the basic calopterygid-type (Fig. 13.2). The lestid-type is characterised by one clearly defined field on each side of the head (Lestidae, Chlorolestidae, Hemiphlebiidae, Megapodagrionidae, Euphaeidae, Diphlebiidae, Dicteriastidae), whereas a clearly-limited field of micropapillae is present on the head of Diphlebiidae. The latter condition is intermediate between Calopterygidae and Lestidae. The representatives of Lestoidea and Hemiphlebioidea are quite similar to each other. All species studied from these superfamilies have the lestid-type of arrester design. Our data supports hypotheses about the close relationship between Lestidae and Chlorolestidae (Tillyard, 1917), and about the close relationship between Lestidae and Megapodagrionidae (Fraser, 1957).

The second type (coenagrionid-type) is characterised by two clearly defined fields at each side of the head (Protoneuridae, Platycnemididae, Coenagrionidae, Pseudostigmatidae) (Fig. 13.2). The dorsal field is smaller, and usually contains shorter microtrichia than the ventral field. Additionally, the rear surface of the head is characterised by the cuticular comb located mediad to the fields of microtrichia. The comb is often covered by microtrichia and presumably serves as a support for the SPC. Representatives of Coenagrionoidea are rather similar in the design of the arrester. The Protoneuridae, Platycnemidinae, Coenagrionidae, and

Pseudostigmatidae show differences in microtrichia structure; however, the general design of the arrester system remains similar.

The fixation mechanism consists of hook-like and mushroom-like microtrichia in *Sympecma* (Lestidae), and plate-like microtrichia on both surfaces (Coenagrionoidea). Mace-like microtrichia on both surfaces presumably acts like an elastic snap (Lestinae) (Fig. 13.1 C) (Gorb, 1999a).

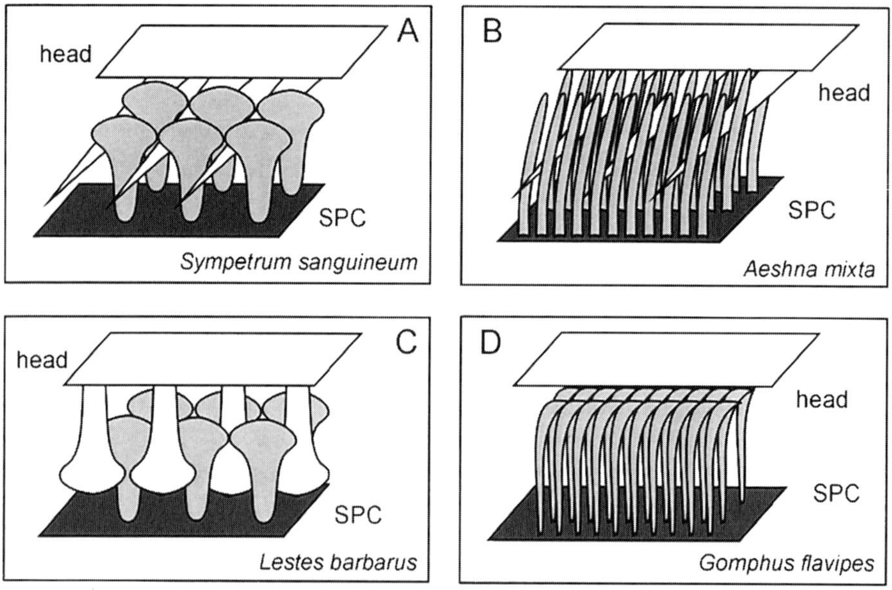

Fig. 13.1. Diagrams of corresponding frictional surfaces existing in the odonate head-arrester system. A. *Spines - plates system* of libellulids. B. *Spines - filaments system* of aeshnids. C. *Clavate MT - clavate MT system* of lestids. D. *Leaves - smooth surface system* of gomphids.

Morphological and functional stages, intermediate between the calopterygid-type and the coenagrionid-type, exist in representatives of the Hetaerininae (Calopterygidae). In *Hetaerina* spp., areas of longer micropapillae exist in the medial part of the head's rear surface. A further stage in the field structure and micropapilla elongation exists in Polythoridae, Chlorocyphidae, and Rimanellidae. In the series of taxa Calopterygidae, Hetaerininae, Chlorocyphidae, and Coenagrionidae, many intermediate conditions of head characters exist. In most of the Coenagrionoidea, the characters of the calopterygid-type, such as micropapillar coverage of the head, are lost. Greatly reduced coverage was found in the species of genus *Copera* (Platycnemididae, Calicnemidinae).

Our data supports the hypothesis about the close relationship between the Pseudostigmatidae and Coenagrionidae + Protoneuridae + Platycnemididae (Tillyard, 1917). Thus, the common trend in arrester evolution is the loss of micropapillar coverage of the head's rear surface, and the development of clearly defined fields of long microtrichia with rather diverse shapes.

The Calopterygoidea is the most diverse superfamily. Considerable differences, concerning the arrester system, are found in representatives of different families. Some calopterygoid families are very similar to families from the Lestoidea or Coenagrionoidea. This is in contrast with the hypothesised monophyly of the group Caloptera (Zalessky, 1932; Belyshev, 1973). The head-characters in the Euphaeidae, similar to those in the Lestoidea, may support the point-of-view that these taxa are more closely related, as is stated in the most common phylogenetical reconstructions. Data on the morphometry of the head-characters (Steinmann, 1967) and on behaviour (Heymer, 1975), supports this conclusion.

1.2 Anisozygoptera and Anisoptera

The head is nearly oval, and medio-laterally widened. There are three basic types of MFH in Anisoptera (Fig. 13.2). The gomphid-type is characteristic for the Epiophlebiidae, Gomphidae, and Petaluridae. The shape of the head varies from medio-laterally widened (*Epiophlebia, Gomphus, Sinogomphus*) to dorso-ventrally lengthened (*Phenes, Uropetala*). Lateral to the occiput there is a cuticular surface sparsely covered by microtrichia and hair sensilla. Clear-cut MFHs are absent. The libelluloid-type is characteristic of Cordulegastridae, Aeshnidae and Libelluloidea. There is one clearly isolated MFH on each side of the head, lateral to the occiput, on the level of its ventral edge. The anactin-type was found only in representatives of the aeshnid tribe Anactini. The head-capsule is similar to that in the Aeshnidae. There are two MFHs on each side of the head.

Fixation mechanisms consist of cone-shaped and thin, seta-like microtrichia in Aeshnidae (Fig. 13.1 B), and cone-shaped and mushroom-like microtrichia in Corduliidae and Libellulidae (Fig. 13.1 A). Smooth or slightly sculptured surfaces on the head of Gomphidae, Petaluridae, and Epiophlebiidae correspond to compressed microtrichia of the SPC (Fig. 13.1 D) (Gorb, 1999a).

Since specialised MFHs are absent in most insects, it can be suggested that arrester systems have evolved from a smooth surface, or one sparsely covered with micropapilla, to the formation of microtrichia fields. If this is correct, the gomphid-type represents the groundplan type of the arrester system. This condition is found in representatives of Epiophlebiidae, Gomphidae, Neopetaliidae, Petaluridae, and Chlorogomphinae. Species within these groups have fields covered with trichoid sensilla and sparsely-distributed microtrichia. It is likely that the postcervical sclerites (SPCs) in their lateral position, can touch these fields when the head is in the most posterior position. It is also suggested that the compressed, leaf-like microtrichia of the SPC, which were found in these taxa, may adhere to the smooth surface of the head. An additional adaptation for the fixation of SPCs to the smooth surface of the head is the particular shape of SPCs, which are elongated along the dorso-ventral axis.

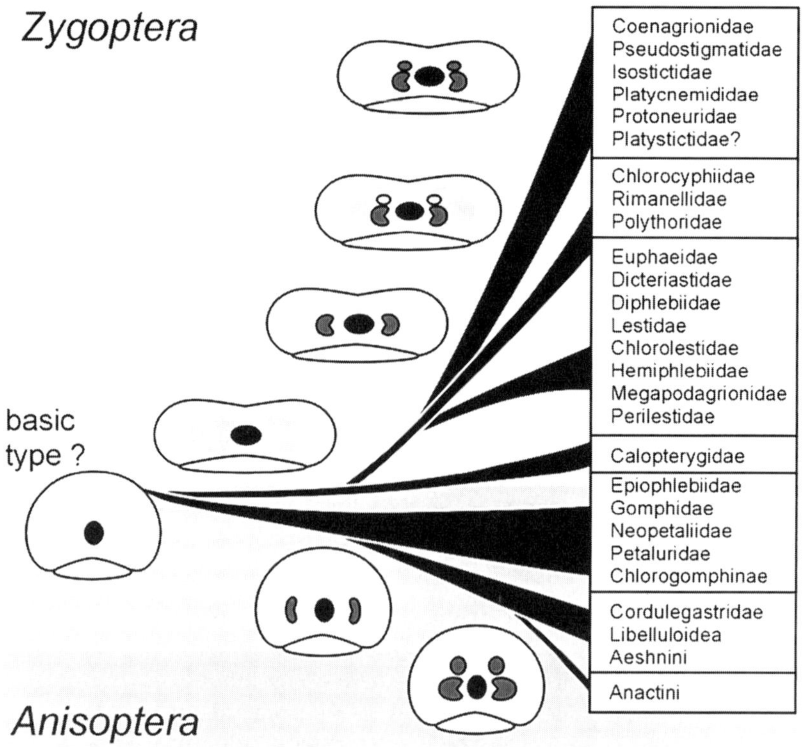

Fig. 13.2. Diagram of possible evolutionary changes of the arrester system. It is hypothesised that the smooth posterior side of the head is ancestral. Sparsely-distributed micropapillae covering large areas of the head, were transformed into compact fields densely covered with microtrichia. In some families of Anisoptera and Zygoptera, each field convergently divided into a dorsal and a ventral area.

The Chlorogomphidae is a transitional group between the Gomphidae and the Cordulegastridae. The cone-shaped microtrichia on the fields of the posterior head-surface are more densely arranged than in representatives of the Gomphidae. The cuticular folds, which are very common among representatives of the Gomphidae, and well-developed MFHs that are characteristic of the Aeshnidae and Libelluloidea, are presented on the head of Cordulegastridae. Libelluloidea (Corduliidae + Libellulidae) are characterised by a rather uniform pattern of arrester-characters. The cone-shaped microtrichia, which have been found on the head fields of most studied representatives of this superfamily, resemble those of the Cordulegastridae. An additional similarity between Corduliidae and Libellulidae is the comma-like shape of MFHs.

It seems that the arrester system as such, is an autapomorphy of the Odonata. The proposed evolutionary path of this character complex in the suborders Anisoptera + Anisozygoptera is largely congruent with the evolutionary scheme of the chromosome complexes (Kiauta, 1968), and is easily compatible with Fraser's hypothesised phylogeny of Odonata (Fraser, 1957).

Correlations between different arrester-designs and odonate behaviour are difficult to interpret at present. The use of high-speed videorecordings may help to comparatively analyse the flight behaviour amongst different odonates. Although even without this information source, the present results demonstrate that morphological analysis of the arrester system can usefully contribute to phylogenetic reconstruction of the relationships within the Odonata. It is likely that arrester evolution is closely connected to the evolution of the odonate copulative process, which is one of the most unusual among Insecta due to specialised, secondary male copulative organs and tandem postures during copulation. Additional selective pressure may have resulted from the specialised habits of dragonflies and damselflies. They are active, flying predators that often feed on the wing. The arrester system contributes to the prevention of head disturbance, and to the gaze stabilisation.

2. OVIPOSITOR COVERAGE

Functional significance of the microtrichia coverage of the inner surface of odonate ovipositor valvulae was briefly discussed in Chapter 5. Similar cuticular outgrowths of variable types were also described for females of other groups of insects (Mickoleit, 1973a; Austin and Browning, 1981; Field and Austin, 1994). It is difficult to resolve a possible evolutionary path of the structures described, because all types of the microstructures have many transitional forms. It may be hypothesised that the evolution of the microstructures in the odonate ovipositor has been directed from the single, sparse microtrichia, through the microtrichia with shallowly-divided crests, to the comb-like microtrichia pattern (Fig. 13.3).

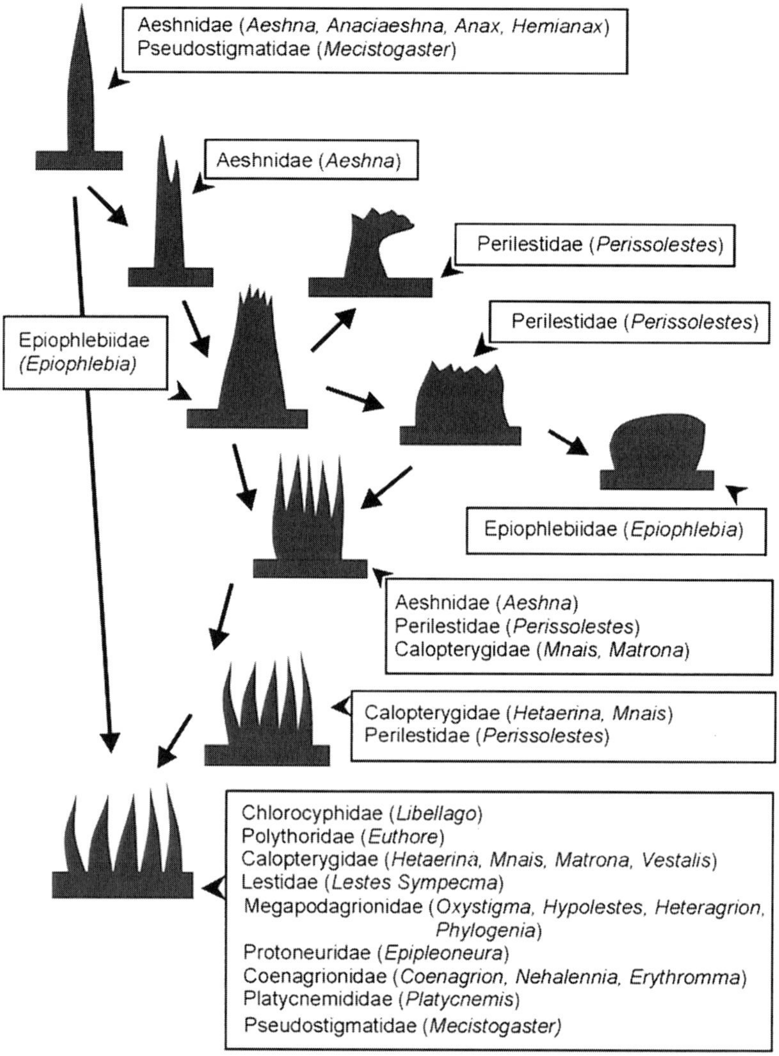

Fig. 13.3. Hypothetical diagram of the possible course of evolution of microstructures on the inner surface of odonate ovipositor valvulae.

The reconstruction of a possible evolutionary scenario is impeded by the existence of many transitional forms. It is likely to assume that the plate-like microsculpture of the upper valvulae of *Epiophlebia superstes* (Anisozygoptera, Epiophlebiidae), has evolved independently from the comb-like microtrichia existing in most Zygoptera. However, the proposed scenario of evolution of the ovipositor coverage (Fig. 13.3) may not reflect species evolution.

3. WING-LOCKING DEVICES

Diversity of designs of the wing-locking devices described above (Chapter 8) let us presume that these systems appeared independently in many insect lineages. A model species for the understanding of possible mechanisms of evolution of cuticular outgrowths is the water bug *Ranatra linearis*, in which the mesothoracic fold serves as the wing-locking device. The cuticle is covered with elongated microtrichia, which hold an underwater air layer on almost all parts of the body surface (plastron) (Chapter 3). However, the lateral wall of the fold, where the wing is locked in the resting position, is covered with plate-like outgrowths that seem to be mechanically more stable than the plastron structures (Fig. 13.4 A, B). The SEM study showed that plate-like microtrichia are fused, plastron outgrowths, because transitional states are present at the border between the two distinctive types of structures. Such transitional outgrowths are thicker and less densely distributed on the surface (Fig. 13.4 C). Presumably, different functional requirements of both surfaces resulted in the divergence of an initially uniform cuticle coverage, into two distinct types of outgrowths (Fig. 13.4 D).

In beetles, a number of frictional structures are known to be involved in securing the elytra on the resting beetles' body (Chapter 8). Flightless-correlated reduction of frictional surfaces in the elytra-to-body binding-mechanism is a phenomenon currently described in tenebrionid beetles. It is well-known that many species have, secondarily, lost their ability to fly. This loss of ability to fly correlates with the appearance of other, non-releasable mechanisms of elytra binding to the body, and the loss of frictional patches common for most of Coleoptera.

The flightless tenebrionid *Tabarus* sp., is restricted to the rainforests of Papua New Guinea. The investigation of the body-to-elytra binding mechanism of this species shows that a system of interlocking macrostructures has evolved as a permanent elytra-to-body attachment mechanism. These macrostructures are located between the two elytra in the midline, between the anterior edge of the elytra and the mesonotum, as well as between the elytral epipleuron, pterothorax and abdominal ventrites (Bouchard and Gorb, in press).

Firstly, the metathoracic elytra-locking device, which usually secures the antero-median parts of the elytra into longitudinal depressions on the metanotum (Beutel and Haas, 2000), is lost. The complete loss of this locking device has only been reported previously in the tenebrionid species *Asida corsica*, a flightless, ground-dwelling desert beetle with a true subelytral cavity. In addition, the frictional patches of microtrichia between the ventral surface of the elytra and the body, found in winged and flightless tenebrionids, are completely absent in *Tabarus* sp. The basal patches (TAL-EAL), which are part of the coleopteran groundplan and which were previously thought to be absent only in beetles with or without reduced elytra (Samuelson, 1994; 1996), are also absent in *Tabarus* sp..

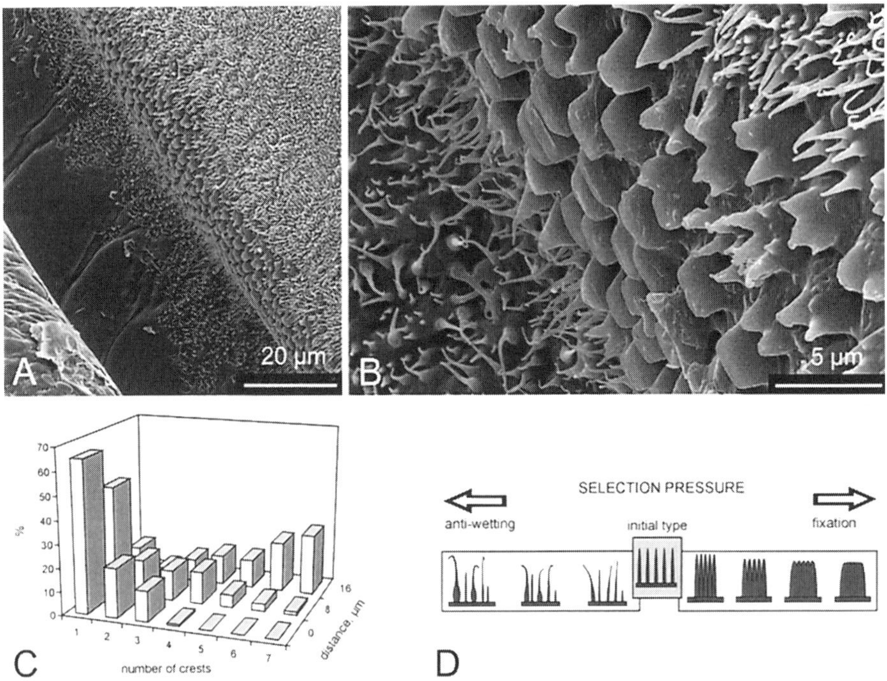

Fig. 13.4. Water stick insect *Ranatra linearis* (Heteroptera, Nepidae); mesothoracic fold serving as the wing-locking device. A, B. Cuticular outgrowths supplementing the fold. C. Number of single crests per area (x-axis), which corresponds to a single cell, at a certain distance (y-axis) from the midline of plate-like outgrowths. D. Hypothetical diagram of the divergence of initial cuticle coverage into two distinct types of outgrowths.

It is obvious that the correlated reduction of certain structures has occurred many times, independently. Nevertheless, it may turn out that other, specific ways of coping with lost flying abilities, are unique to larger or smaller monophyletic units. However, a broad, comparative study is required to gain evidence for this hypothesis.

4. EVOLUTION OF PAD STRUCTURES IN THE BACKGROUND OF HEXAPOD PHYLOGENY

Morphological and ultrastructural studies of the cuticle of hexapod adhesive pads in representatives of most hexapod orders were shortly summarised in Chapter 9. The principal goal of this subsection is to provide an evolutionary interpretation, based on a cladistic analysis, of the distribution of tarsal attachment structures, and of a broad spectrum of other characters of adults and immature stages (Fig. 13.5). Leg structures specialised mainly for prey-capture and/or copulation, were excluded.

The most interesting result of our literary investigation and comparative ultrastructural study is that diversity of attachment, or adhesive structures, of hexapods is generally based on only two basic mechanisms: hairy surfaces or smooth flexible pads. We suggest that a fibrous composite material, such as hexapod cuticle (Neville, 1975), is preadaptative and may provide only limited options for design of attachment systems.

Leg attachment devices are adaptive structures. Their construction and properties may correlate with the preferred substrata normally used by particular hexapod species. Hexapod attachment pads probably evolved to facilitate walking on plant surfaces. Plant surfaces have a wide range of textures: they may be smooth, hairy, and covered with waxes or with moist secretions. As with any integument, plant cuticle is a functional organ reflecting response to a variety of environmental pressures. This results in the particular design, ultrastructure, and chemistry of the plant surface. During the long period of co-evolution between flowering plants and hexapods, plants have not only developed structures attracting pollinators, but also a wide variety of structural and chemical attributes of their surfaces related to defense against herbivores. The co-evolution of plant surfaces and hexapod pads results from a competition between hexapod attachment systems and plant anti-attachment surfaces (Chapter 14).

In spite of the adaptability of pad characters, the phylogenetic analysis demonstrates some general evolutionary trends in the pad structures (Beutel and Gorb, 2001). Arolium, euplantulae, and hairy tarsomeres have evolved independently in several lineages (Fig. 13.5). Nevertheless, some of them support monophyletic lineages (e.g. Embioptera + Dermaptera; Paurometabola excl. Dermaptera + Embioptera; Isoptera + Blattodea; Neuropterida + Strepsiptera + Coleoptera; Hymenoptera + Mecopterida). Other structures such as claw-pads (Ephemeroptera), balloon-shaped, eversible pads (Thysanoptera), or fossulae spongiosae (Reduviidae) are unique to members of larger or smaller monophyletic units. It is plausible to assume that the evolution of flight and the correlated necessity to cling to vegetation, or other substrates, was a major trigger for the evolution of adhesive pads. Groups with the potential to evolve a great variety of designs in adhesive pads are Heteroptera and Diptera.

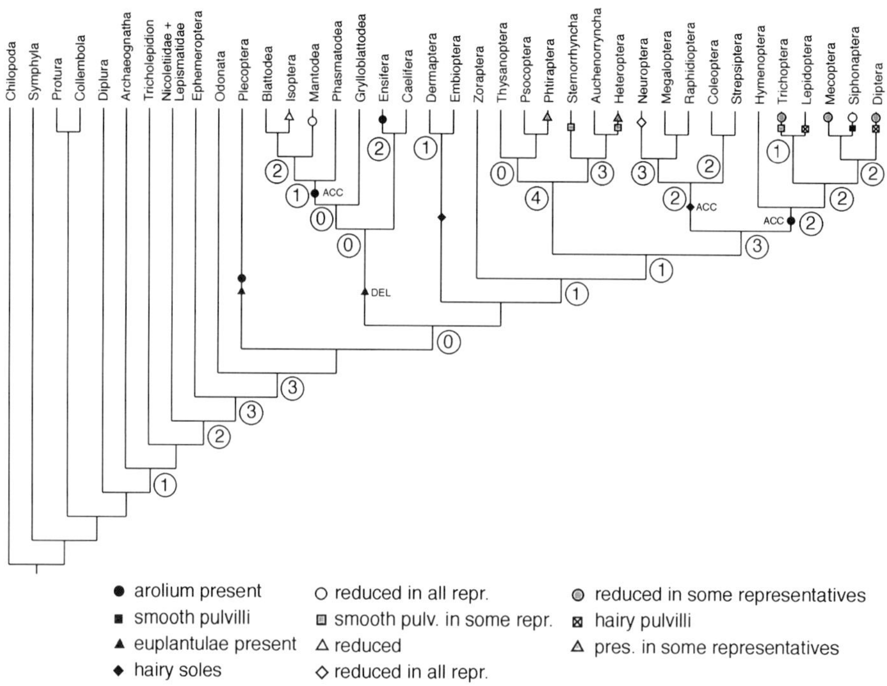

Fig. 13.5. Selected tree of 24 minimal length trees, adhesive pad characters are mapped on cladogram; numbers left of branches are branch support values for selected clades (Beutel and Gorb, 2001). ACC (character transformation accelerated) and DEL (character transformation delayed) are alternative options to the character state optimisation in PAUP.

If representatives of other animal phyla are taken into account, the insect cuticle might reveal other biological fibrous composites, such as skin in amphibians (Ernst, 1973a; b; Emerson and Diehl, 1980; Green, 1981; Linnenbach, 1985; Green and Carson, 1988; Hanna and Barnes, 1990; Emerson, 1991) and mammals (Schliemann, 1970; 1971; 1974; 1975; Schliemann and Rehn, 1980; Thewissen and Etnier, 1995). Vertebrate skin often forms attachment structures similar to smooth and hairy pads of insects

(Braun, 1878; Dellit, 1934; Ruibal and Ernst, 1965; Ernst and Ruibal, 1967; Hiller, 1968; Stork, 1983a; Bauer and Good, 1986; Röll, 1995; Irschick et al., 1996). A plausible explanation for the existence of these two principal designs is that leg pads are primarily involved in and adapted to facilitate locomotion. This means that the ability to easily detach from a surface is of equal importance to fast attachment on a variety of surfaces. Other mechanisms such as suckers, hooks, or glue cannot work well, because of their inability to detach rapidly (Nachtigall, 1974; Gorb, 1998c).

An analysis of many characters was carried out, principally to allow a more reliable interpretation of the evolution of adhesive pad characters. Trees obtained, are largely congruent with or without consideration of attachment structures. However, a better resolution is obtained with the full character set. Inclusion of attachment pad characters in the analysis does not imply circularity in evolutionary interpretation (Luckow and Bruneau, 1997).

Ephemeropterans are characterised by an unusual, autapomorphous condition of the distal leg: the presence of claw-pads. A balloon-shaped, eversible structure between the claws is a unique autapomorphy of Thysanoptera. Distally-broadened adhesive setae on the claws have evolved in Psocoptera. A claw-pad is present in phthirapterans, probably as an adaptation to ectoparasitism. An arolium and smooth pulvilli have evolved independently in several lineages of Heteroptera. Euplantulae are present in some mallophagans and mirines (Heteroptera, Miridae), but are absent from other heteropterans. A fossula spongiosa is present in piratines (Heteroptera, Reduviidae).

The development of different types of adhesive structures plays an important role in the diversification of several orders. Hymenopterans have retained an arolium, which is ascribed to the groundplan of a clade comprising Hymenoptera + Mecopterida, and representatives of non-apocritan families («Symphyta») possess an additional, unique type of attachment structure derived from tarsal thorns (plantar lobes). Most Mecopterida have also retained the arolium, however, it is secondarily absent from the ground-living boreids and from notiothaumids (Mickoleit, 1971). Most dipterans posses a plate-like empodium (Tabanoidea, Nematocera excl. Tipulidae) and hairy pulvilli. The latter character state was found in all taxa examined with the exception of Tipulidae (Hennig, 1973). The presence of an arolium in members of this family is probably a plesiomorphic condition.

Adhesive structures are not found in entognathous hexapods, which live mainly in soil and leaf litter. However, rather complicated claws and associated structures have evolved in diplurans, collembolans, and proturans. Subdivision of the tarsus and reduction of the ring-shaped pretarsus, still more or less distinct in Diplura and Ellipura, are autapomorphies of Insecta. These changes, especially adaptive flexibility of the tarsus, are probably preconditions for further modification and the appearance of different attachment devices.

Chapter 14

Ecological implications

Insect attachment pads did not evolve to enable flies to hold onto windows, but mainly as adaptations for efficient locomotion on plant surfaces. However, plant surfaces have a wide range of textures. They may be smooth, hairy, or covered with waxes or moist secretions. As any integument, plant cuticle, with related structures, is a functional organ reflecting a number of environmental pressures. These pressures resulted in the particular design, ultrastructure, and chemistry of the plant surface. During the long period of co-evolution between flowering plants and insects, plants have developed not only structures attracting pollinators, but also a wide variety of structural and chemical attributes of their surface related to defence against herbivorous predators. Actually, the evolution of plant surfaces and insect pads is a competition between insect attachment systems and plant anti-attachment surfaces. However, mechanisms of insect attachment to plant surfaces remain somewhat understudied. This chapter deals with ecological implications of insect attachment pads and shows the role of plant surfaces in preventing insect attachment.

1. PLANT SURFACES

The plant cuticle is not homogenous, but is organised in different layers or structures (Fig. 14.1). It consists of three main zones: (1) the cuticle proper, consisting of a lamellate layer; (2) the reticulate region, traversed by cellulose microfibres; and (3) the pectinaceous layer (Jeffree, 1986). The cuticle morphology in many plant species shows, however, a substantial divergence from such a general scheme. An important character of the plant surface is an external layer of epicuticular waxes that generally coats the cuticle surface.

1.1 Wax layer

The wax layer widely varies in thickness. It is an extremely thin layer on the cuticles of aquatic plants, whereas substantial crusts appear as pruinescence on the fruits, stems and leaves. During the last ten years, several comparative studies have been carried out on the microsculpture of cuticular waxes of flowering plants (Barthlott and Ehler, 1977; Barthlott and Wollenweber, 1981; Barthlott, 1998). The wax layer has a crystalline structure. The most common types of crystal shapes are tubes, solid rodlets, filaments, plates, ribbons, and granules. Several examples of wax coverage are given in Figure 14.2. Current classification of plant epicuticular waxes, based on high resolution scanning electron microscopy of 13,000 plant species, distinguishes 23 types of wax (Barthlott et al., 1998b).

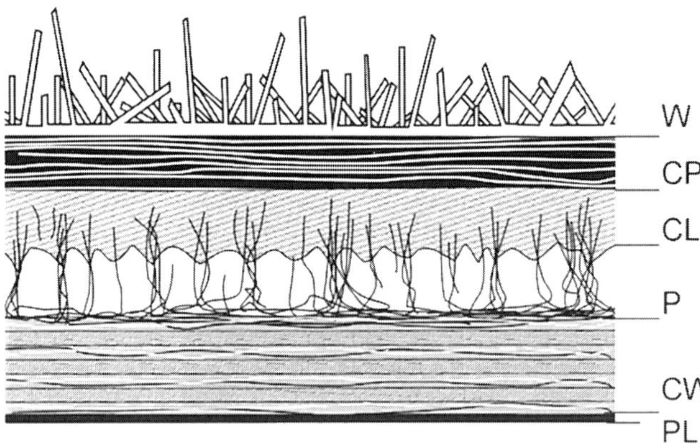

Fig. 14. 1. The generalised structure of a plant cuticle. W, epicuticular wax; CP, cuticle proper; CL, cuticular layer of reticulate region traversed by cellulose microfibrils; P, pectinaceous layer and middle lamella; CW, cell wall; PL, plasmalemma.

Plant waxes are a mixture of different chemical compounds, such as hydro-carbons, wax esters, primary and secondary alcohols, fatty acids, aldehydes, ketones, β-diketones, triterpenoids, and flavonoids (Holloway, 1969; Baker, 1982; Walton, 1990; Bianchi, 1995). Interestingly, the wax ultrastructure correlates with its chemical composition. For example, it has been previously shown that aldehydes play a major role in the production of filamentous wax types, characteristic of the surfaces of *Pisum sativum*. Similarly, the ketone fractions seems to be responsible for the crenate plate morphology of the waxes in *Allium porrum* and *Brassica oleracea* (Jeffree, 1986).

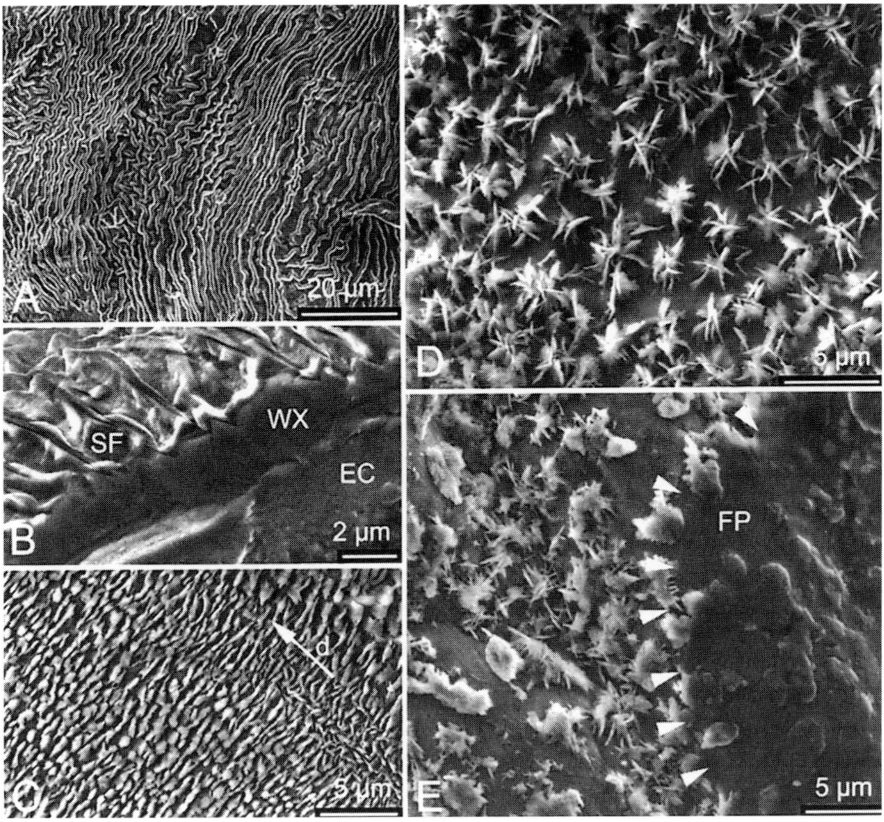

Fig. 14.2. Plant waxes. A, *Acer platanoides*. B. Same, cross section of the leaf. C. *Iris palustris*. D. *Lactuca serriola*. E. Same, with the footprint of a fly *Calliphora erythrocephala*. d, distal direction of the leaf; EC, epicuticle; FP, footprint; SF, leaf surface; WX, wax layer.

Waxes are an essential structural element of the surface, and of fundamental functional and ecological importance for the interaction between plants and their environment. The epicuticular wax on some plant surfaces, once abraded, is not replaced. However, many species, such as *Eucalyptus*, produce wax continuously, thus keeping the surface in a fully functional state (Juniper, 1995).

1.2 Trichomes

Trichomes are hair-like protuberances extending from the epidermis of aerial plant tissues (Levin, 1973). Their walls are about as thick as the surrounding epidermis, or can be somewhat thicker. Some mature trichomes lack cytoplasm, whereas it is present in others. The diversity of structural types of trichomes is immense. They may be unicellular or multicellular, glandular or non-glandular, straight, spiral-shaped, hooked, unbranched or stellate. Specialised types are known as stinging emergences of *Urtica spp.*, insect-trapping glandular trichomes of *Drosera spp.* and *Pinguicula*, and nectar secreting hairs of nectaries (Jeffree, 1986). Trichomes seem to be highly adaptive structures. In *Arabidopsis thaliana*, which is currently a model plant of developmental biology, trichomes are single cells with 3 to 4 branches (Hulskamp and Schnittger, 1998; Hulskamp et al., 1999). In the wild type, the trichomes are regularly distributed on the leaf (Fig. 14.3 A, C). New trichomes are initiated at the basal zone of the leaf with a distance of approximately 2-3 cells. Due to leaf growth, the trichomes are distributed on mature leaves at a distance of 30-40 cells. In plants, overexpressing a positive regulator of trichome fate (maize R gene) results in appearance of new trichomes between older ones (Fig. 14.3 B). In such plants, the density of trichomes is much higher on mature leaves than on those of wild type plants. There are also several mutations of trichomes with more than three branches (Fig. 14.3 D).

Trichome functions are improvement of plant climbing abilities, an increase in hydrophobicity of the surface, a decrease in wind velocity, preventing water loss, etc. There are some examples of trichomes responsible for trapping insects and small animals in carnivorous plants, such as *Sarracenia purpurea, Genlisea,* and *Darlingtonia* (Jeffree, 1986). Presumably, felted trichome layers provide a physical barrier against insect predators, protecting young leaves (Curtis and Lersten, 1978). Insects can also be penetrated by hooked trichomes through the intersegmental membranes of the limbs and abdomens. Trichomes are also involved in chemical defence mechanisms. In some plants of the families Fabaceae and Solanaceae, glandular trichomes secrete adhesive and toxic fluids, which trap and kill insects (Jeffree, 1986).

2. PLANT SURFACE AND ATTACHMENT PADS

The ability to attach to smooth substrates made it possible for insects to adapt to a wide variety of ecological niches on the plant surfaces, which are often vertical and smooth. If compared with related species living in litter, the arboreal carabid beetles developed a wide variety of adaptations for niches on plant surfaces (Stork, 1987).

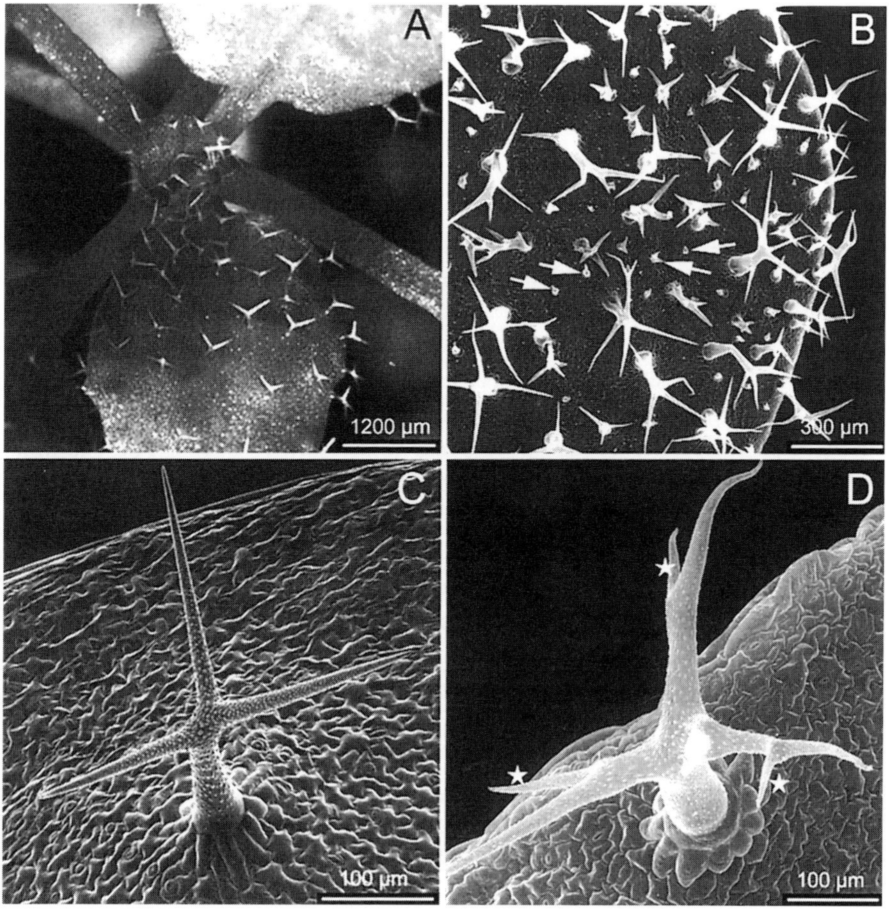

Fig. 14.3. Trichomes in *Arabidopsis thaliana*. A. Light microscopical picture of rosette leaves in the wild type. B. Scanning electron micrograph (SEM) of the plant overexpressing a positive regulator of trichome fate. C. SEM micrograph of a wild type trichome. D. SEM micrograph of a try mutant with trichomes with more than 3 branches. Courtesy of Arp Schnittger, University of Tübingen.

Plant surfaces structured with trichomes and/or covered with waxes made it impossible for some pest-insects to walk on the plant surface. In the evolution of herbivorous insects, structures and surfaces have developed as adaptations to attachment to particular plant surfaces or to a variety of them.

Also, attachment structures on legs of eryophyid mites are adapted to the plant surfaces (Shevchenko, 1970). At the same time, plants improved their surfaces to protect them against the attachment of herbivores.

2.1 The role of wax crystals in preventing insect attachment

Many insects have excellent attachment abilities on a variety of smooth surfaces. For example, the chrysomelid beetle *Chrysolina polita,* adheres well, independently of the molecular properties of the surface material (Stork, 1980b). However, it has been previously suggested that wax crystals on the flowering shoots of plants, such as *Salix* spp, *Hypenia*, and *Eriope* are adaptations to prevent crawling insects from robbing nectar and other resources (Juniper et al., 1989; Harley, 1991; Juniper, 1995; Eigenbrode, 1996). The slippery surfaces, or other types of surface coverage, are plant adaptations called *the greasy pole syndrome* (Harley, 1991). For example, the lower internodes of the South American representatives of the genera *Eriope* and *Hypenia* (Lamiaceae), are densely pubescent; the upper internodes are densely covered by wax. Ants of all sizes and species are unable to walk on these stems and rob the nectar (Juniper, 1995). At the ground level, ants are repelled by hairs. Ants invading from neighbouring vegetation are unable to climb the waxy stems. In the North American fern plant *Ceratopteris triangularis*, a powdery wax-like substance covers the underside of the fronds, particularly areas surrounding the sori. These epicuticular secretions have been interpreted as detachable structures, preventing insect herbivores from reaching the reproductive region of the plant (Juniper, 1995).

Heavy glaucous coatings on the ventral parts of plants, such as commonly occur on the leaf-sheaths of Poaceae, may present an obstacle to the ascent of insects to the leaves. Plant individuals with a waxy bloom are less strongly attacked by insect herbivores. An example is the crassulaceous plant *Dudleya britonii,* which is resistant against the pyralid moth *Rhagea stigmella*.

Although it is known that most wax crystals are easily detached structures and that dust contamination reduces proper functioning of attachment pads, only few studies have experimentally shown that the wax crystal structure directly influences insect attachment. Dust is used, for example, in an application as a material contaminating the adhesive organs of the mite *Varroa jacobsoni,* to prevent their attachment to the body surface of honey bee *Apis mellifera* (Ramires, 1989). It has also been experimentally demonstrated that aphids partly lose their attachment abilities after having walked on talcum powder (Dixon et al., 1990).

Adhesion of the coccinellid beetle *Hippodamia convergens,* to glass can be greatly reduced by a deposit of crystalline tetracosanoic (C_{24}) acid, which is a common constituent of plant wax (Eigenbrode, 1996). Crystal density may be an important factor in preventing attachment. It has been reported that wax crystal density in *Brassica* mutants is negatively correlated to the occurrence of the flea beetles (Fig. 14.9 A). Effects of crystal size and shape on insect attachment have yet to be tested.

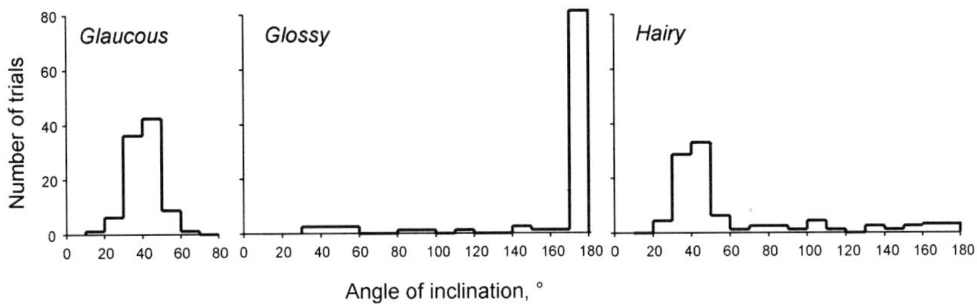

Fig. 14. 4. Histogramms showing the inclination angles at which the beetle *Phaedon cochleariae* (Chrysomelidae), fell off three different leaf surfaces. Data from Stork (1980c)

Juvenile leaves of *Eucalyptus* are usually covered by a glaucous bloom, resulting in a silvery-grey appearance. In some species, wax also occurs on the adult foliage, and on the fruits and buds. Wax coverage on both surfaces of the juvenile leaves resembles a network of thin rods. It has been observed that the adult foliage of *E. nitens* had been intensely grazed by the eucalyptus tortoise beetle, *Paropsis charybdis* Stål (Chrysomelidae), whereas the waxy juvenile leaves on the same trees remained untouched (Edwards, 1982). These beetles can cling to almost any surface, including smooth glass and adult foliage of *Eucalyptus*. However, adults placed on waxy juvenile leaf surfaces repeatedly slipped off. Contact of beetles with the juvenile leaves was followed by grooming of the tarsi, because of wax contamination (Edwards, 1982). These observations show that wax coverage provides physical protection for plants. Mechanical removal of the wax crystals prevented slipping from juvenile leaves. The density of the *P. charybdis* beetle and other chrysomelids was higher on eucalypt species with less wax coverage (Edwards and Wanjura, 1990).

In *Brassica* spp., epicuticular waxes contaminate attachment pads of the mustard beetles *Phaedon cochleariae*, preventing their attachment (Stork, 1980c; 1986). This has been verified with the scanning electron microscope. Beetles placed on leaves of waxy plants fell off when these were raised to a slope of about 40° (Fig. 14.4). On smooth leaves, they remained attached at angles up to 180°. Since these insects can withstand wind speeds up to 100 mph before they are swept from smooth surfaces, these experiments showed the effectiveness of the anti-attachment systems of certain glaucous plants. Mechanical removal of the wax coverage from the leaves of *Brassica napus* increased settling and feeding by the flea beetle *Phyllotreta cruciferae* (Chrysomelidae) (Bodnaryk, 1992). Additionally, the role of waxy coverage of *Brassica* crops in protection against herbivore insects has been supported by other studies (Anstey and Moore, 1954; Way and Murdie, 1965; Stoner, 1990; Eigenbrode and Espelie, 1995; Eigenbrode and Kabalo, 1999).

However, there are several insect groups with exceptional adaptations for attachment to plant surfaces, even to those covered with wax crystals. The cabbage aphid *Brevicoryne brassicae,* walks better on wax blooms of *Brassica* than on smooth surfaces (Åhman, 1990). This can be explained by the long legs of this species, which enables the insect to apply its tarsi nearly perpendicular to the plant surface, and reduce lateral force vectors that could reduce attachment (Southwood, 1986).

Lacewing larvae possess a so-called anal adhesive organ producing a mucilaginous secretion (Spiegler, 1962). Interestingly, the force required to detach the abdominal organ from waxy plants is at least three times greater compared to smooth plant surfaces. This suggests that the secretion is able to wet (or probably dissolve) wax crystals and incorporate them into the adhesive matrix, or colloid. Similar abilities have been reported for some larval syrphids (Diptera) (Eigenbrode, 1996). A disadvantage of this method of attachment is that an insect may have difficulties breaking the abdominal attachment, since its tarsi lose contact first, and an additional supporting point is absent. However, there is evidence that adhesive forces of first instar larvae of *Chrysoperla plorabunda* (Chrysopidae), are 20 to 200-fold greater on glossy types of *Brassica oleracea* plants than on waxy types (Eigenbrode et al., 1999).

Other insects use locomotory «tricks» to attach to waxy surfaces. The chrysomelid beetle *Paropsis aegrota,* feeding on waxy *Eucalyptus* leaves usually walks on the leaf edge (Edwards and Wanjura, 1990). This behaviour allows the mechanical advantage of opposing the tarsi by applying more force perpendicular to the surface.

2.2 Trichomes and insect-plant interactions

The structure and density of plant trichomes vary greatly in angiosperms in general, and in different mutants of the same plant species. Trichomes play an important role in plant resistance against insect herbivores. The lime aphid *Eucallipterus tiliae,* is inhibited from feeding by stellate trichomes densely covering leaves of some lime species. The potato leaf-hopper

Empoasca fabae, is unable to orient to a proper feeding position or to move effectively on a pubescent variety of soybean (Lee et al. 1986). Trichome density is an important factor in soybean resistance to leaf-hoppers (Fig. 14.5 A). However, it is not clear whether these examples of trichome-mediated plant defence are effects of decreased attachment abilities. Wheat resistance to the cereal leaf beetle *Oulema melanopus* (Chrysomelidae), may, at least in part, be based on the trichomes preventing the tarsi and their setae from contacting the underlying leaf surface, thus preventing adhesion (Ringlund and Everson, 1968).

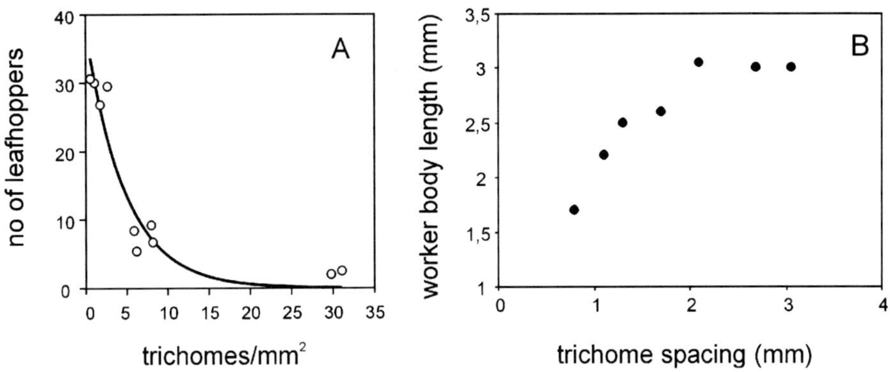

Fig. 14.5. Effect of trichome properties on insects. A. Occurrence of the potato leaf-hopper *Empoasca fabae,* on the soybean *Glycine max* as a function of the density of trichomes. Data from (Levin, 1973). B. Worker body lengths of ant species as a function of trichome spacing. Each point represents a mean value obtained from a pair of species, ant and plant. Data from Davidson et al. (1989).

Larvae and adults of the flea beetle *Altica subplicata,* show a strong preference for feeding on the young leaves of the willow *Salix cordata.* Its leaves are densely pubescent compared to other willow species. In *S. cordata*, pubescence density decreases with leaf age (Gannon et al., 1994). The tarsi of larvae are totally different from those of adults. In larvae, an attachment organ of each leg consists of a smooth fleshy arolium connected to a single claw. The tarsus of the adult beetle is composed of tarsomeres, ventrally covered with clusters of adhesive setae.

Attachment tests performed under different wind conditions have shown that larvae and adults could attach equally well to smooth and hairy leaves (Fig. 14.6). Wind caused difficulty for larval and adult attachment and movement (Gannon et al., 1994). Larvae fell off hairy surfaces more easily.

This has been explained by smooth, fleshy attachment devices probably being better adapted to smooth surfaces.

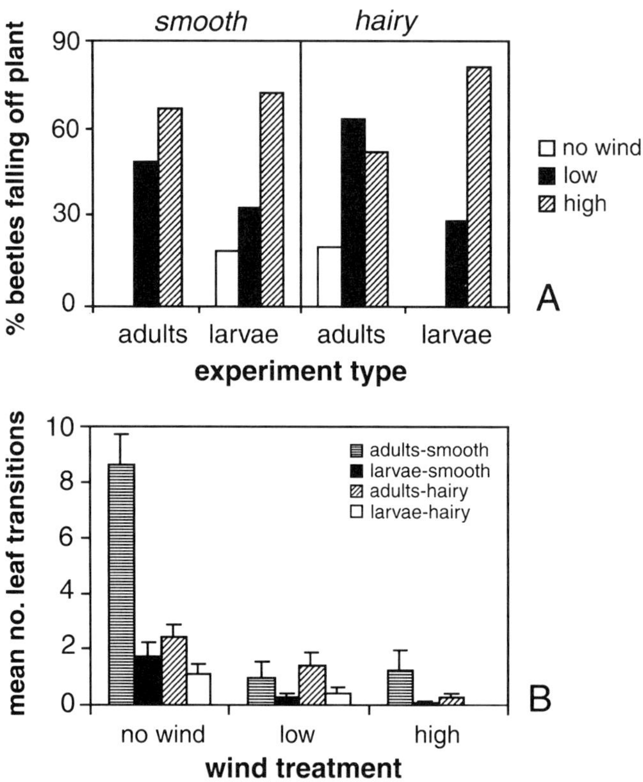

Fig. 14. 6. Attachment ability of the flea beetle *Altica subplicata* (Chrysomelidae) on sand-dune willows. A. Percentage of larvae and adults that fell off smooth and pubescent (hairy) leaves during three wind treatments. B. Larval and adult movement indicated by the number of leaf transitions (absolute value of leaf number at the end of trial minus leaf number at the beginning of trial) after initial placement on smooth or hairy leaves. Data are means and standard errors for adults and larvae placed on smooth or hairy leaves, exposed to different wind treatments. Data from Gannon et al. (1994).

In oak aphids, correspondence of attachment structures on legs to the host plant surface, is an important aspect of their specialisation to the plant substrata. The aphid *Tuberculoides annulatus*, feeding mainly on leaves of *Quercus robur*, has difficulties walking on the leaves of *Q. ilex*. The aphid

Myzocallis schreiberi, living on *Q. ilex*, walks more easily on the surface of this plant (Kennedy, 1986).

Attachment tests carried out on the chrysomelid beetle *Phaedon cochleariae,* on different mutants of *Brassica* revealed that the trichomes, rather than complementing the effect of the glaucous wax bloom in reducing the adhesion of the adhesive setae, may provide anchorage sites for the tarsal claws (Fig. 14.4) (Stork, 1980c).

3. ANT-PLANTS

Secondarily wingless insects living in the canopy, such as ants, must have outstanding attachment abilities to plant surfaces. When subordinate arboreal ants with restricted territories fall off their tree in the tropical rain forest, they will die, since no odour track will lead them back. This is especially true for host-specific ants.

The Asian weaver ant species *Oecophylla smaragdina,* is unable to walk on the waxy bloom, but shows exceptional attachment forces on the smooth plant wax. Such a performance is needed during the construction of their leaf-tent nests. Before leaves are connected by larval silk, workers draw the leaves together. Doing this, ants are attached to the smooth upper side of a leaf and must sustain the entire pulling force (Federle et al., 2000).

W. Federle with co-workers, carried out a series of experiments on ants that are able to walk on slippery, waxy blooms on *Macaranga* ant-plants (Federle et al., 1997). Many insects, including other ant species, slip off these plant surfaces (Fig. 14.7). Only the symbiotic partner ants of *Macaranga* are capable of climbing on these plants (*Crematogaster (Decacrema) msp.1* and *C. msp. 6*) (Fig. 14.8). For this reason, the wax blooms of the ant-plant seem to be an ecological isolation mechanism for the symbiotic ants. This mechanism is based exclusively on the influence of the ant attachment abilities, but not on the repellent effects of the wax. The comparison of surfaces in different species of the *Macaranga* revealed a high correspondence between the occurrence of wax coverage and obligatory ant associations (50% in ant-plants; 6,7% in non-myrmecophytes).

The mechanism that enables these specialist «wax-runners» to cling to slippery wax blooms, is still unknown. However, it appears that smaller ant species are, as a rule, better climbers. Interestingly, superior «wax-runners» do not attach any better to a smooth perspex surface. On the contrary, they attach worse than closely related species that are unable to climb up waxy stems (Federle et al., 2000).

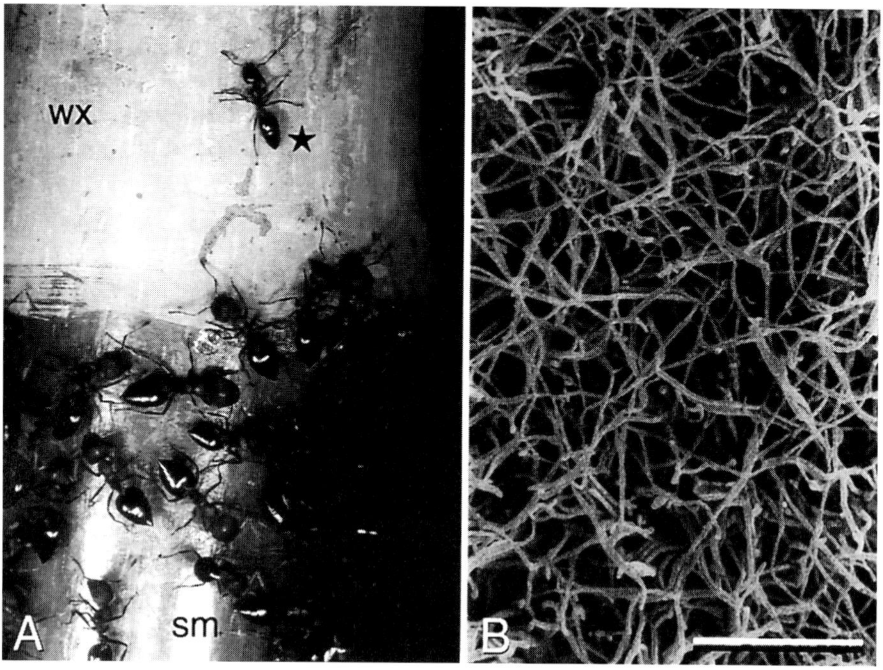

Fig. 14. 7. Protective barrier-effect of slippery epicuticular wax crystal layers on *Macaranga*. A. The wax bloom (*wx*) in the lower part of the twig had been wiped off (*sm*), so that worker ants of the aggressive generalist *Crematogaster cf. artifex* could gain access. The specialist «wax-runner» *Crematogaster* (*Decacrema*) (asterisk) is apparently well protected by the barrier (Federle et al., 2000), with permission of the Company of Biologists Ltd. B. SEM-picture of the wax coverage on the stem. Fine structure of the slippery epicuticular wax crystals on the stem surface of *Macaranga lamellata* (Markstädler et al., 2000), with permission of Birkhäuser Verlag AG, Basel. Scale bar = 5 μm.

In some cases, ant behaviour suggests a need from the plant for protection from other ant species. Plant surfaces covered by trichomes can also influence ant associations by affecting locomotion of some species. Worker body sizes of the numerically predominant ant species are positively and significantly correlated with the mean distances between stem trichomes (Fig. 14.5 B). However, the latter fact is explained not by the reduction of ant attachment, but rather by the cutting of the pheromon trail of large ant species (Davidson et al., 1989). As a result of a strong competition between ant species to occupy myrmecophytic host-plants and use their resources, selection leads to an appearance of plant adaptations, which favours the ant partner that provides benefit to the plant (anti-herbivore protection). These adaptations play the role of an ecological filter against non-symbiotic ants, which may use plant resources but do not provide substantial benefits to the plant.

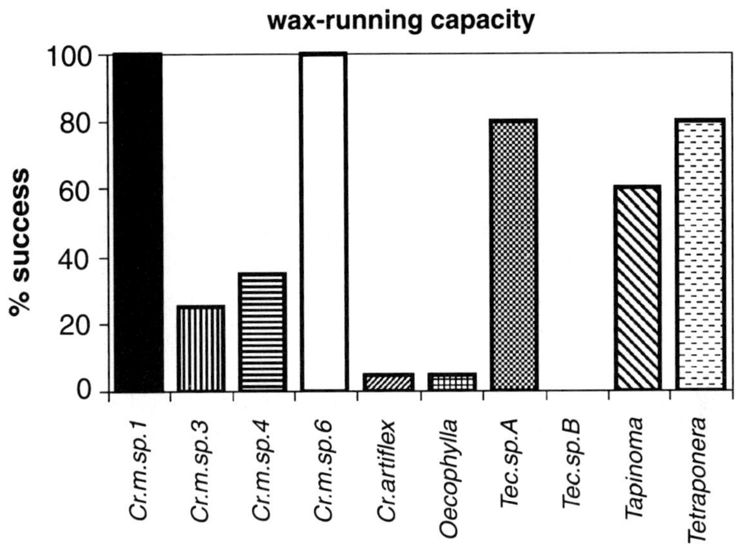

Fig. 14.8. Running capacity of the ten ant species on vertical glaucous twigs of *Macaranga pruinosa,* of 15 mm diameter (n=20 ant workers per species). All ants that successfully walked a stretch of 5 cm, upward or downward along the twig, have been classified as «successful». Ant species tested are *Crematogaster (Decacrema) msp. 1, 3, 4, 6, C. cf. artiflex, Oecophylla smaragdina, Technomyrmex sp. A, B, Tapinoma sp. A, Tetraponera sp. A.* Data from Federle et al. (1997).

4. TRAPS OF CARNIVOROUS PLANTS

Wax coverage on the inner surface of *Nepenthes* pitchers resembles specialised plates about 1 μm long, which overlap like roof tiles (Fig. 14.9). Since they are attached only by a fragile stem, they break off readily, forming a slippery surface for the feet of insects, preventing their escape (Martin and Juniper, 1970). Thus, wax layers can prevent insects from gaining a foothold, since the claws break through the wax network and there is also no contact between the adhesive setae of the insect and the pitcher surface (Juniper and Burras, 1962). If the loose crystalline material on a *Nepenthes* plant is removed, insects do not slip off the pitcher surface. However, the properties of this surface can be easily restored with the covering of talcum powder (Knoll, 1914).

The beetle *Paropsis charybdis* (Chrysomelidae), is able to walk on smooth surfaces using its attachment pads, but slipped off the inner surface of the pitcher (Edwards, 1982). After contact with the waxy area of the pitcher, beetles usually clean their pads of the waxy contamination.

A similar structure of waxes has also been reported from other carnivorous plants. Detachable wax systems exist in *Darlingtonia* and in certain insect-trapping Bromeliaceae (Juniper, 1995). Carnivorous plant systems remain the best-known wax-based mechanism preventing insect attachment. However, there are still many unresolved questions. Some of them are: (1) which adaptations enable some specialised insects, especially dipterans, to use *Nepenthes* pitchers as their habitats (Shinonaga and Beaver, 1979; Beaver, 1983)? (2) which pad design is more resistant to contamination, the hairy one or the smooth one? (3) which force contributing to overall attachment, friction or adhesion, is actually reduced by dusty contamination?

5. EFFECT OF ATTACHMENT ABILITIES ON MULTITROPHIC INTERACTIONS

In some cases, a higher degree of resistance against herbivores has evolved in non-waxy forms of plants. As previously reported, glossy forms of eight crops are resistant against 17 insect species (Eigenbrode, 1996). This can be explained by specific attachment abilities of entomophagous insects. In other words, more predators, such as coccinellid larvae and lacewing larvae exist on glossy plants (Way and Murdie, 1965), because these insects can adhere better to smooth surfaces. In this respect, most entomophagous insects are similar to the herbivores.

Larvae of the lacewing *Chrysoperla carnea,* cannot hold onto the waxy Brussel sprout, but are able to attach to less waxy leaves of *Brassica chinensis* (Arzet von, 1973). Larvae of the coccinellid beetle *Adalia bipunctata,* does not feed on waxy *Brassica* plants, but spends about 40% of the time feeding on *B. chinensis* (Shah, 1982). 45% of *Coccinella septempunctata* and 32% of *Hippodamia variegata,* fell off pea plants within a three minute period (Kareiva and Sahakian, 1990).

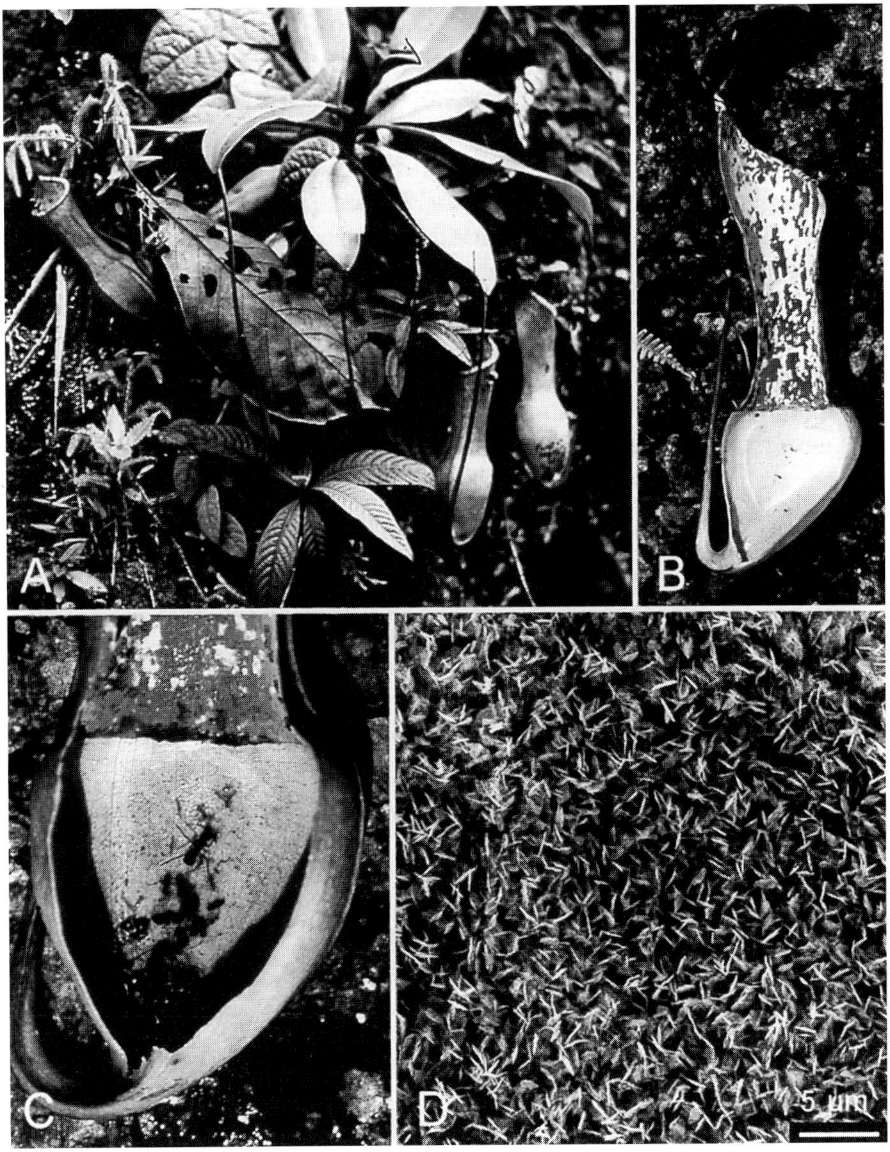

Fig. 14.9. Pitchers of the carnivorous plant *Nepenthes eustachya*. A. The plant under natural conditions. B. Inner part of the pitcher. C. Victims in the lower fluid-filled part of the pitcher. D. Wax-crystals in the upper part. The upper-interior pitcher surface above the liquid is covered with densely packed plates of wax crystals, which detach very easily from the supporting surface. Courtesy of Heiko Rischer, University of Würzburg.

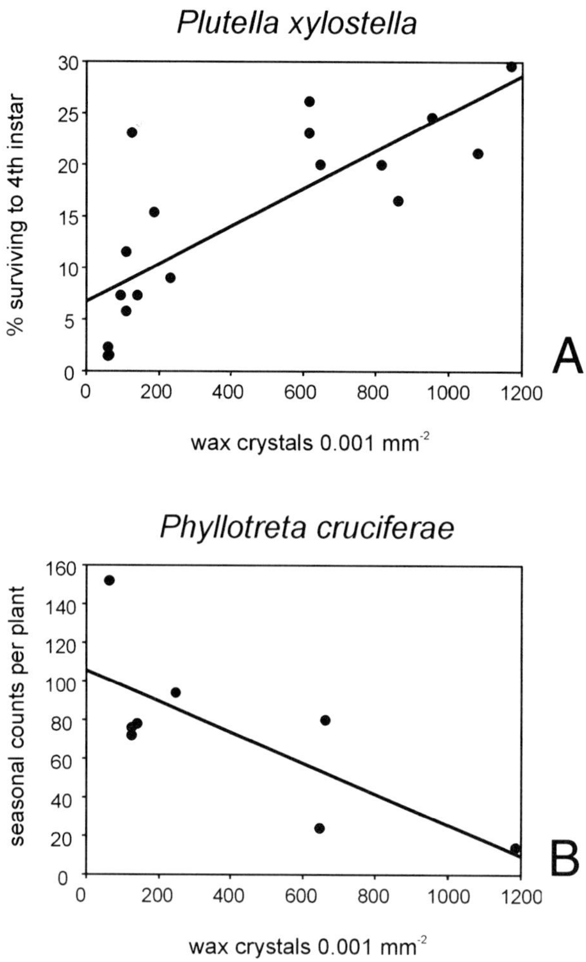

Fig. 14.10. Relationships between crystal densities of *Brassica* lines and survival of the larvae of the moth *Plutella xylostella,* and infestation levels by the flea beetle *Phyllotreta cruciferae*. Data from Eigenbrode (1996)

The aphid parasitoid *Diaretiella rapae,* was more commonly observed on glossy *Brassica* plants. The aphelenid parasitoid *Eretmocerus sp*., parasitised more nymphs of the whitefly *Bemisia sp.* on the glossy cabbage plants (Eigenbrode, 1996). Predatory wasps *Polistes dominulus* (Hymenoptera, Vespidae), slip significantly more frequently from the leaf surface of the waxy mutants of the *Brassica oleracea* plants, than from glossy mutants (Eigenbrode et al., 2000). However, the specific mechanism for these effects still remains unexplained.

Predators as the coccinellid beetle *Hippodamia convergens*, the anthocorid bug *Orius insidiosus,* and the larva of the lacewing *Chrysoperla carnea,* are more mobile on smooth mutants of *Brassica*, if compared with a standard variety (Eigenbrode, 1996). Therefore, these predators are more effective at reducing populations of larvae of the diamondback moth *Plutella xylostella,* on smooth plants than on those with surface waxes (Eigenbrode et al., 1995) (Fig. 14.10). Surfaces of smooth plant mutants appear advantageous for herbivores, however, predators and parasitoids can also attach more easily to these varieties. Thus, control of populations of herbivores seems to be better on the smooth plants.

Since smaller insects have generally better attachment abilities on different surfaces, including those covered with wax crystals, we may expect that only some predatory and parasitic insects exist on the waxy plant mutants. By the amount of crystalline waxes, the plant may mediate the presence of herbivores and the effectiveness of particular predators.

6. ECOLOGICAL SIGNIFICANCE OF BIOLOGICAL SURFACES

As mentioned previously, locomotion always deals with friction and adhesion. Insects usually walk on plant substrata, and therefore many aspects of insect-plant relationships deal with both mechanical and chemical surface interactions. Presumably, physical and chemical aspects of such interactions are different in a variety of ecological groups of insects, such as herbivores and parasites, specialist and generalist phytophages, and plant-ants. From the plant perspective, completely different functions of the surface profile and coverage, such as insect trapping function and selective defence against herbivores may involve similar general mechanisms.

This chapter demonstrated that attachment-mediated insect-plant interaction is a broad field of research. However, there are only a few model systems that have been studied in some detail. The results, in most cases, cannot be transferred to other insect-plant systems because of lack of knowledge on the surface structure, both plant wax and insect secretion chemistry, insect locomotory behaviour, forces contributing to attachment, etc. There is currently no concept explaining which type of attachment device performs better on a particular plant surface. The scaling effects also remain vaguely understood. Only interdisciplinary studies integrating contributions from morphology, physics, chemistry, and ecology may explain the exact role of plant waxes and trichomes, together with insect attachment devices in the complex interactions in natural networks.

Chapter 15

Nature's design as a basis for biomimetics

The so called theory of inventive problem solving is based on three main principles in engineering: (1) all patents are based on about 40 inventive principles; (2) all technology trends are predictable; (3) important inventions come from outside the industry within which they are applied. Throughout evolution, nature has constantly been called upon to act as an engineer in solving technical problems. Organisms have evolved an immense variety of shapes and structures. Although often intricate and fragile, they can nonetheless deal with extreme mechanical loads. Many functional solutions are based on a variety of ingenious structural solutions. This book has demonstrated a variety of biomechanical systems of insects, adapted for attachment of parts of the body to each other, or to attach the organism to a substrate. Understanding these is of major scientific interest, since we can learn about their use as structural elements and their biological role and function. This knowledge is also highly relevant for technical applications.

Many of the greatest challenges for today's engineering science is miniaturisation. Insects and other animals have solved many problems correlated with extremely small size, during their evolution. Previous chapters showed that zoologists and morphologists have collected a huge amount of information about the structure of such living micromechanical systems. This information can be utilised to mimic them for industrial applications. There are three main areas, where nature's solutions of attachment problems may be applied: (1) precise mechanics, (2) gluing technology, and (3) material science of surface-active composite materials. Possible innovations may also appear at the boundaries of the named areas. Also, knowledge about the properties of natural systems might be useful for pest control, by modifying plant surface. Only a few selected examples are discussed in this chapter. More information about the use of biological principles for engineering developments may be obtained from special literature on biomimetics and bionics (Nachtigall, 1998).

1. PRECISE MECHANICS

MEMS is a new word in engineering science meaning micro electro-mechanical systems. Since many electro-mechanical devices become smaller and smaller, all systems, known from the existing macroscale devices, have to be miniaturised in different ways. Attachment of parts of the MEMS can be achieved by gluing parts together, but sometimes releasable attachment fasteners are also required. Natural attachment systems may be a rich source for surface patterning in the macro scale as, for example, in the tyre industry.

1.1 Frictional systems

The hook-and-loop principle was transferred into industrial applications from nature. In 1948, a Swiss mountaineer George de Mestral, during his walk in the field, was quite frustrated by the hooking fruits clinging to his clothes. While picking them off, he realised that this principle could be used to make a fastener of a completely new type. Together with a weaver at a textile plant in France, he designed, from cotton, a prototype called «locking tape». Later they came up with the idea of using nylon to fabricate hooks. Nylon, when sewn under infrared light, forms rigid hooks. The hooks compose the rigid part of the fastener. The corresponding surface is made of a textile tape consisting of thin, flexible loops (Fig. 15.1 A, B, D-F). Presently, there are two main types of such velcro fasteners: (1) the hook-loop principle mentioned above, and (2) the mushroom-and-loop principle, in which hooks are replaced with mushroom-shaped structures (Fig. 15.1 C, G).

Patent databases contain a huge number of ideas dealing with applications of existing fasteners. They are almost universally used: in textile industry, medicine, sports, transport, etc. However, all of these applications use the same types of available hook-like tapes (Fig. 15.1 F, G). Natural systems provide a variety of microscale surface patterns, which may serve as a source for future prototyping of novel types of releasable fasteners, and microfasteners (Fig. 15.1 H). Since forces in the contact areas of most natural systems have not been previously measured, the first step in this direction should be to identify the interesting properties. This requires the knowledge of biologists and measuring techniques used in material science. It is necessary to study the mechanical properties of natural material, of which systems are made, and the range of forces holding contacting surfaces together.

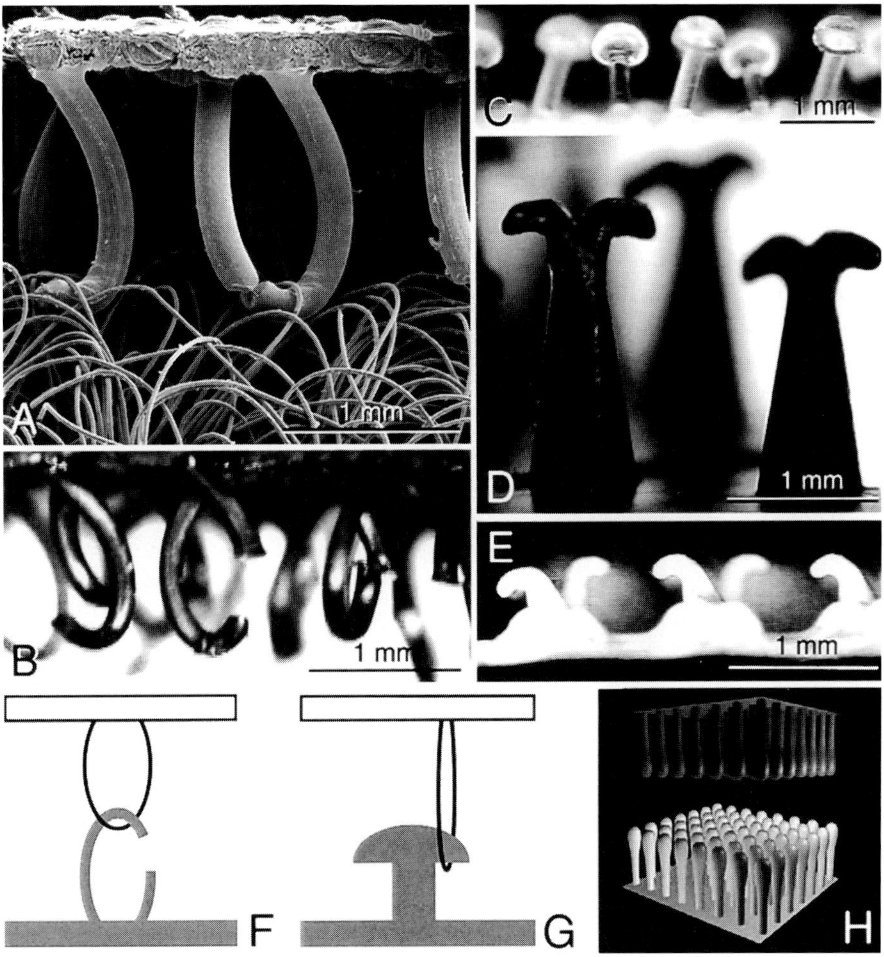

Fig. 15.1. Velcro fasteners. A, B, D, E, F. Hook-and-loop principle. C, G. Mushroom-and-loop principle. H. Hypothetical microlocking mechanism made of two identical surfaces covered by clavate structures. A. SEM micrograph of the most used velcro fastener in action (courtesy of J. Berger, MPI of Developmental Biology, Tübingen). C-E. Diversity of the rigid part of industrial fasteners. F-G. Diagram of two main types of fasteners.

An engineering approach, applied after detailed studies on the natural system, would be most promising. However, engineers can also, simply copy the surface shape of a variety of scales and materials using available technologies of chemistry and processing. Both approaches may run parallel for some time and possibly converge later. A short description of modern technologies, which may be applied for prototyping of diverse surface microsculpture is given in subsection 1.2.

Walking machines usually use suckers to hold onto vertical surfaces, and under a surface. A primary disadvantage of this attachment principle is that a very smooth substrate surface is required. The future goal should be walking robots, able to walk on a variety of surfaces. Insects can walk rather well on smooth and structured substrata, on inclines, vertical surfaces, and some of them even on the ceiling. Hairy and smooth leg attachment pads (Chapter 9) are promising candidates to mimic into robot soles adapted for locomotion. Similar principles can be applied for design of microgripper mechanisms, with an ability to adapt to a variety of surface profiles (see subsection 1.3).

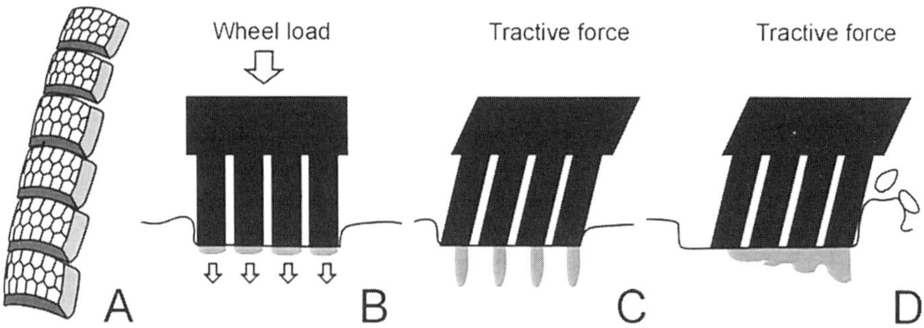

Fig. 15.2. Honeycomb tyre. A. Diagram of the pattern of the tyre surface. B-D. How tyre sipes work in the snow. B. Even ground compacts the snow. C. Tilted edges produce a pressure peak, which results in further penetration into the snow. D. Sliding of rubber on snow, and snow on snow.

One would hardly expect a real innovation to improve, for example, existing tyre profiles. However, a leader of the tyre industry, Continental®, has developed a winter tyre with honeycomb profiles similar to those existing on the attachment pads of the grasshopper *Tettigonia viridissima* (Chapter 9). The company promises enhanced wear performance on the dry road, less aquaplaning and better braking on wet roads, substantially improved lateral guidance, better grip, and more traction on ice. Fine grooves and longitudinal sipes in the individual tread blocks, provide lateral guidance (Fig. 15.2). These are intersected by lateral sipes which provide maximum traction. This optimisation ensures maximum safety on wintry roads. As lateral forces are generated in bends, the honeycomb sipes provide more gripping edges than conventional sipes, thus considerably improving trackholding when cornering.

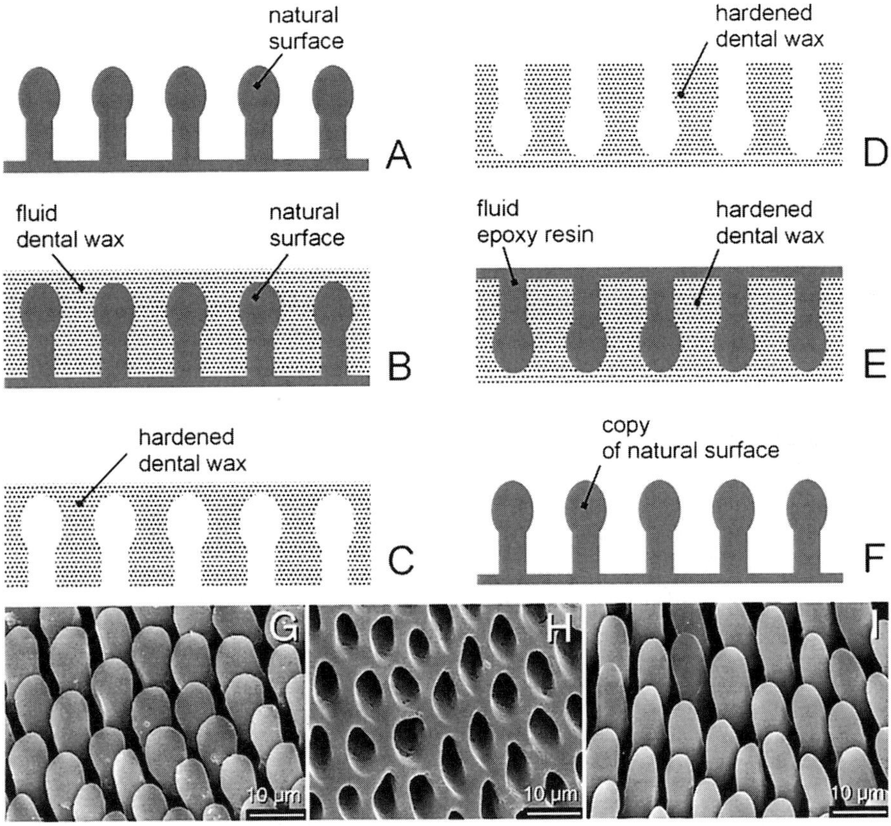

Fig. 15.3. Dental wax cast method for surface replicas. A-F. Diagram of the method (explanations in text). G. Ventral microtrichia field of the head-arresting system of a damselfly, *Pyrrhosoma nymphula*, air dried. H. «Negative» replica of the surface shown in G. I. «Positive» replica of the surface shown in G by the use of the «negative» shown in H.

1.2 Microfabrication technology

There are many difficulties in testing mechanical properties with very small pieces of living tissues. Surfaces of the living tissues, which are cut off the body, dry out very fast. Measurement of frictional forces in a fluid will give completely different results from experiments done in the air. It might be possible to test the contribution of shape of surface structures first, without the influence of material properties specific for biological tissues. The dental wax cast technique can be applied to produce surface casts of an epoxy resin (Fig. 15.3). This method uses two-component dental wax to obtain «negative» casts from a living surface at room temperature.

Wax, applied on the biological surface, spreads very quickly over the sample and fills surface irregularities (Fig. 15.3 A-B). After hardening (2-5 min), the biological sample can be easily removed without any damage to the «negative» wax cast or the biological sample, because of a high elasticity of the hardened wax (Fig. 15.3 C-D). Then the wax casts are filled with Spurr's resin (Spurr, 1969), or other low viscosity epoxy resin, which after polymerisation at 60-70°C for 24h, become relatively stiff and again can be removed from the «negative» without damage (Fig. 15.3 E-F). The «negative» replica can be used again to fabricate a number of resin «positives». This method delivered good results with relatively long cuticle outgrowths (10-30 μm), as well as with small structures (0.02 μm) (Fig. 15.3 G-I). Even structures of about 0.2 μm in diameter can be resolved.

Such replicas can be used to test attachment properties of surfaces defined by shape. By varying the resin composition, polymerised material can demonstrate a variety of material properties, from the very hard to quite soft. By using samples with the same shape and different mechanical properties of material, dependence of attachment forces on material properties at a given surface profile can be tested. The main disadvantage of this method of prototyping is that the area of the replica corresponds exactly to the original area of the biological sample. For industrial prototypes larger areas are usually needed. In addition, if an interesting structure pattern is located on the wavy surface, its shape will be exactly repeated in the cast.

Since diverse morphological properties of the surface, such as dimensions of outgrowth and their pattern on the surface, can be quantified by the use of scanning electron microscopy, profiliometry, and atomic force microscopy, another method of the surface microstructuring can be applied. So-called lithographic technique allows the fabrication of surface shape from different artificial materials. This technique uses the shadow projection of an absorber relief, into a radiation-sensitive, resist material (Fig. 15.4 A). Both the exposure duration and further development of the irradiated area depend on the type of resist material, and on the three-dimensionality of the required resulting structure (Fig. 15.4 B). With the use of negative resist, an inversion of the structure is possible. To create the positive casts of such structures from a wide range of materials, such as metal, glass, polymers, and ceramics, or to allow a mass manufacturing process, the electroforming-moulding technique is used (Fig. 15.4 C-F). The resist structure is filled by electroplating with a metal, for use as a mould for further injection moulding.

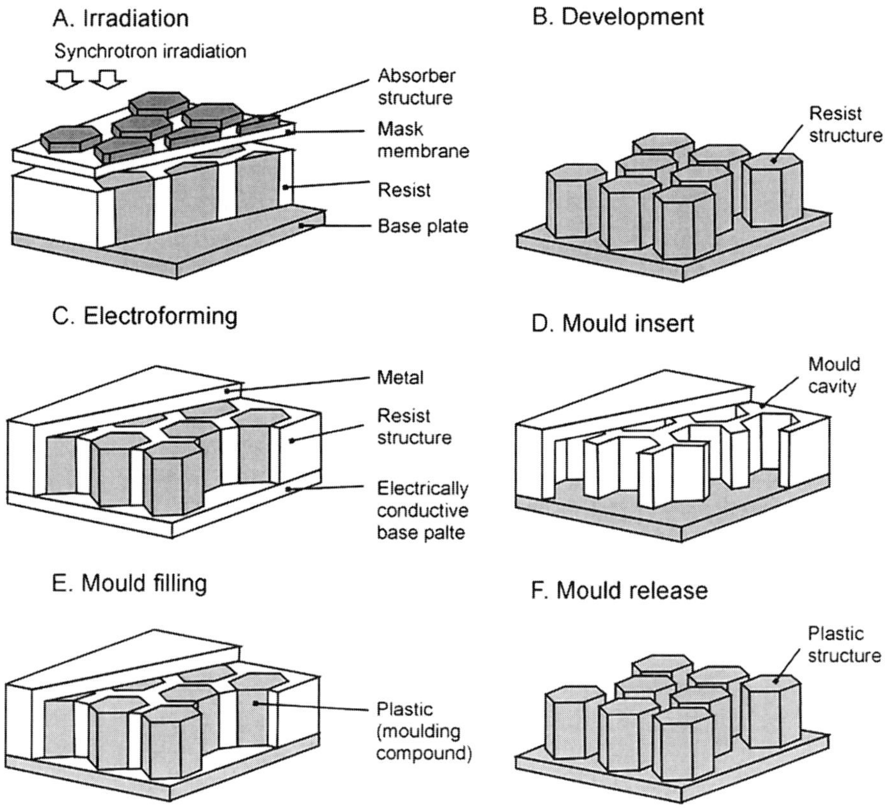

Fig. 15.4. Lithographic technology (LIGA) of surface microstructuring (Institut für Mikrotechnik Mainz GmbH).

1.3 Anti-adhesive and anti-frictional systems

It is sometimes important to decrease adhesive or frictional forces rather than to increase them. This problem becomes even more important with small object dimension, when adhesive and frictional forces are relatively high, because of different scaling properties with the mass. In many micromechanical systems, greater adhesive forces may reduce precision of the mechanism. Below, an example of a microgripper mechanism is discussed. An additional topic for biomimetics of insect microsystems may be lubrication in micro-joints, which still remains vaguely studied.

Object manipulation on the millimetre scale (for instance the handling of MEMS parts) is usually performed with mechanical grippers. Since the mechanical contact between gripper and object is necessary, adhesion becomes increasingly important with decreasing object size. The gripping force can be adjusted to an appropriate value. The release of the micropart, however, is commonly accomplished by the action of gravity after opening the gripper jaws.

As the dimensions decrease, adhesion decreases by the power of two, while gravity decreases by a power of three. The fact that the gravitational force decreases much faster than the adhesion force causes problems for the disconnecting process, as well as for accompanying manipulation accuracy. Adhesion is known as the sum of several contributions all arising from atomic or molecular interactions. The major contribution to adhesion arises from capillary forces due to adsorbed water (Scherge et al., 1999). The water forms capillary necks holding both parts together. One way to control capillary action is to generate hydrophobic, i.e. non-polar surfaces. These surfaces do not wet, thus no water can be adsorbed. Another way is to reduce the area of contact. This can be achieved by microstructuring of the gripper jaws. Since adhesion is a function of the contact area, attracting forces can be reduced significantly. From biological systems, we can learn some principles to optimise MEMS, where mass to force relationships are different from the macroscale mechanical systems (Scherge and Gorb, 2000b). Insect attachment pads are analogous systems for technical microgrippers.

First of all, the attachment force (gripping force) in living systems is optimised for the weight of the mobile system. The use of visco-elastic materials ensures maximum gripping force at minimal contact areas due to the surface adaptability (Gorb et al., 2000; Scherge and Gorb, 2000b). Such an adaptability has the additional advantage of ability to attach to a variety of surfaces, independent of their surface sculpture. Hydro-carbon coverage is probably able to suppress the action of adsorbed water and prevent capillarity.

2. COMPOSITE MATERIALS

The new century is often called the age of materials. It is expected that nowadays technology will improve all types of existing materials and create new ones. Again, biological systems may provide many innovations for material design and their surface structuring. Insect cuticle has fascinated material scientists over decades, because of the possibility of combining in one particular structure, such as a wing or leg, a wide range of mechanical properties from stiff sclerites and veins, to soft membranes and elastic resilin-bearing joints. If one is going to create attachment systems based on principles similar to biological systems, it is necessary to evaluate which materials would be the best candidates, (1) to mimic such a wide range of properties of natural materials, and (2) to allow similar possibilities of combining subunits with different material properties into a structural continuum with local mechanical differences.

Among thousands of existing polymers, polyurethanes are possibly the most appropriate class of materials for this purpose. Polyurethanes have a wide range of material properties depending on their chemical composition and further foaming. Furthermore, technology of local foaming provides the possibility of varying material properties in the same piece of material. Presumably, polyurethane foams can be used to mimic biological attachment devices with visco-elastic properties.

3. GLUES

The industry of adhesives is presently following different goals (Hennemann, 2000): (1) an increase in the reliability of glued contact; (2) mimicking of natural, environment-friendly glues; (3) development of mechanisms for application of a minute amount of glue to the surface. An additional challenge is the use of substances, which (1) allow multiple attachment and detachment, and (2) enable attachment to a variety of surfaces.

Many insect attachment devices correspond to at least some of these requirements. One such example is the hairy surface of the leg pulvillus in flies. This system uses a secretion enabling hairs to attach and detach to diverse substrata very quickly. The hair design includes a mechanism that delivers the secretion, in extremely small amounts, directly to the contact area, and only then when the contact to the substrate is achieved (Chapter 9). In the case of natural systems, the mechanical and chemical properties of one surface are already predetermined. This can limit possible adhesive substances optimised for this particular surface. Needless to say, the goal of biomimetics is not to imitate the entire system, but to use interesting principles or design details for new industrial applications, or for improvement of existing ones. However, it is a long way from the knowledge of natural systems, to real engineering innovations.

4. PEST CONTROL

Most arthropod attachment systems cannot attach to surfaces covered by tiny dust particles. This phenomenon was already used to prevent attachment of the parasitic tick *Varroa jacobsoni,* to its host, the honey bee (Ramires, 1989). As was previously mentioned in Chapter 14, attachment abilities of leg pads in many insects are reduced on plant surfaces covered by wax crystals. Additionally, in a series of experiments on the chrysomelid beetle *Gastrophysa viridula*, it was shown that surface roughness strongly influences attachment of insects with the hairy pad system. Minimum attachment ability was observed at a roughness ranging from 0.3 μm to 3 μm (Fig. 15.5).

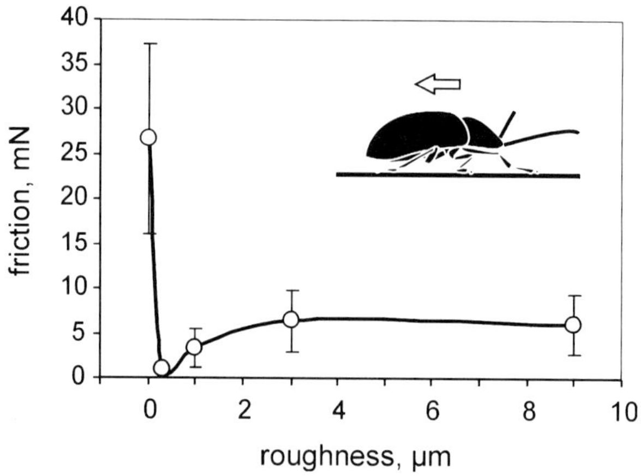

Fig. 15.5. Attachment abilities of the beetle *Gastrophysa viridula* (Chrysomelidae), on substrates with different roughness; centrifugal experiments.

Knowledge about the structural properties of insect leg attachment devices and plant surfaces, together with experimental data on attachment abilities of insect pests on a variety of structured substrata, might be useful for pest control. Directional changes of plant sorts, using genetic technology or a selective process by man, may result in plant surfaces preventing or reducing attachment of particular insect pests. Development of such a method of biological control requires, however, not only structural and experimental data on plant hosts and insect pests, but also additional complex studies on the genetic background of surface pattern formation in plants, and on ecology of food chains connected to the particular plant species.

References

Abrol, D.P. (1986) Flight range and significance of wing-hooks in *Megachile femorata* Smith (Hymenoptera, Megachilidae). J. Anim. Morphol. Physiol. *3*:107-112.

Aiken, R.B. and A. Khan (1992) The adhesive strength of the palettes of males of a boreal water beetle, *Dytiscus alaskanus* J. Balfour Browne (Coleoptera: Dytiscidae). Can. J. Zool. *70*:1321-1324.

Akimov, I.A., I.A. Starovir, A.V. Yastrebtsov, and V.G. Gorgol (1988) *Varroa mite - the agent of the bee varroatosis*. Kiev: Naukova Dumka.

Altner, H. and L. Prillinger (1980) Ultrastructure of invertebrate chemo, thermo and hygroreceptors and its functional significance. Int. Rev. Cytol. *67*:69-139.

Andersen, S.O. (1963) Characterization of a new type of cross-linkage in resilin, a rubber-like protein. Biochem. Biophys. Acta *69*:249-262.

Andersen, S.O. (1964) The cross-links in resilin identified as dityrosine and trityrosine. Biochem. Biophys. Acta *93*:213-215.

Andersen, S.O. (1966) Covalent cross-links in a structural protein, resilin. Acta Physiol. Scand. *66 (Suppl. 263)*:1-83.

Andersen, S.O. (1979) Biochemistry of insect cuticle. Annu. Rev. Entomol. *24*:29-61.

Andersen, S.O. and F.M. Barrett (1971) The isolation of ketocatechols from insect cuticle and their possible role in sclerotization. J. Insect Physiol. *17*:69-83.

Andersen, S.O. and T. Weis-Fogh (1964) Resilin, a rubber-like protein in arthropod cuticle. Adv. Insect Physiol. *2*:1-65.

Anderson, T.F. and A.G. Richards (1942a) An electron microscope study of some structural colors in insects. J. Appl. Phys. *13*:748-758.

Anderson, T.F. and A.G. Richards (1942b) Nature through the electron microscope. Sci. Monthly *55*:187-192.

Andreev, K. (1988) Ecology of mites foresing on the beetles of the genus *Necrophorus* (Silphidae). Nauch. Dokl. Vyssh. Shkol. Biol. Nauki *1988*:44-47.

Anstey, T.H. and J.F. Moore (1954) Inheritance of glossy foliage and cream petals in Green Sprouting Broccoli. J. Heredity *45*:39-41.

Arens, W. (1989) Comparative functional morphology of the mouthparts of stream animals feeding on epilithic algae. Arch. Hydrobiol. Suppl. *83*:253-354.

Arnold, J.W. (1974) Adaptive features on the tarsi of cockroaches (Insecta: Dictyoptera). Int. J. Insect Morphol. Embryol. *3* :317-334.

Arnquist, G. (1988) Sexual selection in a water strider: the function, nature of selection and heritability of a male grasping apparatus. In: Abstr. and Author Index. Proc. 18th Int. Congr. Entomol. 3-9 July 1988, Vancouver, Vancouver: pp. 24-24.

Arun Rachel, G. (1988) Bursal glands of Reduviidae (Insecta; Heteroptera). Proc. Ind. Acad. Sci. Animal Sci. *97*:211-222.

Arzet von, H.R. (1973) Suchverhalten der Larven von *Chrysopa carnea* Steph. (Neuroptera: Chrysopidae). Z. Angew. Entomol. *74*:64-79.

Asahina, S. (1954) *A morphological study of a relic dragonfly Epiophlebia superstes Selys (Odonata; Anisozygoptera).* Tokyo: The Japan Society of the promotion of Science.

Aspöck, H. and U. Aspöck (1971) Raphidioptera (Kamelhalsfliegen). In M. Beier (ed.): Handbuch der Zoologie, Berlin, New York: W. de Gruyter, pp. 1-48.

Ass, M.J. (1963) Predatory legs in Arthropoda: their origination and function. Zool. Zhurn. *42*:1346-1362.

Austin, A.D. and T.O. Browning (1981) A mechanism for movements of eggs along insect ovipositors. Int. J. Insect Morphol. Embryol. *10*:93-108.

Autumn, K., Y.A. Liang, S.T. Hsieh, W. Zesch, W.P. Chan, T.W. Kenny, R. Fearing, and R.J. Full (2000) Adhesive force of a single gecko foot-hair. Nature *405*:681-685.

Åhman, I. (1990) Plant-surface characteristics and movements of two *Brassica*-feeding aphids, *Lipaphis erysimi* and *Brevicoryne brassicae*. Symp. Biol. Hung. *39*:119-125.

Baehr, M. (1980) Zur Funktionsmorphologie und evolutiven Bedeutung der elytralen Sperrmechanismen der Scaritini (Coleoptera: Carabidae). Ent. Gen. *6*:311-333.

Baker, E.A. (1982) Chemistry and morphology of plant epicuticular waxes. In D.F. Cutler, K.L. Alvin and C.E. Price (eds.): The plant cuticle, London: Academic Press, pp. 139-165.

Baker, G.T., A. Chandrapatya, and H.H.J. Neshitt (1987) Morphology of several types of cuticular suckers in mites (Arachnida, Acarina). Spixiana *10*:131-137.

Balderson, J. (1991) Mantodea (Praying mantids). In CSIRO (ed.): The Insects of Australia, Ithaca, New York: Cornell University Press, pp. 348-359.

Barbier, R. (1985) Morphogenese et evolution de la cuticle et des crochets des fauses-pattes, au cours du developpement larvaire de la *Galleria mellonella* L. (Lepidoptera, Pyralidae). Bull. Soc. Zool. France *110*:205-221.

Barth, F.G. (1970) Die Feinstruktur des Spinneninteguments. II. Die räumliche Anordnung der Mikrofasern in der lamellierten Kutikula und ihre Beziehung zur Gestalt der Porenkanäle (*Cupiennius salei* Keys., adult, häutungsfer, Tarsus). Z. Zellforsch. Mikrosk. Anat. *104*:87-106.

Barth, F.G. (1986) Zur Organisation sensorischer Systeme: die cuticularen Mechanoreceptoren der Arthropoden. Verh. Dtsch. Zool. Ges. *79*:69-90.

Barthlott, W. (1998) Scanning electron microscopy of the epidermal surface in plants. In D. Claugher (ed.): Scanning electron microscopy in taxonomy and functional morphology, Oxford: Clarendon, pp. 69-94.

Barthlott, W. and N. Ehler (1977) Raster-Elektronenmikroskopie der Epidermisoberflächen von Spermatophyten. Trop. Subtrop. Pflanzenwelt *19*:367-467.

Barthlott, W. and C. Neinhuis (1997) Purity of the sacred lotus or escape from contamination in biological surfaces. Planta *202*:1-8.

Barthlott, W. and C. Neinhuis (1998a) Lotusblumen und Autolacke: Ultrastruktur pflanzlicher Grenzflächen und biomimetische unverschmutzbare Werkstoffe. In W. Nachtigall and A. Wisser (eds.): 4. Bionik - Kongress, München 1998, Stuttgart, Jena, Lübeck, Ulm: Gustav Fischer Verlag, pp. 281-293.

Barthlott, W., C. Neinhuis, D. Cutler, F. Ditsch, I. Meusel, I. Theisen, and H. Wilhelmi (1998b) Classification and terminology of plant epicuticular waxes. Bot. J. Linn. Soc. *126*: 237-260.

Barthlott, W. and E. Wollenweber (1981) Zur Feinstruktur, Chemie und taxonomischen Signifikanz epikutikularer Wachse und ähnliche Sekrete. Trop. Subtrop. Pflanzenwelt *32*:7-67.

Bauchhenss, E. (1979a) Die Pulvillen von *Calliphora erythrocephala* Meig. (Diptera, Brachycera) als Adhäsionsorgane. Zoomorphologie *93*:99-123.

Bauchhenss, E. (1979b) *Die Pulvillen von Calliphora erythrocephala Meig. als Adhäsionsorgane*. PhD Theses, München.

Bauchhenss, E. and M. Renner (1977) Pulvillus of *Calliphora erythrocephala* Meig. (Diptera; Calliphoridae). Int. J. Insect Morphol. Embryol. *6*:225-227.

Bauer, A.M. (1998) Morphology of the adhesive tail tips of carphodactyline geckos (Reptilia: Diplodactilidae). J. Morphol. *235*:41-58.

Bauer, A.M. and D.A. Good (1986) Scaling of scansorial surface area in the genus *Gekko*. In Z. Rocek (ed.): Studies in Herpetology, Prague: pp. 363-366.

Bauer, T. and M. Kredler (1988) Adhesive mouthparts in a ground beetle larva (Coleoptera, Carabidae, *Loricera pilicornis* F.) and their function during predation. Zool. Anz. *221*:145-156.

Baur, F. and S.N. Gorb (2001) How the bee releases its leg attachment devices. Biona Report *in press*:

Beament, J.W.L. (1955) Wax secretion in the cockroach. J. Exp. Biol. *32*:514-538.

Beaver, R.A. (1983) The communities living in *Nepenthes* pitcher plants: fauna and food webs. In J.H. Frank and L.P. Lounibos (eds.): Phytotelmata: terrestrial plants as hosts for aquatic insect communities, New Jersey: Plexus Publishing, Inc., pp. 129-159.

Beckemeyer, R. (1997) Functional morphology of tandem flight in Odonata. Argia *9*:13-17.

Beier, M. (1968) Mantodea (Fangheuschrecken). In M. Beier (ed.): Handbuch der Zoologie, Berlin, New York: W. de Gruyter, pp. 1-47.

Beier, M., U. Gerhardt, A. Kästner, and H.G. Vitzthum (1941) Chelicerata. In T. Krumbach (ed.): Handbuch der Zoologie, Berlin: W. de Gruyter.

Belyshev, B.F. (1973) *Dragonflies of Siberia* . Novosibirsk: Nauka (in Russian).

Bennet-Clark, H.C. (1975) The energetics of the jump of the locust *Schistocerca gregaria*. J. Exp. Biol. *63*:53-83.

Bennet-Clark, H.C. and E.C.A. Lucey (1967) The jump of the flea: a study of the energetics and a model of the mechanism. J. Exp. Biol. *47*:59-76.

Berlese, A. (1909) *Gli Insetti* . Milano.

Bertram, D.S. and R.G. Bird (1961) Studies on mosquito-borne viruses in their vectors. I. The normal fine structure of the mid-gut epithelium of adult female of *Aedes aegypti* (L.) and the functional significance of its modification following a blood meal. Trans. Roy. Soc. Trop. Med. Hyg. *55*:404-423.

Betz, O. (1996) Function and evolution of the adhesion-capture apparatus of *Stenus* species (Coleoptera, Staphylinidae). Zoomorphology *116*:15-34.

Betz O. (1998) Comparative studies on the predatory behaviour of *Stenus* spp. (Coleoptera : Staphylinidae): the significance of its specialized labial apparatus. J.Zool., Lond. *244*:527-544.

Beutel, R. and S.N. Gorb (2001) Ultrastructure of attachment specializations of hexapods (Arthropoda): evolutionary patterns inferred from a revised ordinal phylogeny. Journal of Zoological Systematics and Evolutionary Research *in press*.

Beutel, R.G. and F. Haas (2000) Phylogenetic relationships of the suborders of Coleoptera (Insecta). Cladistics *16*:1-39.

Bey-Bienko, G.J. (1964) *Key of insects of the European part of the USSR*. Moscow and Leningrad: Nauka.

Bey-Bienko, G.J. (1966) *General entomology*. Moscow: Vysshaja Shkola.

Bianchi, G. (1995) Plant waxes. In R.J. Hamilton (ed.): Waxes: chemistry, molecular biology and functions, Dundee: The Oily Press, pp. 177-222.

Blunck, H. (1912) Beitrag zur Kenntnis der Morphologie und Physiologie der Haftscheiben von *Dytiscus marginalis* L. Z. Wiss. Zool. *100*:459-492.

Bocharova-Messner, O.M. and A.Z. Dmitriev (1984) Morphological and functional analysis of the wing venation in Odonata according to the data of the scanning electron microscopy. In: IX Congress of All-Union Entomological Society. Abstr. Theses., Kiev: Naukova Dumka, pp. 65-65.

Bodnaryk, R.P. (1992) Leaf epicuticular wax, an antixenotic factor in Brassicaceae that affects the rate and pattern of feeding in flea beetles, *Phyllotreta cruciferae* (Goeze). Can. J. Plant. Sci. *72*:1295-1303.

Bouligand, Y. (1965) Sur une architecture torsadee repandue dans de nombreuses cuticles d'Arthropodes. C. R. Hebd. Seanc. Acad. Sci. Paris *260*:3665-3668.

Bouligand, Y. (1972) Twisted fibrous arrangements in biological materials and cholesteric mesophases. Tiss. Cell. *4*:189-217.

Bowden, F.P. (1957) Adhäsion und Reibung. Endeavour *16*:5-18.

Bowden, F.P. and D. Tabor (1986) *The friction and lubrication of solids*. Oxford: Clarendon Press.

Boyde, A. and S.J. Jones (1996) Scanning electron microscopy of bone: instrument, specimen, and issues. Microsc. Res. Techn. *33*:92-120.

Brainerd, E.L. (1994) Adhesion force of ants on smooth surfaces. Amer. Zool. *34*:128-128.

Braun, M. (1878) Zur Bedeutung der Cuticularborsten auf den Haftlappen der Geckotiden. Arb. Zool. -Zoot. Inst. Würzburg *4* :231-237.

Bregetova, N.G., E.M. Bulanova-Zachvatkina, and V.I. Volgin (1955) *Mites of Rodents of the fauna of USSR*. Moscow and Leningrad: Izdatelstvo AN SSSR.

Breidbach, O. (1980) Die Tarsen von Insekten - ein schones Beispiel von Konvergenz. Mikrokosmos *69*:200-201.

Bremond, J. (1974) Remarques sur le phenomene de convergence des membres prehensiles chez la mante religieuse et le crustace *Squilla mantis*. Entomologiste *30*:183-188.

Briceno, R.D. and W.G. Eberhard (1988) Functional morphology of cerci in male earwigs. In: Proc. 18th Int. Congr. Entomol. Vancouver. July 3-rd-9th. 1988. Abstr. and Author Index, Vancouver: pp. 208-208.

Brodsky, K. (1930) Zur Kenntnis der Wirbellosenfauna der Bergströme Mittelasiens. II. *Deuterophlebia mirabilis* Edw. Z. Morphol. Ökol. Tiere *18*:289-321.

Büttiker, W., H.W. Krenn, and J.F. Putterill (1996) The proboscis of eye-frequenting and piercing Lepidoptera (Insecta). Zoomorphology *116*:77-83.

Calvert, P.P. (1920) The Costa Rican species of *Epigomphus* and their mutual mating adaptions. Transactions of the American Entomological Society *46*:323-354.

Carver, M., G.F. Gross, and T.E. Woodward (1991) Hemiptera. In: The insects of Australia, New York: Cornell University Press, pp. 429-510.

Chu, H., D.M. Norris, and S.D. Carlson (1975) Ultrastructure of the compound eye of the diploid female beetle, *Xyleborus ferrugineus*. Cell Tiss. Res. *165*:23-26.

Clausen, M.B. and A.G. Richards (1951) Studies on arthropod cuticle. 6. Intracellular and extracellular formation of cuticle in the developing posterior spiracular chamber of a blowfly larva (*Phormia regina*). J. Morphol. *89*:199-215.

Common, I.F.B. (1969) A wing locking or stridulatory device in Lepidoptera. J. Austral. Entomol. Soc. *8*:121-125.

Compère, P. and G. Goffinet (1987a) Elaboration and ultrastructural changes in the pore canal system of the mineralized cuticle of *Carcinus maenas* during the moulting cycle. Tiss. Cell. *19*:859-875.

Compère, P. and G. Goffinet (1987b) Ultrastructural shape and three-dimensional organization of the intracuticular canal systems in the mineralized cuticle of the green crab *Carcinus maenas*. Tiss. Cell. *19*:839-857.

Conde-Boytel, R., E.H. Erickson, and S.D. Carlson (1989) Scanning electron microscopy of the honey-bee, *Apis mellifera* L.(Hymenoptera: Apidae) pretarsus. Int. J. Insect Morphol. Embryol. *18*:59-69.

Condé, B. and J. Pagés (1991) Diplura. In CSIRO (ed.): The Insects of Australia, Ithaca, New York: Cornell University Press, pp. 269-271.

Cook, M. (1970) Composition of mussel and barnacle deposits at the attachment interface. In R.S. Manly (ed.): Adhesion in Biological Systems, New York and London: Academic Press, pp. 139-150.

Cooper, M.A. (1987) A remarkable secondary sexual structure of the male of *Anacrabro fritzi* Leclercq (Hymenoptera; Sphecidae). Entomol. Mon. Mag. *123*:223-224.

Costello, W.J. and C.K. Govind (1984) Contractile proteins of fast and slow fibers during differentiation of lobster claw muscle. Dev. Biol. *104*:434-440.

Courtney, G. (1990) Cuticular morphology of larval mountain midges (Diptera, Deuterophlebiidae): implications for the phylogenetic relationships of Nematocera. Can. J. Zool. *68*:556-578.

Cox, S.C. (1989) The electrical and mechanical properties of the proleg retractor muscle of the Chinese oak silkmoth larva. Physiol. Entomol. *14*:265-272.

Crampton, G.C. (1926) A comparison of the neck and prothoracic sclerites throughout the orders of Insects. Trans. Amer. Ent. Soc. *52*:192-248.

Cribb, B.W. and M.K. Jones (1995) Reappraisal of the pore channel system in the grooved pegs of *Aedes aegypti*. Tiss. Cell. *27*:47-53.

Curtis, J.D. and N.R. Lersten (1978) Heterophylly in *Populus grandidentala* (Salicaceae) with emphasis on resin glands and extrafloral nectaries. Am. J. Bot. *65*:1003-1010.

D'Andrea, M. and S. Carfi (1988) Spines on the wing veins in Odonata. 1. Zygoptera. Odonatologica *17*:313-335.

Dalingwater, J.E. (1975a) SEM observations on the cuticle of some decapod crustaceans. Zool. J. Linn. Soc. *56*:327-330.

Dalingwater, J.E. (1975b) The reality of arthropod cuticular laminae. Cell Tiss. Res. *163*:411-413.

Dashman, T. (1953) Terminology of the pretarsus. Ann. Entomol. Soc. Am. *46*:56-62.

Davidson, D.W., R.R. Snelling, and J.T. Longino (1989) Competition among ants for myrmecophytes and the significance of plant trichomes. Biotropica *21*:64-73.

De Meijere, J.C. (1901) Über das letzte Glied der Beine bei den Arthropoden. Zool. Jb. Anat. *14*:417-476.

Dellit, W. (1934) Zur Anatomie und Physiologie der Geckozehe. Jen. Z. Naturwiss. *68*:613-656.

Dethier, V.G. (1963) *The physiology of insect senses*. New York: Wiley.

Dethier, V.G. (1976) *The hungry fly. A physiological study of the behavior associated with feeding*. Cambridge and London: Harvard University Press.

Dewitz, H. (1883) Die Befestigung durch einen klebenden Schleim beim Springen gegen senkrechte Flächen. Zool. Anz. *6*:273-274.

Dewitz, H. (1884) Über die Fortbewegung der Tiere an senkrechten glatten Flächen vermittels eines Secretes. Archiv Ges. Physiol. *33*:440-481.

Dixon, A.F.G., P.C. Croghan, and R.P. Gowing (1990) The mechanism by which aphids adhere to smooth surfaces. J. Exp. Biol. *152*:243-253.

Dlussky, G.M. and E.B. Fedoseeva (1988) Origin and early stages of the evolution of ants (Hymenoptera, Formicidae). In A.P. Rasnitsyn (ed.): Cretaceous biocoenotical crisis and insect evolution, Moscow: Nauka.

Dolynskaja, I.V. (1993) Characteristics of the notodondid moth pupae (Lepidoptera, Notodontidae). Communication 1. Key to genera. Vestn. Zool. *1993*:39-43.

Domagala, P. and H. Chiradella (1984) Structure and function of the terminal abdominal appendages (pygopodia) of the photurid firefly larvae. Biol. Bull. *166*:299-309.

Dougherty, W.J. (1990a) Barnacle adhesion: reattachment of the adult barnacle *Chthamalus fragilis* Darwin to polystyrene surfaces followed by centrifugational shearing. J. Crustac. Biol. *10*:469-478.

Dougherty, W.J. (1990b) SEM observations on the interfacial surface of the cement of the adult barnacle, attached to natural and synthetic adherends. Tiss. Cell. *22*:463-470.

Driscoll, C.A. and M.A. Condon (1994) Labellar modifications of *Blepharoneura* (Diptera: Tephritidae): Neotropical fruit flies that damage and feed on plant surfaces. Ann. Entomol. Soc. Am. *87*:448-453.

Dubinin, V.B. (1951) *Mites of Birds. Fauna of the USSR*. Moscow: Nauka.

Dunkle, S.W. (1991) Head damage from mating attempts in dragonflies (Odonata: Anisoptera). Entomological News *120*:37-41.

Edwards, J.S. and M. Tarkanian (1970) The adhesive pads of Heteroptera: a re-examination. Proc. Roy. Ent. Soc. London A *45*:1-5.

Edwards, P.B. (1982) Do waxes of juvenile *Eucalyptus* leaves provide protection from grazing insects? Aust. J. Ecol. *7*:347-352.

Edwards, P.B. and W.J. Wanjura (1990) Physical attributes of eucalypt leaves and the host range of chrysomelid beetles. Symp. Biol. Hung. *39*:227-236.

Eigenbrode, S.D. (1996) Plant surface waxes and insect behaviour. In G. Kerstiens (ed.): Plant cuticles - an integral functional approach, Oxford: BIOS, pp. 201-222.

Eigenbrode, S.D. and K.E. Espelie (1995) Effects of plant epicuticular lipids on insect herbivores. Ann. Rev. Entomol. *40*:171-194.

Eigenbrode, S.D., S. Moodie, and T. Castagnola (1995) Generalist predators mediate resistance to a phytophagous pest in cabbage with glossy leaf wax. Entomol. Exp. Appl. *77*:335-342.

Eigenbrode, S.D. and N.N. Kabalo (1999) Effects of *Brassica oleracea* waxblooms on predation and attachment by *Hippodomia convergens*. Entomol. Exp. Appl. *91*:125-130.

Eigenbrode, S.D., N.N. Kabalo, and K.A. Stoner (1999) Predation, behavior, and attachment by *Chrysoperla plorabunda* larvae on *Brassica oleracea* with different surface waxblooms. Entomol. Exp. Appl. *90*:225-235.

Eigenbrode, S.D., L. Rayor, J. Chow, and P. Latty (2000) Effects of wax bloom variation in *Brassica oleracea* on foraging by a vespid wasp. Entomol. Exp. Appl. *97*:161-166.

Eisner, T. and D.J. Aneshansley (2000) Defense by foot adhesion in a beetle (*Hemisphaerota cyanea*). Proc Nat Acad USA *97*:6568-6573.

Elzinga, R.J. and A.B. Broce (1986) Labellar modifications of Muscomorpha flies (Diptera). Ann. Entomol. Soc. Am. *79*:150-209.

Emeljanov, A.F. (1987) Phylogeny of Cicadina (Homoptera, Cicadina) according to data on comparative morphology. Trudy VEO *69* :19-109.

Emerson, S.B. (1991) The ecomorphology of Bornean tree frogs (family Rhacophoridae). Zool. J. Linn. Soc. *101*:337-357.

Emerson, S.B. and D. Diehl (1980) Toe pad morphology and mechanisms of sticking in frogs. Biol. J. Linn. Soc. *13*:199-216.

Ernst, K.D. (1969) Die Feinstruktur von Riechsensillen auf der Antenne des Aaskäfers *Necrophorus*. Z. Mikrosk. Anat. *94*:72-102.

Ernst, K.D. (1972) Die Ontogenie der basiconischen Riechsensillen auf der Antenne von *Necrophorus*. Z. Mikrosk. Anat. *129*:217-236.

Ernst, V. (1973a) The digital pads of the tree frog, *Hyla cinerea*. I. The epidermis. Tiss. Cell. *5* :83-96.

Ernst, V. (1973b) The digital pads of the tree frog, *Hyla cinerea*. II. The mucous glands. Tiss. Cell. *5*:96-104.

Ernst, V. and R. Ruibal (1967) The structure and development of the digital lamellae of lizards. J. Morphol. *120*:233-266.

Erri Babu, D. and K. Anger (1987) The structure and modification of integumental tissues in *Pagurus bernhardus* (L.) (Decapoda; Anomura). J. Exp. Mar. Biol. Ecol *112*:267-281.

Evans, M.E.G. (1973) J. Zool. Lond. *169*:181-194.

Federle, W., U. Maschwitz, B. Fiala, M. Riederer, and B. Hölldobler (1997) Slippery ant-plants and skilful climbers: selection and protection of specific ant partners by epicuticular wax blooms in *Macaranga* (Euphorbiaceae). Oecologia *112*:217-224.

Federle, W., K. Rohrseitz, and B. Hölldobler (2000) Attachment forces of ants measured with a centrifuge: better «wax-runners» have a poorer attachment to a smooth surface. J. Exp. Biol. *203*:505-512.

Ferguson, A.H. (1955) The proventriculus of immature Anisoptera with reference to its use in taxonomy. Doct. Diss. StateUniv. Louisiana *15*:

Field, L.H., A. Evens, and D.L. Macmillan (1987) Sound production and stridulatory structures in hermit crabs of the genus *Trizopagurus*. J. Mar. Biol. Assoc. U. K. *67*:89-110.

Field, S.A. and A.D. Austin (1994) Anatomy and mechanics of the telescopic ovipositor system of *Scelio* Latreille (Hymenoptera: Sceleonidae) and related genera. Int. J. Insect Morphol. Embryol. *23*:135-158.

Filshie, B.K. and D.S. Smith (1980) A proposed solution to a fine-structural puzzle: the organization of gill cuticle in a crayfish *(Panulirus)*. Tiss. Cell. *12*:209-226.

Filshie, B.K. and D.F. Waterhouse (1969) The structure and development of a surface pattern on the cuticle of the green vegetable bug *Nezara viridula*. Tiss. Cell. *1*:367-385.

Foelix, R. (1982) *The biology of spiders*. Massachusetts and London: Harvard University Press.

Foelix, R., R. Jackson, A. Henksmeyer, and S. Hallas (1984) Tarsal hairs specialized for prey-capture in the salticid *Portia* . Rev. Arachnol. *5*:329-334.

Fraenkel, G. and K.M. Rudall (1940) A study of the physical and chemical properties of the insect cuticle. Proc. Roy. Soc. London B *129*:1-35.

Fraenkel, G. and K.M. Rudall (1947) The structure of insect cuticles. Proc. Roy. Soc. London B *134*:111-143.

Francouer, A. and R. Loiselle (1988) Evolution du strigile chez les formicides (Hymenopteres). Natur. Can. *115*:333-335.

Frantsevich, L.I. (1998) The coxal articulation of the striking leg: a comparative study. J. Morphol. *236*:127-138.

Frantsevich, L.I. and L.A. Frantsevich (1999) The midcoxal prong in flies (Diptera: Brachycera). Ent. Gen. *23*:233-250.

Frantsevich, L.I. and S.N. Gorb (1998) The probable purpose of the mid-coxal prong in Brachycera (Diptera). Naturwissenschaften *85*:31-33.

Fraser, F.C. (1957) A reclassification of the order Odonata. Roy. Zool. Soc. N. S. W. 1-133.

Frazier, S.F., G.S. Larsen, D. Neff, L. Quimby, M. Carney, R.A. DiCaprio, and S.N. Zill (1999) Elasticity and movements of the cockroach tarsus in walking. J. Comp. Physiol. A *185*:157-172.

Freeland, J., R.H. Crotizier, and J. Marc (1982) On the occurence of arolia in ant feet. J. Austral. Entomol. Soc. *21*:257-262.

Furth, D.G. and K. Suzuki (1990) Comparative morphology of the tibial flexor and extensor tendons in insects. Syst. Entomol. *15*:443-441.

Furth, D.G., W. Traub, and I. Harpaz (1983) What makes *Blepharida* jump? A structural study of the metafemoral spring of a flea beetle. J. Exp. Biol. *227*:43-47.

Gannon, A.J., C.E. Bach, and G. Walker (1994) Feeding patterns and attachment ability of *Altica subplicata* (Coleoptera: Chrysomelidae) on sand-dune willow. Great Lakes Entomologist *27*:89-101.

Gasc, J.-P., S. Renous, and A. Diop (1982) Structure microscopique de l'epiderme palmaire du Saurien *Coleodactylus amazonicus* (Andersson, 1918) (Sphaerodactylinae), comparee a celle de l'epiderme des feuilles de la litiere, substrat locomoteur de l'animal. C. R. Acad. Sc. Paris *294*:169-174.

Ghasi-Bayat, A. (1979) Zur Oberflächenstruktur der tarsalen Haftlappen von *Coreus marginatus* (L.) (Coreidae,Heteroptera). Zool. Anz. *203*:345-347.

Ghasi-Bayat, A. and I. Hasenfuss (1980a) Die Oberflächenstrukturen der Prätarsus von *Elasmucha ferrugata* (Fabricius) (Acanthosomatidae, Heteroptera). Zool. Anz. *205*:76-80.

Ghasi-Bayat, A. and I. Hasenfuss (1980b) Über den Transportweg der Haftflüssigkeit der Pulvilli bei *Coptosoma scutellatum* (Geoffr.) (Plataspididae, Heteroptera). Nachrichtenblatt der Bayerischen Entomologen *1980*:58-58.

Ghasi-Bayat, A. and I. Hasenfuss (1980c) Zur Herkunft der Adhäsionsflüssigkeit der Tarsalen Haftlappen bei den Pentatomidae (Heteroptera). Zool. Anz. *204*:13-18.

Ghiradella, H. (1989) Structure and development of iridescent butterfly scales: lattices and laminae. J. Morphol. *202*:69-88.

Ghiradella, H. (1994) Structure of butterfly scales: patterning in an insect cuticle. Microsc. Res. Techn. *27*:429-438.

Gibbs, D.G. (1967) The proventriculus of some trichopterous larvae. J. Zool. ,Lond. *152*:245-256.

Gilbert, C., E. Bauer, and M. Burger (1995) Proprioceptive reflex loop controlling head posture about the roll axis in flies. In: Abstr. Thes. 4th ICN, Cambridge: pp. 185-185.

Gilby, A.R. (1980) Chemical methods (Lipids). In T.A. Miller (ed.): Cuticle techniques in Arthropods, New York, Heidelberg and Berlin: Springer Verlag, pp. 217-252.

Giles, E.T. (1963) The comparative external morphology and affinities of the Dermaptera. Trans. R. Ent. Soc. Lond. *115*:95-164.

Gillett, J.D. and V.B. Wigglesworth (1932) The climbing organ of an insect, *Rhodnius prolixus* (Hemiptera, Reduviidae). Proc. Roy. Soc. London B *111*:364-376.

Goel, S.C. (1972) Notes on the structure of the unguitractor plate in Heteroptera (Hemiptera). J. Entomol. *46*:167-173.

Gogala, M. (1984) Vibration producing structures and songs of terrestrial Heteroptera as systematic character. Biol. Vestn. *32*:19-36.

Gorb, S.N. (1989) Functional morphology of the arrester-system in Odonata. Vestn. Zool. *1989*:62-67.

Gorb, S.N. (1990a) Inner morphology of the arrester-system in the damselfly *Erythromma najas* Hansemann (Zygoptera, Coenagrionidae). Vestn. Zool. *1990*:59-62.

Gorb, S.N. (1990b) Microsculpture of the head-fixation system of dragonflies in scanning electron microscope. Zool. Zhurn. *69*:148-154.

Gorb, S.N. (1990c) Morphology of the head-fixation system in calopterygoid damselflies (Odonata, Zygoptera, Calopterygoidea). Zool. Zhurn. *69*:37-45.

Gorb, S.N. (1991a) Flight reflexes connected to the head-fixation system in dragonflies (Odonata). Zh. Evol. Bioch. Physiol. *27*:46-51.

Gorb, S.N. (1991b) Sensilla in the odonate arrester system (Odonata). Vestn. Zool. *1991*:61-68.

Gorb, S.N. (1993) The skeleton-muscle organization of the head fixation system in odonates and its evolutionary implications: a comparative study. Ann. J. SGSPO *1*:3-20.

Gorb, S.N. (1994) Central projections of ovipositor sense organs in the damselfly *Sympecma annulata* (Zygoptera, Lestidae). J. Morphol. *220*:139-147.

Gorb, S.N. (1995a) Design of the predatory legs of water bugs (Hemiptera: Nepidae, Naucoridae, Notonectidae, Gerridae). J. Morphol. *223*:289-302.

Gorb, S.N. (1995b) Scanning electron microscopy of pruinosity in Odonata. Odonatologica *24*:225-228.

Gorb, S.N. (1995c) The dragonfly head arrester: A sensory organ connected to flight reflexes. In: Proc. 23rd Göttingen Neurobiology Conference, Stuttgart, New York: Georg Thieme Verlag, pp. 193-193.

Gorb, S.N. (1996a) Design of insect unguitractor apparatus. J. Morphol. *230*:219-230.

Gorb, S.N. (1996b) Initial stage of tandem contact in *Platycnemis pennipes* Pall. (Zygoptera; Platycnemididae). Odonatologica *25*:371-376.

Gorb, S.N. (1997a) Armored cuticular membranes in Brachycera (Insecta, Diptera). J. Morphol. *234*:213-222.

Gorb, S.N. (1997b) Porous channels in the cuticle of the head-arrester system in dragon/damselflies (Insecta: Odonata). Microsc. Res. Techn. *37*:583-591.

Gorb, S.N. (1997c) The jumping mechanism of cicada *Cercopis vulnerata*: skeletal structures, musculature, frictional surfaces and kinematic model of leg movements. In R. Blickhan (ed.): Abstract Papers I. International Conference on Motion Systems, Jena: Innovationskolleg Bewegungssysteme, pp. 121-122.

Gorb, S.N. (1997d) Ultrastructural architecture of the microtrichia of the insect cuticle. J. Morphol. *234*:1-10.

Gorb, S.N. (1998a) Frictional surfaces of the elytra to body arresting mechanism in tenebrionid beetles (Coleoptera: Tenebrionidae): design of co-opted fields of microtrichia and cuticle ultrastructure. Int. J. Insect Morphol. Embryol. *27*:205-225.

Gorb, S.N. (1998b) Functional morphology of the head-arrester system in Odonata. Zoologica *148*:1-132.

Gorb, S.N. (1998c) Reibungssysteme bei Insekten. In W. Nachtigall and A. Wisser (eds.): Technische Biologie und Bionik. 4. Bionik - Kongress, München 1998, Stuttgart, Jena, Lübeck, Ulm: Gustav Fisher Verlag, pp. 185-189.

Gorb, S.N. (1998d) The design of the fly adhesive pad: distal tenent setae are adapted to the delivery of an adhesive secretion. Proc. Roy. Soc. London B *265*:747-752.

Gorb, S.N. (1999a) Evolution of the dragonfly head-arresting system. Proc. Roy. Soc. London B *266*:525-535.

Gorb, S.N. (1999b) Ultrastructure of the thoracic dorso-medial field (TDM) in the elytra-to-body arresting mechanism in tenebrionid beetles (Coleoptera: Tenebrionidae). J. Morphol. *240*:101-113.

Gorb, S.N. (2000) Ultrastructure of the neck membrane in dragonflies (Odonata). J. Zool. Lond. *250*:479-494.

Gorb, S.N. and F.G. Barth (1996) A new mechanosensory organ on the anterior spinnerets of the spider *Cupiennius salei* (Araneae, Ctenidae). Zoomorphology *116*:7-14.

Gorb, S.N., Gorb, E.V. and V. Kastner (2001) Scale effects on the attachment pads and friction forces in syrphid flies (Diptera, Syrphidae). J. Exp. Biol. *204*:1421-1431.

Gorb, S.N., Y. Jiao, and M. Scherge (2000) Ultrastructural architecture and mechanical properties of attachment pads in *Tettigonia viridissima* (Orthoptera Tettigoniidae). J. Comp. Physiol. A *186*:821-831.

Gorb, S.N. and M. Scherge (2000) Biological microtribology: anisotropy in frictional forces of orthopteran attachment pads reflects the ultrastructure of a highly deformable material. Proc. Roy. Soc. London B *267*:1239-1244.

Gracham-Smith, G.S. (1930) Further observations on the anatomy and function of the proboscis of the blow-fly, *Calliphora erythrocephala* L. Parasitology *22*:47-115.

Grassé, P.-P. (1949) Insectes: Paléontologie, Géonémie, Aptérygotes, Ephéméroptères, Odonaptères, Battoptéroides, Orthoptéroides, Dermaptéroides, Coléotères. In: Traité de Zoologie: Anatomie, Systématique, Biologie., Paris: Masson,

Green, D.M. (1981) Adhesion and the toe-pads of tree frogs. Copeia *1981*:790-796.

Green, D.M. and J. Carson (1988) The adhesion of tree frog toe-pads to glass: cryogenic examination of a capillary adhesion system. J. Nat. Hist. *22*:131-135.

Grodnicky, D.L. (1988) Structure and function of the scale coverage of the wings in butterflies (Lepidoptera/Hesperioidea, Papilionoidea). Entomol. Rev. *67*:251-256.

Grodnicky, D.L. and M.V. Kozlov (1989) Structural and functional organization of the scale coverage of the wings in butterflies (Papilionida- Lepidoptera). Uspeh. Sovrem. Biol. *107*:446-457.

Gronenberg, W. (1996a) Fast actions in small animals: springs and click mechanisms. J. Comp. Physiol. A *178*:727-734.

Gronenberg, W. (1996b) The trap-yaw mechanism in the dacetine ants *Daceton armigerum* and *strumigenys* sp. J. Exp. Biol. *199*:2021-2033.

Günther, K. and K. Herter (1974) Dermaptera (Ohrwürmer). In M. Beier (ed.): Handbuch der Zoologie, Berlin, New York: W. de Gruyter, pp. 1-158.

Haas, F. (1995) The phylogeny of the Forficulina, a suborder of the Dermaptera. Syst. Entomol. *20*:85-98.

Haas, F., S.N. Gorb, and R.J. Wootton (2000) Elastic joints in dermapteran hind wings materials and wing folding. Arthropod Structure and Development *29*:135-143.

Haas, F. and R.J. Wootton (1996) Two basic mechanisms in insect wing folding. Proc. Roy. Soc. London B *263*:1651-1658.

Hackman, R.H. (1975) Expanding abdominal cuticle in the bug *Rhodnius* and in the tick *Boophilus*. J. Insect Physiol. *21*:1613-1623.

Hackman, R.H. and M. Goldberg (1987) Comparative study of some expanding arthropod cuticles: the relation between composition, structure and function. J. Insect Physiol. *33*:39-50.

Halcrow, K. and C.V.L. Powell (1992) Ultrastructural diversity in the pore canal systems of amphipod crustaceans. Tiss. Cell. *24*:417-436.

Hammond, P.M. (1989) Wing-folding mechanism of beetles, with special reference to investigations of adephagan phylogeny (Coleoptera). In T. Ervin, G.E. Ball and D.R. Whitehead (eds.): Carabid beetles: their evolution, natural history, and classification, Boston, London: Junk Publishers: The Hague, pp. 113-180.

Hanna, G. and W.-J.P. Barnes (1990) Adhesion and detachment of the toe pads of tree frogs. J. Exp. Biol. *155*:103-125.

Harley, R. (1991) The greasy pole syndrome. In C.R. Huxley and D.E. Cutler (eds.): Ant-plant interactions, Oxford: Oxford University Press, pp. 430-433.

Hasenfuss, I. (1977a) Die Herkunft der Adhäsionsflüssigkeit bei Insekten. Zoomorphology *87*:51-64.

Hasenfuss, I. (1977b) Woher stammt die Adhesionsflüssigkeit, die bei Insekten ein Haften an glatten Flächen ermöglicht? Verh. Dtsch. Zool. Ges. *277*:277-277.

Hasenfuss, I. (1978) Über das Haften von Insekten an glatten Flächen - Herkunft der Adhäsionsflüssigkeit. Zool. Jb. Anat. *99*:115-116.

Hasenfuss, I. (1999) The adhesive devices in larvae of Lepidoptera (Insecta, Pterygota). Zoomorphology *119*:143-162.

Heckmann, C.W. (1983) Comparative morphology of arthropod exterior surfaces with capability of binding a film of air underwater. Int. Rev. Ges. Hydrobiol. *68*:715-736.

Heie, O.E. (1987) Morphological structure and adaptations. In: Aphids: biology, natural enemies and control, Amsterdam et al.: pp. 393-400.

Heilig, S. and K. Sander (1986) Zahnradsektoren zur Koordination der Sprungbeine - eine larvale Synapomorphie der fulgoromorphen Zikaden (Homoptera, Cicadina, Fulgoroidea). Zool. Jb. Syst. *113*:307-317.

Heinzeller, J., P. Seifert, and B. Aschauer (1989) Architektur des Unguitraktor-Apparatus bei *Chironomus thummi* (Diptera). Verh. Dtsch. Zool. Ges. *82*:260-260.

Heming, B.S. (1970) Functional morphology of the thysanopteran pretarsus. Can. J. Zool. *49*:91-108.

Heming, B.S. (1972) Functional morphology of the pretarsus in larval Thysanoptera. Can. J. Zool. *50*:751-766.

Heming, B.S. (1973) Metamorphosis of the pretarsus in *Frankliniella fusca* (Hinds) (Thripidae) and *Haplothrips verbasci* (Osborn) (Phlaeothripidae) (Thysanoptera). Can. J. Zool. *51*:1211-1234.

Hengstenberg, R. (1991) Gaze control in the blowfly *Calliphora:* a multisensory, two-stage integration process. The Neurosciences *3*:19-29.

Hennemann, O.-D. (2000) Kleben von Kunststoffen. Anwendung, Ausbildung, Trend. Kunststoffe *90*:184-188.

Hennesey, R.D. (1981) At-rest setal wing coupling and restraining mechanisms in the Encyrtidae and Aphelinidae (Hymenoptera; Chalcidoidea). Ann. Entomol. Soc. Am. *74*:172-176.

Hennig, W. (1973) Diptera (Zweiflügler). In M. Beier (ed.): Handbuch der Zoologie, Berlin, New York: W. de Gruyter, pp. 1-335.

Henning, B. (1974) Morphologie und Histologie der Tarsen von *Tettigonia viridissima* L. (Orthoptera, Ensifera). Z. Morphol. Tiere *79*:323-342.

Hepburn, H.R. (1969) The proventriculus of Mecoptera. J. Georgia Entomol. Soc. *4*:159-167.

Hepburn, H.R. (1971) Proboscis extension and recoil in Lepidoptera. J. Insect Physiol. *17*:637-656.

Hepburn, H.R. (1976) *The insect integument.* Amsterdam: Elsevier.

Hepburn, H.R. (1985) Structure of the integument. In G.A. Kerkut and L.I. Gilbert (eds.): Comprehensive insect physiology biochemistry and pharmacology, Oxford et al.: Pergamon Press, pp. 1-58.

Hepburn, H.R. and A. Ball (1973) On the structure and mechanical properties of beetle shells. J. Mat. Sci. *8*:618-623.

Hepburn, H.R. and H.D. Chandler (1976) Material properties of arthropod cuticles: the arthrodial membranes. J. Comp. Physiol. A *109*:177-198.

Hepburn, H.R. and H.D. Chandler (1978) Tensile mechanical properties and transconformational changes of chitins. In R.A. Muzzarelli and E.R. Parisier (eds.): Proceedings of the first international conference on chitin/chitosan, Cambridge, Mass.: Massachusetts Institute of Technology, pp. 124-143.

Hepburn, H.R. and I. Joffe (1976) Locust solid cuticle - a time sequence of mechanical properties. J. Insect Physiol. *20*:497-506.

Hermann, H.R., M.A. Mullen, and I.B. Wallace (1975) Suction discs of *Blepharocera separata*. J. Georgia Entomol. Soc. *10*:145-150.

Hertz, H. (1881) Über den Kontakt elastischer Körper. J. Reine Angew. Math. *92*:156-156.

Hesse, M. (1981) Auf welche Weise transportieren Insecten den Blütenstaub? Linz. Biol. Beitr. *13*:50-50.

Heymer, A. (1975) Der stammesgeschichtliche Aussagewert der Pterostigma-Oberflächenstruktur bei Odonaten. Z. Zool. Syst. Evolutionsforsch. *13*:81-91.

Highley, K., M. Coomey, M. DenBeste, and L. Wolfram (1977) Frictional properties of skin. J. Invest. Dermatol. *69*:303-305.

Hill, D.E. (1978) Function of the pretarsus in living *Phidippus regius*. Peckhamia *1*:70-71.

Hiller, U. (1968) Untersuchungen zum Feinbau und zur Funktion der Haftborsten von Reptilien. Z. Morphol. Tiere *62*:307-362.

Hiller, U. and R. Blaschke (1967) Zum Haftproblem der Gecko-Füsse. Naturwissenschaften *54*:344-345.

Hilton, D.F.J. (1983) Mating isolation in two species of *Nehalennia* (Zygoptera; Coenagrionidae). Odonatologica *12*:375-379.

Hinton, H.E. (1970) Some little-known surface structures. In A.C. Neville (ed.): Insect ultrastructure, Oxford and Edinburgh: Blackwell Scientific Publishers, pp. 41-58.

Hinton, H.E. (1976) The fine structure of the pupal plastron of simulid flies. J. Insect Physiol. *22*:1061-1070.

Hintzpeter, U. and T. Bauer (1998) The antennal setal traps of the Ground beetle *Loricera pilicornis*: a specialisation for feeding on Collembola. J. Zool. ,Lond. *208*:605-630.

Hlavac, T.F. (1974) *Merope tuber* (Mecoptera): a wing-body interlocking mechanism. Psyche *81*:303-306.

Hofender, H. (1927) Über die larven der Blepharoceriden und ihren merkwürdigen Anheftungsapparat. Verh. Zool. -Botan. Ges. Wien *77*:82-98.

Holloway, P.J. (1969) Chemistry of leaf waxes in relation to wetting. J. Sci. Food Agric. *20*:124-128.

Holway, T.R. (1935) Preliminary note on the structure of the pretarsus and its possible phylogenetic significance. Psyche *42* :1-124.

Homann, H. (1957) Haften Spinnen an einer Wasserhaut? Naturwissenschaften *44*:318-319.

Hölldobler, B. and E.O. Wilson (1990) *The Ants*. Cambridge Massachusetts: Harvard University Press.

Hulskamp, M. and A. Schnittger (1998) Spatial regulation of trichome formation in *Arabidopsis thaliana*. Semin. Cell Dev. Biol. *9*:213-220.

Hulskamp, M., A. Schnittger, and U. Folkers (1999) Pattern formation and cell differentiation: Trichomes in *Arabidopsis* as a genetic model system. Int. Rev. Cytol. *186*:147-178.

Hunter, P. and R.M.T. Rosario (1988) Associations of Mesostigmata with other arthropods. Ann. Rev. Entomol. *33*:393-417.

Hyder, D.E. and C.Y. Oseto (1989) Structure of the stridulatory apparatus and analysis of the sound produced by *Smicronyx fulvus* and *Smicronyx sordidus* (Coleoptera, Curculionidae, Erirrhininae, Smicronychini). J. Morphol. *201*:69-84.

Ilychev, V.D. (1987) *Biological damages*. Moscow: Vysshaja Shkola.

Irschick, D.J., C.C. Austin, K. Petren, R.N. Fisher, J.B. Losos, and O. Ellers (1996) A comparative analysis of clinging ability among pad-bearing lizards. Biol. J. Linn. Soc. *59*:21-35.

Ishii, S. (1987) Adhesion of a leaf feeding ladybird *Epilachna vigintioctomaculata* (Coleoptera: Coccinellidae) on a vertically smooth surface. Appl. Ent. Zool. *22*:222-228.

Israelachvili, J. (1992) *Intermolecular & Surface Forces*. London: Academic Press.

Janitschek, H. (1970) Protura (Beintastler). In M. Beier (ed.): Handbuch der Zoologie, Berlin, New York: W. de Gruyter, pp. 1-72.

Jeffree, C.E. (1986) The cuticle, epicuticular waxes and trichomes of plants, with references to their structure, functions and evolution. In B.E. Juniper and T.R.E. Southwood (eds.): Insects and the plant surface, London: Edward Arnold, pp. 23-64.

Jelinski, M. (1988) An attempt to estimate the degree of arolium adherence of female *Varroa jacobsoni* Oudemans, 1904. Przeglad Zoologiczny *32*:95-97.

Jensen, M. and T. Weis-Fogh (1962) Biology and physics of locust flight. V. Strength and elasticity of locust cuticle. Phil. Trans. Roy. Soc. London B *245*:137-169.

Jiao, Y., S.N. Gorb, and M. Scherge (2000) Adhesion measured on the attachment pads of *Tettigonia viridissima* (Orthoptera, Insecta). J. Exp. Biol. *203*:1887-1895.

Joffe, I., H.R. Hepburn, and S.O. Andersen (1975) On the mechanical properties of the *Limulus* solid cuticle. J. Comp. Physiol. A *101*:147-160.

Johnson, B. and P. Talbot (1987) Ultrastructural analysis of the pleopod tegumental glands in male and female lobsters, *Homarus americanus*. J. Crustac. Biol. *2*:288-301.

Juniper, B.E. (1995) Waxes on plant surfaces and their interactions with insects. In R.J. Hamilton (ed.): Waxes: chemistry, molecular biology and functions, Dundee: Oily, West Ferry, pp. 157-174.

Juniper, B.E. and J.K. Burras (1962) How pitcher plant trap insects. New Scientist *269*:75-77.

Juniper, B.E., R.J. Robins, and D.M. Joel (1989) *The carnivorous plants*. London: Academic Press.

Kabata, Z. (1968) Copepoda parasitica on Australian fishes. V11. *Shiinoa occlusa* gen. et sp. nov. J. Nat. Hist. *2*:497-504.

Kabata, Z. (1979) *Parasitic Copepoda of British fishes*. London: The Royal Society.

Kabata, Z. and B. Cousens (1972) The structure of the attachment organ of Lernaeopodidae (Crustacea; Copepoda). J. Fish. Res. Bd. Canada *29*:1015-1023.

Kaltenbach, A. (1978) Mecoptera (Schnabelhafte, Schnabelfliegen). In M. Beier (ed.): Handbuch der Zoologie, Berlin, New York: W. de Gruyter, pp. 1-111.

Karasev, V.P. (1989) Scale coverage of the curculionid beetles of the genus *Tichius* Germar (Coleoptera, Curculionidae). In: Dynamics of zoocoenozes and animal conservation in Belorussia, Minsk: pp. 85-85.

Kareiva, P. and R. Sahakian (1990) Tritrophic effects of a simple architectural mutation in pea plants. Nature *345*:433-434.

Karner, M. (1997) Bifunctional muscles and interlocking mechanism in *Cantharis:* how not to move the wings. In R. Blickhan (ed.): Abstract Papers II. Biomechanic Workshop of the Studygroup Morphology, Jena: Innovationskolleg Bewegungssysteme, pp. 191-192.

Kendall, U.D. (1970) The anatomy of the tarsi of *Schistocerca gregaria* Forskål. Z. Zellforsch. *109*:112-137.

Kennedy, C.E. (1986) Attachment may be a basis for specialization in oak aphids. J. Ecol. Entomol. *11*:291-300.

Kevan, P.G. and M.A. Lane (1985) Flower petal microtexture is a tactile cue for bees. Proc. Nat. Acad. Sci. USA *85*:4750-4752.

Kiauta, B. (1968) Evolution of the chromosome complement in Odonata. Entomologische Berichte *28*:97-100.

Kinzelbach, R.K. (1971) Strepsiptera (Fächerflügler). In M. Beier (ed.): Handbuch der Zoologie, Berlin, New York: W. de Gruyter, pp. 1-73.

Knitter, H. (1998) *In situ* fixation of intact tandem linkages of Zygoptera in the field. Odonatologica *27*:383-390.

Knoll, F. (1914) Über die Ursache des Ausgleitens der Insektenbeine an wachsbedekten Pflanzenteilen. Jahrb. Wiss. Bot. *54* :448-497.

Kosaki, A. and R. Yamaoka (1996) Chemical composition of footprints and cuticula lipids of three species of lady beetles. Jpn. J. Appl. Entomol. Zool. *40*:47-53.

Kovoor, J. and H.M. Peters (1988) The spinning apparatus of *Polenecia producta* (Araneae,U loboridae): structure and histochemistry. Zoomorphology *108*:47-59.

Kozlov, M.V. (1988) Palaeontology of butterflies and problems in Papilionida order phylogeny. In A.G. Ponomarenko (ed.): Cretaceous biocoenotic crisis and insect evolution, Moscow: Nauka, pp. 16-69.

Kölsch, G. and O. Betz (1998) Ultrastructure and function of the adhesion-capture apparatus of *Stenus* species (Coleoptera, Staphylinidae). Zoomorphology *118*:263-272.

Krenn, H.W. (1990) Functional morphology and movements of the proboscis of Lepidoptera (Insecta). Zoomorphology *110*:105-114.

Kullberg, A. (1988) The case, mouthparts, silk and silk formation of *Rheotanytarsus muscicola* Kieffer (Chironomidae; Tanytarsini). Aquatic Insects *10*:249-255.

Lacombe, D. and V.R. Liguori (1969) Comparative histological studies of the cement apparatus of *Lepas anatifera* and *Balanus tintinnabulum*. Biol. Bull. *137*:170-180.

Lee, Y.I., M. Kogan, and J.R.J. Larsen (1986) Attachment of the potato leaf-hopper to soybean plant surfaces as affected by morphology of the pretarsus. Entomol. Exp. Appl. *42*:101-107.

Lees, A.D. and C.H. Waddington (1942) The development of bristles in normal and some mutant types of *Drosophila melanogaster* . Proc. Roy. Soc. London B *131*:87-110.

Lees, A.M. and J. Hardie (1988) The organs of adhesion in the aphid *Megoura viciae*. J. Exp. Biol. *136*:209-228.

Levin, D.A. (1973) The role of trichomes in plant defence. Q. Rev. Biol. *48*:3-15.

Ligouri, V.R. (1970) Histological and histoenzymological characteristics of the cement apparatus in barnacles. In R.S. Manly (ed.): Adhesion in Biological Systems, New York and London: Academic Press, pp. 123-137.

Linnenbach, M. (1985) Zum Feinbau der Haftscheiben von *Hyla cinerea* (Schneider, 1799) (Salientia: Hylidae). Salamandra *21*:81-85.

Liske, E. (1982) Proprioceptive control of head position and head movement in the praying mantis. Naturwissenschaften *69*:452-452.

Liske, E. (1984) Head movements of the praying mantis with particular reference to visual and proprioceptive information. Int. Congr. Entomol. Proc. *17*:195.

Liske, E. (1989) Neck hair plate sensilla of the praying mantis: central projections of the afferent neurones and their physiological responses to imposed head movement in the yaw plane. J. Insect Physiol. *35*:677-687.
Liske, E. and W.J. Davis (1986) Behavioral suppression of head grooming in the male praying mantis during mating. Naturwissenschaften *73*:333-334.
Liske, E. and W. Mohren (1984) Saccadic head movements of the praying mantis, with particular reference to visual and proprioceptive information. Physiol. Entomol. *9*:29-38.
Locke, M. (1960a) The cuticle and wax secretion in *Calpodes ethlius* (Lepidoptera, Hesperidae). Quart. J. Mic. Sci. *101*:333-338.
Locke, M. (1960b) The cuticular pattern in an insect - the intersegmental membranes. J. Exp. Biol. *37*:398-406.
Locke, M. (1961) Pore channels and related structures in insect cuticle. J. Biophys. Biochem. Cytol. *10*:589-618.
Locke, M. (1964) The structure and formation of the integument in insects. In M. Rockstein (ed.): The Physiology of Insecta, New York: Academic Press, pp. 123-213.
Locke, M. (1965) Permeability of insect cuticle to water and lipids. Science *147*:295-298.
Locke, M. (1966) The structure and formation of the cuticulin layer in the epicuticle of an insect, *Calpodes ethlius* (Lepidoptera, Hesperiidae). J. Morphol. *118*:461-494.
Locke, M. (1976) The role of plasma membrane plaques and Golgi complex vesicles in cuticle deposition during the moult/intermoult cycle. In H.R. Hepburn (ed.): The insect integument, Amsterdam: Elsevier, pp. 237-258.
Lopatin, I.K. (1980) *Zoogeography*. Minsk: Vysshaja shkola.
Loxton, R.G. and I. Nicholls (1979) The functional morphology of the praying mantis forelimb (Dictyoptera: Mantodea). Zool. J. Linn. Soc. *66*:185-203.
Luckow, M. and A. Bruneau (1997) Circularity and independence in phylogenetic tests of ecological hypotheses. Cladistics *13*:145-152.
Maderson, P.F.A. (1964) Keratinized epidermal derivatives as an aid to climbing in gekkonid lizards. Nature *203*:780-781.
Maki, T. (1936) Studies on the skeletal structure, musculature and nervous system of the alder fly *Chauliodes formosanus* Petersen. Mem. Fac. Sci. Agric. Taihoku Imp. Univ. *16*:117-243.
Malicky, H. (1999) Trichoptera (Köcherfliegen). In M. Beier (ed.): Handbuch der Zoologie, Berlin, New-York: W. de Gruyter, pp. 1-114.
Mallach, N. (1936) Über Bau und Funktion der Cenchri am Tenthredinidenthorax. Berlin. Inaugural Dissertation, pp. 1-59.
Markl, H. (1962) Borstenfelden an der Gelenken als Schwersinnesorgane bei Ameisen und anderen Hymenopteren. Z. Vergl. Physiol. *45*:475-569.
Markstädler, C., W. Federle, R. Jetter, M. Riederer, and B. Hölldobler (2000) Chemical composition of the slippery epicuticular wax blooms on *Macaranga* (Euphorbiaceae) ant-plants. Chemoecology *10*: 33-40.
Martin J.T. and B.E. Juniper (1970) *The cuticle of plants*. London: Edward Arnold.
Martynov, A.V. (1924) Wing venation and tracheation in Odonata and Ephemeroptera. Russ. Ent. Obozr. *18*:145-174.
Mathew, G. (1984) Occurrence of an unusual type of frenulum in *Pycnarmon caberalis* Guen (Lepidoptera, Pyraustidae). Entomon *9*:223-223.
Matthes, D. (1970) Der Meloide *Cerocoma schafferi* (L.). Entomologische Blätter für Biologie und Systematik der Käfer *66*:33-40.
Matushkina, N.A. and S.N. Gorb (1997) Skeleton-muscle organisation of the endophytic ovipositor in Odonata. Vestn. Zool. *31*:57-70.

McCafferty, W.P. and D.W. Bloodgood (1989) The female and male coupling apparatus in *Tortopus* mayflies. Aquatic Insects *11*:141-146.

Memmi, M.P. and A.S. Elfimov (1987) Functional morphology of the cuticle in crustaceans of the genus *Dosima* (Cirripedia, Lepadomorpha). Vid. Medd. Dan. Naturalist Foren. *141*:81-96.

Messner, B. (1983) Die Georyssiden-Imagines (Insecta, Coleoptera) kleben ihre tarnende Erdkruste mit einem Drusensekret fest. Entomologische Nachrichten und Berichte *27*:233-234.

Messner, B. (1986a) Die Plastronatmung bei Insekten. Wiss. Z. E. M. Arndt-Univ. Greifswald. Math Naturwiss. R. *35*:30-34.

Messner, B. (1986b) Zur Plastronatmung bei Blepharoceriden - Puppen der Gattung *Liponeura* (Insecta, Diptera). Zool. Jb. Anat. *114*:503-510.

Messner, B. (1988) Vorschlag fur die Neufassung des Begriffes 'plastron' bei den Arthropoden. Deutsche Entomologische Zeitschrift *35*:379-389.

Messner, B. and J. Adis (1987) Die Plastronstrukturen der submers lebenden neotropischen Barenraupe von *Palustra laboulbeni* Bar, 1873 (Lepidoptera: Arctiidae). Zool. Jb. Anat. *115*:531-538.

Messner, B. and J. Adis (1988) Die Plastronstrukturen der bisher einzigen sumbers lebenden Diplopodenart *Gonographis adisi* Hoffman 1985 (Pyrgodesmidae, Diplopoda). Zool. Jb. Anat. *117*:277-290.

Messner, B. and L. Dieckmann (1987) Die Plastronstukturen des submers an Schilf lebenden Russelkafers *Dicranthus elegans* (Fabricius, 1801) (Coleoptera, Curculionidae). Zool. Jb. Anat. *115*:115-125.

Messner, B., I. Groth, U. Messner, and T. Geisel (1987) Die Plastronstrukturen der larve, der puppe und des submers lebenden Weibchens von *Acentria nivea* (Olivier, 1791) (Lepidoptera, Pyralidae). Zool. Jb. Anat. *115*:163-180.

Mickoleit, G. (1971) Das Exoskelet von *Notiothauma reedi* McLachlan. Ein Beitrag zur Morphologie und Phylogenie der Mecoptera (Insecta). Z. Morphol. Tiere *69*:318-362.

Mickoleit, G. (1973a) Über den Ovipositor der Neuropteroidea und Coleoptera und seine phylogenetische Bedeutung (Insecta, Holometabola). Z. Morphol. Tiere *74*:37-64.

Mickoleit, G. (1973b) Zur Anatomie und Funktion des Raphidiopteren-Ovipositors (Insecta, Neuropteroidea). Z. Morphol. Tiere *76*:145-171.

Miller, G.L., P.R. Miller, and A.R. Brady (1988) Adhesive hairs in lycosid spiders of various life styles, including the occurrence of claw tufts in *Lycosa hentzi* Banks. Bull. Br. Arachnol. Soc. *7*:213-216.

Mitchell, R. (1962) The structure and evolution of water mite mouthparts. J. Morphol. *100*:41-49.

Mittelstaedt, H. (1950) Physiologie des Gleichgewichtsinnes bei fliegenden Libellen. Z. Vergl. Physiol. *32*:442-463.

Morgan, S.G. (1989) Adaptive significance of spination in estuarine crab zoeae. Ecology *70*:464-482.

Mutvei, H. (1974) SEM studies on arthropod exoskeleton. Part 1: Decapod crustaceans, *Homarus gammarus* L. and *Carcinus maenas* (L.). Bull. Geol. Inst. Univ. Uppsala *4*:73-80.

Müller, H.G. (1981) Haftorgane bei Spinnen. Mikrokosmos *70*:381-382.

Müller, H.G. and S. Heimer (1989) Die Kopffatsätze der Zwergspinnenmännchen und ihre biologische Bedeutung. Mikrokosmos *78*:214-216.

Müller, H.J. (1941) Über Bau und Funktion des Legeapparates der Zikaden (Homoptera Cicadina). Z. Morphol. Ökol. Tiere *38*:534-629.

Nachtigall, W. (1974) *Biological mechanisms of attachment*. Berlin, Heidelberg, New York: Springer-Verlag.

Nachtigall, W. (1998) *Bionik. Grundlagen und Beispiele für Ingenieure und Naturwissenschaftler*. Berlin et al.: Springer.

Naldrett, M.J. (1992) Adhesives. In J.F.V. Vincent (ed.): Biomechanics - materials: A practical approach, Oxford, New York, Tokyo: IRL Press, pp. 219-240.

Naldrett, M.J. (1993) The importance of sulphur cross-links and hydrophobic interactions in the polymerization of barnacle cement. J. Mar. Biol. Assoc. U. K. *73*:689-702.

Naumann, I.D. (1991) Hymenoptera. In: The insects of Australia, New York: Cornell University Press, pp. 916-1000.

Neal, J.W. (1982) Significance of opposing abdominal tergal spines on the pupae of the bagworm *Thyridopteryx ephemeraeformis* (Lepidoptera; Psychidae). J. Kans. Entomol. Soc. *55*:605-616.

Neville, A.C. (1965) Circadian organization of chitin in some insect skeletons. Quart. J. Mic. Sci. *106*:315-325.

Neville, A.C. (1967) Chitin orientation in cuticle and its control. In J.W.L. Beament, J.E. Treherne and V.B. Wigglesworth (eds.): Advances in Insect Physiology, London and New York: Academic Press, pp. 213-286.

Neville, A.C. (1970) Cuticle ultrastructure in relation to the whole insect. In A.C. Neville (ed.): Insect ultrastructure, Oxford and Edinburgh: Blackwell Scientific Publishers, pp. 17-39.

Neville, A.C. (1975) *Biology of the arthropod cuticle*. Berlin, Heidelberg, New York: Springer.

Neville, A.C. (1993) *Biology of fibrous composites*. Cambridge: Cambridge University Press.

Neville, A.C. and B.M. Luke (1969a) A two-system model for chitin-protein complexes in insect cuticles. Tiss. Cell. *1*:689-707.

Neville, A.C. and B.M. Luke (1969b) Molecular architecture of adult locust cuticle at the electron microscope level. Tiss. Cell. *1*:355-366.

Neville, A.C., Thomas M.G., and B. Zelazny (1969) Pore canal shape related to molecular architecture of arthropod cuticle. Tiss. Cell. *1*:183-200.

New, T.R. (1991) Neuroptera (Lacewings). In CSIRO (ed.): The Insects of Australia, Ithaca, New York: Cornell University Press, pp. 525-542.

Niederegger, S. and S.N. Gorb (2001) Fly walking: a compromise between attachment and motion? Biona Report *in press*.

Nielsen, E.S. and I.F.B. Common (1991) Lepidoptera. In: The insects of Australia, New York: Cornell University Press, pp. 817-916.

Nutting, W.L. (1963) The hastate setae of certain dermestid beetle larvae: an unusual entangling defence mechanism. J. Arizona Acad. Sci. *2*:189-189.

Obenberger, J. (1964) *Entomologie*. Praha: Nakladatelstvi Ceskoslovenske Akademie Ved.

Olberg, R.M. (1978) Visual and multimodal interneurons in dragonflies. Diss. Abstr. Int. B Sci. Eng. *39*:2175

Olberg, R.M. (1981) Object- and self-movement detectors in the ventral nerve cord of the dragonfly. J. Comp. Physiol. A *141*:327-334.

Olberg, R.M. and R.B. Pinter (1990) The effect of mean luminance on the size selectivity of identified target interneurons in the dragonfly. J. Comp. Physiol. A *166*:851-856.

Olesen, J. (1972) The hydraulic mechanism of labial extension and jet propulsion in dragonfly nymph. J. Comp. Physiol. A *81*:53-56.

Olesen, J. (1979) Prey-capture in dragonfly nymphs (Odonata; Insecta): labial protraction by means of a multi purpose abdominal pump. Vid. Medd. Dan. Naturalist Foren. *141*:81-96.

Palestrini, C., R. Piazza, and M. Zunino (1987) Segnali sonori in the species di Geotrupini (Coleoptera, Scarabaeoidea, Geotrupidae). Boll. Soc. Entomol. Ital. *119*:139-151.

Parker, A.R., Z. Hegedus, and R.A. Watts (1998) Solar-absorber antireflector on the eye of an eocene fly. Proc. Roy. Soc. London B *265*:811-815.

Parry, D.A. (1983) Labial extension in the dragonfly larva *Anax imperator*. J. Exp. Biol. *107*:495-499.

Pasteels, J.J. and J.M. Pasteels (1972) *Les soies cuticulaires des Megachilidae (Hymenoptera; Apoidea) vues au microscope electronique a balayage*. Bruxeles.

Pelletier, Y. and Z. Smilowitz (1987) Specialized tarsal hairs on adult male Colorado potato beetles, *Leptinotarsa decemlineata* (Say), hamper its locomotion on smooth surfaces. Can. Entomol. *119*:1139-1142.

Perez-Miles, F. (1994) Tarsal scopula division in Theraphosinae (Araneae, Theraphosidae): its systematic significance. J. Arachnol. *22*:46-53.

Persson, B.N.J. (1998) *Sliding Friction. Physical Principles and Applications*. Berlin, Heidelberg, New York: Springer.

Peters, H.M. and J. Kovoor (1991) The silk-producing system of *Linyphia triangularis* (Araneae, Linyphiidae) and some comparisons with Araneidae. Structure, histochemistry and function. Zoomorphology *111*:1-17.

Peters, W.L. and I.C. Campbell (1991) Ephemeroptera (Mayflies). In CSIRO (ed.): The Insects of Australia, Ithaca, New York: Cornell University Press, pp. 279-293.

Pfau, H.K. (1971) Struktur und Funktion des sekundären Kopulationsapparates der Odonaten (Insecta; Palaeoptera) ihre Wandlung in der Stammesgeschichte und Bedeutung für die adaptive Enfaltung der Ordnung. Z. Morphol. Tiere *70*:281-371.

Pfau, H.K. (1985) Die eigentümliche Eiablage der *Cordulegaster*-Weibchen. Natur und Museum *115*:35-123.

Pfau, H.K. (1986) Untersuchungen zur Konstruktion, Funktion und Evolution des Flugapparates der Libellen (Insecta; Odonata). Tijdtsch. Ent. *129*:35-123.

Pfau, H.K. (1991) Contributions of functional morphology to the phylogenetic systematics of Odonata. Adv. Odonatol. *5*:109-141.

Pirisinu, Q., G. Spinelli, M.C. Biochierai, G. Guiderelli, and G. Paparo (1988) Stridulatory apparatus and sound production in Coleoptera Palpicornia. In: Proc. 18th Int. Congr. Entomol., Vancouver, July 3rd-9th, 1988: Abstr. and Author Index, Vancouver: pp. 86-86.

Pliske, T.E. and M.M. Salpeter (1971) The structure and development of the hairpencil glands in males of the Queen butterfly, *Danaus gilippus berenice*. J. Morphol. *134*:215-242.

Podczeck, F. and J. Newton (1995) Development of an ultracentrifuge technique to determine the adhesion and friction properties between particles and surfaces. J. Pharm. Sci. *84*:1067-1071.

Pohle, G. (1989) Structure, function and development of setae on gill-grooming appendages and associated mouthparts of pinnoterid crabs (Decapoda: Brachiura). Can. J. Zool. *67*:1690-1707.

Poodry, C.A. (1980) Epidermis: Morphology and Development. In M. Ashburner and T.R.F. Wright (eds.): The Genetics and Biology of Drosophila, London et al.: Academic Press, pp. 443-497.

Popham, E.J. and E. Bevans (1979) Functional morphology of the feeding apparatus in larval and adult *Aeshna juncea* L. (Anisoptera; Aeshnidae). Odonatologica *8*:301-318.

Powell, G.S. and C.L. Hurd (1972) Observations of initial king crab egg attachment. Trans. Amer. Fish. Soc. *101*:557-559.

Pringle, J. (1963) *Insect flight*. Moscow: Izd. Inostr. Literat.

Pritykina, L.N. (1980) Odonatoptera. Trudy PIN AN SSSR *178*:128-134.

Quentin, St.D. (1962) Der Eilegeapparat der Odonaten. Z. Morphol. Ökol. Tiere *51*:165-189.
Rabinowicz, E. (1995) *Friction and Wear of Materials*. New York et al.: Wiley.
Radhakrishnan, V. (1998) Locomotion: dealing with friction. Proc. Nat. Acad. Sci. USA *95*:5448-5455.
Radovic, I.T. (1985) Morphology and adaptive value of the sting apparatus of digger wasps (Hymenoptera; Sphecidae). Acta Entomol. Jugosl. *21*:61-74.
Rambla, M. (1990) Les scopula des Opilions, differences avec les scopula des araignees (Arachnida, Opiliones, Araneae). Bulletin de la Societe Europeenne D'Arachnologie Hors Serie *1990*:293-298.
Ramires, V.B. (1989) Can dusts be used against *Varroa jacobsoni*? Apiacta *24*:2-4.
Rees, B.E. and G.F. Ferris (1939) The morphology of *Tipula reesi* Alex. Microentomology *4*:143-178.
Rentz, D.D.F. (1991) Grylloblattodea (Grylloblattida, Grylloblattaria, Grylloblattodea, Notoptera). In CSIRO (ed.): The Insects of Australia, Ithaca, New York: Cornell University Press, pp. 357-359.
Richards, A.G. (1944) Electron micrographs of mosquito microtrichia. Entomol. News *55*:260-262.
Richards, A.G. (1951) *The integument of arthropods*. Minneapolis: University of Minnesota Press.
Richards, A.G. (1965) The proventriculus of adult Mecoptera and Siphonaptera. Entomol. News *76*:253-256.
Richards, A.G. (1967) Sclerotization and the localization of brown and black colours in insects. Zool. Jb. Anat. *84*:25-62.
Richards, A.G. (1981) The frenulum of moth. Entomol. News *92*:111-114.
Richards, A.G. and T.F. Anderson (1942) Electron microscope studies of insect cuticle. J. Microsc. *71*:135-183.
Richards, A.G. and F.H. Korda (1947) Electron micrographs of centipede setae and microtrichia. Entomol. News *58*:141-145.
Richards, A.G. and P.A. Richards (1979) The cuticular protuberances of insects. Int. J. Insect Morphol. Embryol. *8*:143-157.
Richards, A.G. and S.H. Seilheimer (1977) The proventricular flanges in mosquito larvae. Entomol. News *88*:1-9.
Richards, P.A. and A.G. Richards (1969) Acanthae: a new type of cuticular process in the proventriculus of Mecoptera and Siphonaptera. Zool. Jb. Anat. *86*:158-176.
Richards, P.A., A.G. Richards, and A.C. Hodson (1977) Development and variation in flanges on the proventriculus of larvae of *Aedes aegypti* (Diptera: Culicidae). J. Med. Entomol. *13*:517-523.
Rietschel, P. (1961) Bau, Funktion und Entwicklung der Haftorgane der Blepharoceridenlarven. Z. Morphol. Ökol. Tiere *50*:239-265.
Ringlund, K. and E.H. Everson (1968) Leaf pubescence in common wheat, *Triticum aestivum* L., and resistance to the common leaf beetle, *Oulemma melanopus* L. Crop Science *8*:705-710.
Rinterknecht, E. and P. Levi (1966) Etude au microscope electronique du cycle cuticulaire au cours de 4-eme stade larvaire chez *Locusta migratoria*. Z. Mikrosk. Anat. *72*:390-407.
Robertson, H.M. (1982) Mate recognition and mechanical isolation in *Enallagma* damselflies (Odonata; Coenagrionidae). Evolution *36*:243-250.
Rodova, R.A. (1968) Apparatus of the wing fixation in the resting position in the Chironomidae (Diptera). Entomol. Rev. *47*.

Rodova, R.A. (1971) Larva and female of *Telmatopelopia nemorum* (Gaetghebuer, 1921) (Diptera, Chironomidae). In: Biology and physiology of fresh-water organisms, Leningrad: Nauka, pp. 144-151.

Rodova, R.A. (1974) Chironomid females (Diptera, Chironomidae). In: Biology and productivity of fresh-water invertebrates, Leningrad: Nauka, pp. 154-194.

Rodova, R.A. (1980) Chironomid females of the tribe Tanytarsini (Diptera, Chironomidae). In: Biology, Morphology, and Systematics of Water Invertebrates, pp. 239-255.

Roll, B. (1995) Epidermal fine structure of the toe tips of *Sphaerodactylus cinereus* (Reptilia, Gekkonidae). J. Zool. ,Lond. *235*:289-300.

Rombouts, J.E. (1884) Über die Fortbewegung der Fliegen an glatten Flächen. Zool. Anz. *7*:619-623.

Roscoe, D.T. and G. Walker (1991) The adhesion of spiders to smooth surfaces. Bull. Br. Arachnol. Soc. *8*:224-226.

Ross, E.S. (1991) Embioptera Embidiina (Embiids, web-spinners, foor-spinners). In CSIRO (ed.): The Insects of Australia, Ithaca, New York: Cornell University Press, pp. 405-409.

Roth, L.M. and E.R. Willis (1952) Tarsal structure and climbing ability of cockroaches. J. Exp. Biol. *119*:483-517.

Rothshild, M., Y. Schlein, K. Parker, A.C. Neville, and S. Sternberg (1975) The jumping mechanism of *Xenopsylla cheopis*. 3. Execution of the jump and activity. Phil. Trans. Roy. Soc. London B *271*:499-515.

Rothshild, M. and R. Traub (1971) A revised glossary of terms used in the taxonomy and morphology of fleas. In: Illustrated catalogue of Rothschild collection of fleas in the British Museum (Natural History), London: British Museum (Nat. Hist.), pp. 8-85.

Rovner, J.S. (1978) Adhesive hairs in spiders: behavioral functions and hydraulically mediated movement. Symp. Zool. Soc. Lond. *42*:99-108.

Rovner, J.S. (1980) Morphological and ethological adaptations for prey-capture in wolf spiders (Araneae, Lycosidae). J. Arachnol. *8*:201-215.

Röder, G. (1984a) *Morphologische Untersuchungen an Prätarsen von Diptera and Mecoptera (Insecta).* Nürnberg: Dissertation.

Röder, G. (1984b) Zur Morphologie des Prätarsus der Diptera and Mecoptera. Zool. Jb. Anat. *144*:465-502.

Röll, B. (1995) Epidermal structure of the toe tips of *Sphaerodactylus cinereus* (Reptilia, Gekkonidae). J. Zool. Lond. *235*:289-300.

Rudall, K.M. (1963) The chitin protein complexes in insect cuticle. Adv. Insect Physiol. *1*:257-313.

Ruibal, R. and V. Ernst (1965) The structure of the digital setae in lizards. J. Morphol. *117*:271-294.

Samuelson, G.A. (1994) An elytron to body meshing mechanism of possible significance in the higher classification of Chrysomelidae (Coleoptera). In: Proceedings of the Third International Symposium on the Chrysomelidae, Leiden: Backhuys Publishers, pp. 136-147.

Samuelson, G.A. (1996) Binding sites: elytron-to-body meshing structures of possible significance in the higher classification of Chrysomeloidea. In P.H.A. Jolivet and M.L. Cox (eds.): Chrysomelidae Biology, The Classification, Phylogeny and Genetics, Amsterdam: SPB Academic Publishing, pp. 267-290.

Sandeman, D. and H. Markl (1980) Head movements in flies (*Calliphora*) produced by deflexion of the halteres. J. Exp. Biol. *85*:43-60.

Sander, K. (1956) Bau und Funktion des Sprungapparates von *Pyrilla perpusilla* Walker (Homoptera - Fulgoroidea). Zool. Jb. Anat. *75*:383-388.

Saroyan, J.R., E. Lindner, and C.A. Dooley (1970) Repair and reattachment in the Balanidae as related to their cementing mechanism. Biol. Bull. *139*:333-350.

Schaller, F. (1970) Collembola (Springschwänze). In M. Beier (ed.): Handbuch der Zoologie, Berlin, New York: W. de Gruyter, pp. 1-72.

Schedl, W. (1991) Hymenoptera: Unterordnung Symphyta (Pflanzenwespen). In M. Fischer (ed.): Handbuch der Zoologie, Berlin, New York: W. de Gruyter, pp. 1-117.

Scherge, M. and S.N. Gorb (2000a) Microtribology of biological materials. Tribology Letters *8*:1-7.

Scherge, M. and S.N. Gorb (2000b) Using biological principles to design MEMS. J. Micromech. Microeng. *10*:359-364.

Scherge, M. and S.N. Gorb (2001) *Biological micro- and nanotribology*. Berlin et al.: Springer.

Scherge, M., X. Li, and J.A. Schaefer (1999) The effect of water on friction of MEMS. Tribology Letters *6*:215-220.

Schliemann, H. (1970) Bau und function der Haftorgane von *Thyroptera* und *Myzopoda* (Vespertilionoidea, Microchiroptera, Mammalia). Z. Wiss. Zool. *181*:353-400.

Schliemann, H. (1971) Die Haftorgane von *Thyroptera* und *Myzopoda* (Microchiroptera, Mammalia) - Gedanken zu ihrer Entstehung als Parallelbildungen. Z. Zool. Syst. Evolutionsforsch. *9*:61-80.

Schliemann, H. (1974) Haftorgane bei Fledermäusen. Natur und Museum *104*:15-20.

Schliemann, H. (1975) Über die Entstehung von Haftorganen bei Chiropteren. Mitt. Hamb. Zool. Mus. Inst. *72*:249-259.

Schliemann, H. (1983) Haftorgane - Beispiele für gleichsinnige Anpassungen in der Evolution der Tiere. Funkt. Biol. Med. *2*:169-177.

Schliemann, H. and C. Rehn (1980) Zur Kenntnis der Haftorgane von *Eudiscopus denticulus* (Osgood, 1932) (Mammalia, Microchiroptera, Vespertilionidae). Z. Säugetierkunde *45*:29-39.

Schmidt, K. and W. Gnatzy (1971) Die Feinstruktur der Sinneshaare auf den Cerci von *Grillus bimaculatus* Deg. 2. Die Hautung der Faden und Keulenhaare. Z. Zellforsch. Mikrosk. Anat. *122*:210-226.

Schmitt, M. (1991) Stridulatory devices of leaf beetles (Chrysomelidae) and other Coleoptera. In M. Zunino, X. Belles and M. Blas (eds.): Advances in Coleopterology, Barcelona: Assoc. Eur. Coleopt., pp. 263-280.

Schneider, P. (1981) Gleitkoppelmechanismus bei Blattläusen (Aphididae). Zool. Anz. *206*:178-180.

Schneider, P. and R. Schill (1978) Der Gleitkoppelmechanismus bei vierflügeligen Insekten mit asynchronem Flugmotor. Zool. Jb. Physiol. *82*:365-382.

Schönitzer, K. (1986) Comparative morphology of the antenna cleaner in bees (Apoidea). Z. Zool. Syst. Evolutionsforsch. *24*:35-51.

Schönitzer, K. and G. Lawitzky (1987) A phylogenetic study of the antenna cleaner in Formicidae, Mutillidae and Tiphiidae (Insecta, Hymenoptera). Zoomorphology *107*:273-285.

Schönitzer, K. and M. Penner (1984) The function of the antenna cleaner of the honey-bee (*Apis mellifica*). Apidologie *15*:23-32.

Schrott, A. (1986) Vergleichende Morphologie und Ultrastruktur des Cenchrus-Dornenfeldapparates bei Pflanzenwespen (Insecta: Hymenoptera, Symphyta). Berichte Naturwiss. Med. Ver. Innsbruck *73*:159-168.

Schultz, T.D. and N.F. Hadley (1987) Structural colors of tiger beetles and their role in heat transfer through the integument. Physiological Zoology *60*:737-745.

Schwan, T.G. and D. Corwin (1987) Uropodid mites phoretic on fleas of ground squirrels in California. Proceedings of the Entomological Society of Washington *89*:790-793.

Seibt, U., D. Schneider, and T. Eisner (1972) Duftpinsel, Flügeltaschen und Balz des Tagfalters *Danaus chrysippus*. Z. Tierpsychol. *31*:513-530.

Seidel, B. (1989) Phoresis of *Cyclocypris ovum* (Jurine) (Ostracoda, Podocopida, Cyprididae) on *Bombina variegata variegata* (L.) (Anura, Amphibia) and *Triturus vulgaris* (L.) (Urodela, Amphibia). Crustaceana *57*:171-176.

Seifert, P. and T. Heinzeller (1989) Mechanical, sensory and glandular structures in the tarsal unguitractor apparatus of *Chironomus riparius* (Diptera, Chironomidae). Zoomorphology *109*:71-78.

Shah, M.A.T. (1982) The influence of plant surfaces on the searching behavior of coccinellid larvae. Entomol. Exp. Appl. *31* :377-380.

Shapiro, A.M. and A.H. Porter (1989) The lock-and-key hypothesis: evolutionary and biosystematic interpretation of insect genitalia. Annu. Rev. Entomol. *34*:231-245.

Shevchenko, V.G. (1970) Origin, morphological and functional characteristic of eryophyoid mites (Acarina, Eriophyoidea). In: Studies on the evolutionary morphology of invertebrates, Leningrad: Izd. LGU, pp. 153-183.

Shinonaga, S. and R.A. Beaver (1979) *Pierretia urceola*: a new species of sarcophagid fly found living in *Nepenthes* pitcher plants in west Malaysia. Annales de la Societe Entomologique de France *15*:37-40.

Shvanvich, B.N. (1949) *General entomology*. Moscow, Leningrad: Sovetskaja nauka.

Simmermacher, G. (1884) Untersuchungen über Haftapparate an Tarsalgliedern von Insekten. Zool. Anz. *7*:225-228.

Slifer, E.H. (1950) Vulnerable areas on the surface of the tarsus and pretarsus of the grasshopper (Acrididae, Orthoptera) with special reference to the arolium. Ann. Entomol. Soc. Am. *43*:173-188.

Slifer, E.H. and B.P. Uvarov (1938) Brunner's organ:a structure found on the jumping legs of grasshoppers. Proc. Roy. Soc. London B *13*:111-115.

Slifer, S. (1961) The fine structure of insect sense organs. Int. Rev. Cytol. *11*:125-159.

Smith, E.L. (1972) Biosystematics and morphology of Symphyta. 3. External genitalia of *Euura*. Int. J. Insect Morphol. Embryol. *1*:321-365.

Smith, G.B. and J.A.L. Watson (1991) Thysanura Zygentoma (Siverfish). In CSIRO (ed.): The Insects of Australia, Ithaca, New York: Cornell University Press, pp. 275-278.

Snodgrass, R.E. (1935) *Principles of insect morphology*. New York: McGraw-Hill.

Snodgrass, R.E. (1956) *Anatomy of the honey bee*. New York: Comstock Publishing Associates.

Southwick, E.E. (1985) Bee hair structure and the effect of hair on metabolism at low temperature. Journal of Apicultural Research *24*:144-149.

Southwood, T.R.E. (1986) Plant surfaces and insects - an overview. In T.R.E. Southwood and B.E. Juniper (eds.): Insects and the plant surface, London: Edward Arnold, pp. 1-22.

Spiegler, P.E. (1962) The origin and nature of the adhesive substance in larvae of the genus *Chrysopa* (Neuroptera: Chrysopidae). Ann. Entomol. Soc. Am. *55*:69-72.

Spungis, V. (1990) Ecological approach to the convergence of morphological features in soil-inhabiting gall midge larvae (Cecidomyidae). In: 2nd Int. Congr. Dipterol. Bratislava, Aug. 27 -Sep. 1, 1990. Abstr. Vol., pp. 223-223.

Spurr, A.R. (1969) A low-viscosity epoxy resin embedding medium for electron microscopy. Journal of Ultrastructure Research *26*:31–43.

Starck, J.M. (1985) Stridulationsapparate einiger Spinnen - Morphologie und evolutionsbiologische Aspekte. Z. Zool. Syst. Evolutionsforsch. *23*:115-135.

Starnecker, G. (1998) A safety band prevents falling of the suspended pupa of the butterfly *Inachis io* (Nymphalidae) during moult. A comparison with the girdled pupa of *Pieris brassicae* (Pieridae, Lepidoptera). Zoomorphology *118*:129-136.

Steinmann, H. (1967) A study of the cephalic measurement values on the species of Europaean Odonata families. Dtsch. Entomol. Z. *4*:265-275.

Stern, C. (1954) Two or three bristles. Amer. Sci. *42*:213-247.

Stoner, K.A. (1990) Glossy leaf wax and host-plant resistance to insects in *Brassica oleracea* L. under natural infestation. Environ. Entomol. *19*:730-739.

Stork, N.E. (1980a) A scanning electron microscope study of tarsal adhesive setae in the Coleoptera. Zool. J. Linn. Soc. *68*:173-306.

Stork, N.E. (1980b) Experimental analysis of adhesion of *Chrysolina polita* (Chrysomelidae, Coleoptera) on a variety of surfaces. J. Exp. Biol. *88*:91-107.

Stork, N.E. (1980c) Role of wax blooms in preventing attachment to brassicas by the mustard beetle, *Phaedon cochleariae*. Entomol. Exp. Appl. *28*:100-107.

Stork, N.E. (1981) The structure and function of the adhesive organs on the antennae of male *Harpocera thoracica* (Fallen) (Miridae; Hemiptera). J. Nat. Hist. *15*:639-644.

Stork, N.E. (1983a) A comparison of the adhesive setae on the feet of lizards and arthropods. J. Nat. Hist. *17*:829-835.

Stork, N.E. (1983b) The adherence of beetle tarsal setae to glass. J. Nat. Hist. *17*:583-597.

Stork, N.E. (1986) The form of plant waxes: a means preventing insect attachment? In B.E. Juniper and T.R.E. Southwood (eds.): Insects and the plant surface, London: Edward Arnold, pp. 346-347.

Stork, N.E. (1987) Adaptations of arboreal carabids to life in trees. Acta Phytopathol. Entom. Hungarica *22*:273-291.

Stork, N.E. (1988) Tarsal adhesive mechanisms in Coleoptera and their role in holding onto plant surfaces. In: Proc. 8th Int. Congr. Entomol., Vancouver, July 3rd-9th, 1988. Abstr. and Author Index, Vancouver, pp. 92-92.

Strausfeld, N.J., H.S. Seyan, and J.J. Milde (1987) The neck motor system of the fly *Calliphora erythrocephala*. I. Muscles and motor neurons. J. Comp. Physiol. A *160*:205-224.

Strohmenger, T. and W. Nentwig (1987) Adhesive and trapping properties of silk from different spider species. Zool. Anz. *218* :9-16.

Strümpel, H. (1983) Homoptera (Pflanzensäuger). In M. Fischer (ed.): Handbuch der Zoologie, Berlin, New York: W. de Gruyter, pp. 1-222.

Tanaka, Y. and M. Hisada (1980) The hydraulic mechanism of the predatory strike in dragonfly larvae. J. Exp. Biol. *88*:1-19.

Theischinger, G. (1991) Megaloptera (Alderflies, dobsonflies). In CSIRO (ed.): The Insects of Australia, Ithaca, New York: Cornell University Press, pp. 521-524.

Thewissen, J.G.M. and S.A. Etnier (1995) Adhesive devices on the thumb of vespertilionoid bats (Chiroptera). Journal of Mammalogy *76*:925-936.

Thorpe, W.H. and D.J. Crisp (1947) Studies on plastron respiration. I. The biology of *Aphelocheirus* and the mechanism of plastron retention. J. Exp. Biol. *24*:227-269.

Tillyard, R.J. (1917) *The biology of dragonflies (Odonata or Paraneuroptera)*. Cambridge: University Press.

Tillyard, R.J. (1928) The evolution of the order Odonata. In: Proc. 4-th Intern Congr. Entomol. Ithaca, Hamburg: pp. 543-545.

Tönjes, P. (1989) Zur Spinnfadenbildung und zum Wohnröhrenbau bei der Larve von *Telmatogeton japonicus* Tokunaga, 1933 (Syn. *T. remanei* Remmert, 1963) (Chironomidae, Telmatogetonidae). Zool. Jb. Anat. *118*:117-136.

Törne, O. (1910) Die Saugnäpfe der männlichen Dytiscidae. Zool. Jb. Anat. *29*:415-448.

Tuxen, S.L. (1970) *Taxonomists glossary of genitalia of insects*. Copenhagen: Munksgaard.

Tuzovskij, P.V. (1987) *Morphology and postembryonic development of water mites*. Moscow: Nauka.

Urquhart, F.A. (1970) Mechanism of cremaster withdrawal and attachment in pendant rhopalocerous pupae (Lepidoptera). Can. Entomol. *102*:1579-1582.

Utzeri, C. (1989) Tactile communication through the tandem link in the Odonata and the problem of tandem oviposition in *Sympetrum* (Libellulidae). Opusc. Zool. Flum. *35*:1-6.

Verhoeff, K.W. (1902) Beiträge zur vergleichenden Morphologie des Thorax der Insekten mit Berücksichtigung der Chilopoden. Nova Acta Leopoldina, Halle *81*:64-124.

Vincent, J.F.V. (1981) Morphology and design of the extensible intersegmental membrane of the female migratory locust. Tiss. Cell. *13*:18-31.

Vincent, J.F.V. and S.D.E. Wood (1972) Mechanism of abdominal extension during oviposition in *Locusta*. Nature *235*:167-168.

Waite, J.H. (1983) Adhesion in byssally attached bivalves. Biol. Rev. *58*:209-231.

Walker, G., A.B. Yule, and J. Ratcliffe (1985) The adhesive organ of the blowfly, *Calliphora vomitoria*: a functional approach (Diptera: Calliphoridae). J. Zool. Lond. *205*:297-307.

Wallentin, J., M. Mondon, H. Stadler, S.N. Gorb, and Ch. Ziegler (1999) The secretes of fly secretes: adhesion properties probed by force-distance curves. In D.J. Müller and H.F. Knapp (eds.): Scanning-probe microscopes and organic materials VIII, Workshop, Basel, October 4-6, 1999, Abstract Booklet, Basel: p. 14.

Walther, C. (1969) Zum Verhalten des Krallenbeugersystems bei der Stabheuschrecke *Carausius morosus* Br. Z. Vergl. Physiol. *62*:421-460.

Walton, T.J. (1990) Waxes, cutin and suberin. In J.L. Harwood and J.R. Bowyer (eds.): Lipids, membranes and aspects of photobiology, London: Academic Press, pp. 105-158.

Warth, A.H. (1956) *The chemistry and technology of waxes*. New York: Reinhold.

Warthen, J.D., E.C. Uebel, W.R. Lusby, and V.E. Adler (1981) The cuticular lipids of the walkingstick *Diapheromera femorata* (Say). Insect Biochemistry *11*:467-472.

Watson, J.A.L. (1966) Genital structure and isolating mechanism in Odonata. Proc. Roy. Entomol. Soc. London A *41*:171-174.

Watson, J.A.L. and F.J. Gay (1991) Isoptera (Termites). In CSIRO (ed.): The Insects of Australia, Ithaca, New York: Cornell University Press, pp. 330-347.

Watson, J.A.L. and A.F. O'Farell (1991) Dragonflies and damselflies. In CSIRO (ed.): The Insects of Australia, Ithaca, New York: Cornell University Press, pp. 294-310.

Watson, J.A.L. and G.B. Smith (1991) Archaeognatha Microcoryphia (Bristletails). In CSIRO (ed.): The Insects of Australia, Ithaca, New York: Cornell University Press, pp. 272-274.

Way, M.J. and G. Murdie (1965) An example of varietal resistance of Brussel sprouts. Ann. Appl. Biol. *56*:326-328.

Weber, H. (1930) *Biologie der Hemipteren*. Berlin: Verlag von Julius Springer.

Weele van der, H.W. (1906) Morphologie und Entwicklung der Gonapophysen der Odonata. Tijdtsch. Ent. *49*:99-198.

Weidner, H. (1972) Copeognatha (Staubläuse). In M. Beier (ed.): Handbuch der Zoologie, Berlin, New York: W. de Gruyter, pp. 1-94.

Weis-Fogh, T. (1960) A rubber-like protein in insect cuticle. J. Exp. Biol. *37*:889-907.

Weis-Fogh, T. (1961) Molecular interpretation of the elasticity of resilin, a rubber-like protein. J. Mol. Biol. *3*:648-667.

Weis-Fogh, T. (1970) Structure and formation of insect cuticle. In A.C. Neville (ed.): Insect ultrastructure, Oxford and Edinburgh: Blackwell Scientific Publishers, pp. 165-186.

West, T. (1862) The foot of the fly; its structure and action: elucidated by comparison with the feet of other insects. Trans. Linn. Soc. Lond. *23*:393-421.

Westheide, W. and R. Rieger (1996) *Spezielle Zoologie. I. Einzeller und Wirbelose Tiere.* Stuttgart, Jena, New York: Gustav Fischer Verlag.

Whitten, J.M. (1969a) Coordinated development in footpad of the fly *Sarcophaga bullata* during metamorphosis: changing puffing patterns of the giant cell chromosomes. Chromosoma *26*:215-224.

Whitten, J.M. (1969b) Coordinated development in the fly foot: sequential cuticle secretion. J. Morphol. *127*:73-104.

Whitten, J.M. (1976) Some observations on cellular organization and pattern in flies. In H.R. Hepburn (ed.): The insect integument, New York: Elsevier, pp. 277-297.

Wicksten, M.K. (1978) Attachment of decorating materials in *Loxorhynchus crispatus* (Brachiura, Majidae). Trans. Amer. Microsc. Soc. *97*:217-220.

Wigglesworth, V.B. (1956) The haemocytes and connective tissue formation in an insect, *Rhodnius prolixus*. Quart. J. Mic. Sci. *97*:89-98.

Wigglesworth, V.B. (1973) Haemocytes and basement membrane formation in *Rhodnius*. J. Insect Physiol. *19*:831-844.

Wigglesworth, V.B. (1987) How does a fly cling to the under surface of a glass sheet? J. Exp. Biol. *129*:363-367.

Wildermuth, H. (1984) Drei aussergewohnliche Beobachtungen zum Fortpflanzungsverhalten der Libellen. Mitt. Ent. Ges. Basel *34*:121-129.

Wygodzinsky, P. (1961) On a surviving representative of the Lepidotrichidae (Thysanura). Ann. Entomol. Soc. Am. *54*:621-627.

Yan, W. and Y. Tang (1981) The biochemical composition of the secondary cement of *Balanus reticulatus* Utinomi and *Balanus amaryllis* Darwin. Nanhai Haiyang Kexue Jikan *1981*:145-153.

Zalessky, Yu.M. (1932) The wing venation in dragonflies and their phylogeny. Izvestija AN SSSR. Otd. Matem. Estestv. Nauk *8*:713-733.

Zaytsev, V.F. (1984) Microstructure of the labella of the fly proboscis. II. Pseudotracheal framework; structure and evolution. Entomol. Rev. *63*:35-39.

Zwick, P. (1980) Plecoptera (Sternfliegen). In M. Beier (ed.): Handbuch der Zoologie, Berlin, New York: W. de Gruyter, pp. 1-115.

Index

A

B

C

D

F

G

H

M

N

O

P

Q

T

U

V

W

X

Y

Z